AF411734

STUDIES IN EVOLUTION

THE HISTORY OF PALEONTOLOGY

Advisory Editor

Stephen Jay Gould

Consulting Editor

Thomas J.M. Schopf

See last pages of this volume for a complete list of titles.

STUDIES IN EVOLUTION

CHARLES EMERSON BEECHER

ARNO PRESS

A New York Times Company
New York • 1980

Editorial Supervision: Erin Foley

Reprint Edition 1980 by Arno Press Inc.

Reprinted from a copy in the University of Illinois Library.

THE HISTORY OF PALEONTOLOGY

ISBN for complete set: 0-405-12700-6
See last pages of this volume for titles.

Manufactured in the United States of America

Library of Congress Cataloging in Publication Data

Beecher, Charles Emerson, 1856-1904.
    Studies in evolution.

    (The History of evolution)
    Reprint of the 1901 ed. published by Scribner,
New York, in series: Yale bicentennial publications.
    Includes bibliographies and index.
    1. Evolution. 2. Spines (Zoology) 3. Trilobites.
4. Brachiopoda. 5. Brachiopoda, Fossil. I. Title.
II. Series: History of evolution. III. Series:
Yale bicentennial publications.
QH366.B42 1980   575   79-8324
ISBN 0-405-12704-9

STUDIES IN EVOLUTION

Yale Bicentennial Publications

With the approval of the President and Fellows of Yale University, a series of volumes has been prepared by a number of the Professors and Instructors, to be issued in connection with the Bicentennial Anniversary, as a partial indication of the character of the studies in which the University teachers are engaged.

This series of volumes is respectfully dedicated to

The Graduates of the University

STUDIES IN EVOLUTION

MAINLY REPRINTS OF OCCASIONAL PAPERS
SELECTED FROM THE

*PUBLICATIONS OF THE LABORATORY OF INVERTEBRATE
PALEONTOLOGY, PEABODY MUSEUM
YALE UNIVERSITY*

BY

CHARLES EMERSON BEECHER

NEW YORK: CHARLES SCRIBNER'S SONS
LONDON: EDWARD ARNOLD
1901

UNIVERSITY PRESS · JOHN WILSON
AND SON · CAMBRIDGE, U. S. A.

PREFACE

THE following papers from the publications of the Laboratory of Paleontology have been selected for reprinting on account of their representing more or less closely a distinct line of research; namely, the investigation and study of the development of a number of invertebrate animals. The generalizations resulting from such studies properly belong to the province of organic evolution, while the detailed methods pertain to the observation and interpretation of the stages of growth and decline in the organism.

Nearly all the subject-matter is based upon studies of the remains of fossil animals, many of them coming from the oldest known fossil-bearing rocks. In some instances material representing living species has been introduced for comparison and to illustrate further the problems under investigation.

The first work done in America on the stages of growth of fossil Brachiopoda was a memoir by the present writer, in collaboration with Dr. John M. Clarke, on "The Development of Some Silurian Brachiopoda," published by the University of the State of New York. It seems fitting to include this, in order to complete the work on the development of the Brachiopoda carried on subsequently by the writer. Another joint paper, written in connection with Mr. Charles Schuchert, is also introduced.

The first paper in the present collection, on "The Origin and Significance of Spines," is an attempt to apply the general laws of evolution in the study of a particular structure throughout the animal and vegetable kingdoms, and to discover its true interpretation in terms of ontogeny, phylogeny, and chronology.

The next division of the volume comprises papers on the structure and development of Trilobita. The following section presents developmental studies on the Brachiopoda. It should be stated that this work was undertaken largely in the hope that the results would lead to the principles governing a natural classification of all forms in these two classes. In the brachiopods nothing further than a division into orders and a grouping of the families under the orders was attempted. The elaboration of this classification has been very fully carried out by Mr. Charles Schuchert, in his "Synopsis of American Brachiopoda." For the Trilobita, a new arrangement into orders was suggested, together with a redefining of the families and a grouping of the genera under them.

In the last division are three papers on special problems of development.

The author is greatly indebted to Miss Lucy P. Bush for assistance in arranging the material for this volume, and especially for aid while it was being printed.

Yale University, April 3, 1901.

CONTENTS

CONTENTS

XV

CONTENTS

ILLUSTRATIONS

LIST OF PLATES

FIGURES IN TEXT

I

GENERAL EVOLUTION

1. THE ORIGIN AND SIGNIFICANCE OF SPINES

STUDIES IN EVOLUTION

I

GENERAL EVOLUTION

1. THE ORIGIN AND SIGNIFICANCE OF SPINES

A STUDY IN EVOLUTION *

(PLATE I)

INTRODUCTION

THE presence of spines in various plants and animals is, at times, most obvious to all mankind, and not unnaturally they have come to be regarded almost wholly in the light of defensive and offensive weapons. Their origin, too, is commonly explained as due to the influence of natural selection, resulting in the greater protection enjoyed by spiniferous organisms. But when, upon critical examination, it is seen that some animals are provided with spines which apparently interfere with the preservation of the individual, that other animals develop spines which cannot serve any purpose for protection or otherwise, and that spines themselves are often degenerate or suppressed organs, then it becomes evident that the spinose condition may have other interpretations than the single one of protection.

The object of this article is to make a few observations on spinosity, especially among invertebrate animals, and to endeavor to arrive at some general conclusions relating to the origin and significance of this condition. It is believed that the results have a broader application than is at first apparent, and underlie important laws and principles of organic evolution. In closely related species, the presence or absence of

* *Amer. Jour. Sci.* (4), VI, 1–20, 125–136, 249–268, 329–359, pl. i, 1898.

spines seems in itself a trivial character, indicating at best only specific differences, yet it will be shown that the spines are often the expression of important vital adjustments and conditions, and are not merely external features of the same value as color and many other skin or superficial characters. As will be indicated later on, spines may also arise through the operations of a number of forces and conditions, and it may well be asked, therefore, Do spines have any profound significance? It must be granted at the outset that apart from other characteristics, or when regarded as simple spini-form extensions of certain tissues or organs, they have no such value or meaning. How, then, should they be considered? The reply is evident: Their importance lies not in what they *are*, but in what they *represent*. They *are* simply prickles, thorns, spines, or horns; they *represent*, as will be shown, a stage of evolution, a degree of differentiation in the organism, a ratio of its adaptability to the environment, a result of selective forces, and a measure of vital power.

After studying numerous organisms, the writer is led to believe that in every case no single reason is sufficient to account for this spinose condition. The original cause may not be operative through the entire subsequent phylogeny, so that spines arising from external stimuli and then serving important defensive purposes may at a later period practically lose this function; or spines may become more and more developed simply by increasing diversity of growth forces, or through the multiplicity of effects. In this way causes may follow, overlap, or even coincide with each other; but in interpreting special cases the problems involved may be quite complicated and often obscure.

In reviewing the development of animal life from the earliest Cambrian to the present, one cannot avoid being impressed by the groups of spinose forms which appear here and there throughout geologic time, and give a special phase to contemporary faunas. Tracing these one by one through their geological development, it is noticed that each group began its history in small, smooth, or unornamented species.

As these developed, the spinose forms became more abundant until after the culmination of the group is reached, when this type either became extinct or was continued in smaller and less specialized forms. In applying this principle to any order of plants or animals, several precautions are necessary. The estimate must be based approximately upon the general average of the totality of specific characters, whether a genus, family, order, or even a class is being considered. A short-lived family or genus, or the terminal members of specialized groups, therefore, cannot be taken as representing the developmental status of the larger divisions, because they culminated and disappeared independently of the culmination of the class to which they belong. On a small scale, however, each epitomizes the rise and decline of the larger group, and the principles of correlation commonly applied in ontogeny and phylogeny can likewise be used in the study of spines and spiniferous species, with equally exact results, whenever the principal factors are understood.

Law of Variation.

Before undertaking any general or special examination of the life histories and interpretation of spinose organisms, it is desirable to consider briefly some of the biogenetic principles which are considered to bear directly on the problems here under discussion.

First among these is the law of variation or change, which is so generally recognized as to require but the briefest restatement.

The organic as well as the inorganic world is subject to all the forces of nature, internal and external, molecular and molar, and even a partial stability is gained only through a regulated adjustment. In organisms this change is momentary and persistent, while in most inorganic substances it is slow and intermittent. The results of this continual readjustment constitute modification, which may be progressive or regressive, continuous or discontinuous (in the sense of accelerated, uniform, or retarded). They are everywhere present and the causes always operative. Throughout life the individual changes, and in addition varies from all other individuals. The family also changes with time, and likewise differs from other families. Variation is everywhere present. Moreover, it is generally accepted, and is so taken here, that in its results this variation is not haphazard, but is normally in accordance with certain demands or in harmony with certain surroundings. Whether an organism itself tends to vary in all directions, or is chiefly subject to modifications from external forces, does not alter the preceding statement.

Cope [11] has considered variation as either physico-chemical (molecular) or mechanical (molar). The influence of the first is known as physiogenesis and of the second as kinetogenesis. In the animal kingdom the potency of kinetogenesis is greater as an efficient cause of evolution; while in the vegetable kingdom physiogenesis is apparently of more importance.

The tendency of variation is always in the direction of the establishment of an equilibrium between the organism and its environment. However, the laws of the development of the earth preclude the possibility of a constant environment, and therefore a perfect, permanent, and uniform equilibrium between life and surroundings is unattainable.

The manner of variation is clearly defined as *progressive* and *regressive*. Progressive variation is one of the essential factors of evolution, while regressive variation is towards dissolution. Since the main history of life is told through processes of the former, progressive variation is far greater in importance; while, in general, regressive variation can be applied only to late periods in the history of groups or forms now in their decadence, or to others which in past times have suffered decline and extinction.

The summary of the operation of the law of multiplication of effects, as given by Herbert Spencer,[66] may well be stated here, as it emphasizes one of the principles through which spines have originated.

"It manifestly follows that a uniform force, falling on a uniform aggregate, must undergo dispersion; that falling on an aggregate made up of unlike parts, it must undergo dispersion from each part, as well as qualitative differentiations; that in proportion as the parts are unlike, these qualitative differentiations must be marked; that in proportion to the number of the parts, they must be numerous; that the secondary forces so produced must undergo further transformations while working equivalent transformations in the parts that change them; and similarly with the forces they generate. Thus the conclusions that a part-cause of evolution is the multiplication of effects, and that this increases in geometrical progression as the heterogeneity becomes greater, are not only to be established inductively, but are deducible from the deepest of all truths."

Modification, therefore, may properly include the results of the multiplication of effects. Furthermore, from a knowledge of the life history of the organic world, it is known that this change has been progressive, resulting in the evolution of the higher from the lower, of the complex from the simple, and of the definite from the indefinite.

It must now be asked, Is the amount of variation without limit or is it restricted within bounds which can be determined? As far as can be seen, the limitations of the forms of species of animals and plants end only with the aggregate number of possibilities within the functional scope of the organism. Beyond, in either direction, is death, and a passage from the organic into the inorganic. The restrictions of variation are chiefly those of temperature, pressure, motion, light, space, time, and matter. Within certain limits, these clearly bound the horizon of known possible life. Further, the material constitution of the organic world is naturally subject to ordinary mechanical and chemical laws.

If, instead of the preceding, general, and therefore rather abstract statements of the limits of variation, the subject is considered from the concrete, objective side, the limits between which are found all the variations actually presented by any character or set of characters, in the animal or the vegetable kingdoms, can at once be determined. The fact that the organic world can be divided into kingdoms, sub-kingdoms, classes, orders, etc., and definitions of the divisions given, in itself furnishes sufficient evidence that these have been the limits of organic change, at least under present terrestrial conditions. This does not imply that the phylogenies of groups of animals and plants do not converge and coalesce, and join larger and larger phyla in past ages, so that the gaps between unlike forms are gradually filled by complete series. It does, however, express the definite heterogeneity of the results of development.

For the sake of illustrating an extreme range of variation, it will be granted that the assemblage of characters by which a mammal is now recognized precludes mammalian variation into a cold-blooded, non-vertebrate, lungless animal. Likewise the mammalian skeleton cannot be siliceous or chitinous. Externally mammals may be smooth, hairy, scaly, or plated, but not feathered. There may be found numerous gradations from the smooth to the plated state, and a great range of variation in each type of epidermal structure. In vertebrate animals generally, the hair may vary in length, in fineness, in color and shape; it may form bristles, or spines, or feathers; and, as a skin character, it is related to horn-sheaths, hoofs, nails, claws, scales, and teeth. These constitute the limits of modification in epidermal or exoskeletal growths. The types are few, but the variety in each is almost infinite. The variation may be seen in individuals, but becomes greater in species, and increases still more in larger groups. The gradations are numerous between the hair of a Beaver and the spines of a Porcupine; between the horns of the Giraffe, Rhinoceros, and Antelope; between the nails of Man and the claws of the Carnivora; and between the teeth of a Dog-fish and those of a Tiger.

Definition of Terms.

In the beginning it is well to understand the meaning and extent of the terms included under the comprehensive word spine. In a general sense, *spine* is here used to cover any stiff, sharp-pointed process. A *prickle* is restricted in use to the small, sharp-pointed, conical projections which are purely cuticular; as in the Rose and Blackberry. A *thorn* is a sharp process on plants, usually representing a branch or stem. A *horn* is an excrescence on the head of certain animals, and is properly hollow. An *antler* is a solid bony process, usually deciduous, and generally confined to the male; as in the Deer or Elk. A *spur* is a term applied to the claw-like process on the legs and wings of some birds, and on the hind legs of *Ornithorhynchus* and *Echidna*.

The word spine, therefore, is most comprehensive, and is here intended to include the modified hairs of the *Echidna* and Porcupine; the sharp, prickly scales of the Horned Toad (*Phrynosoma*); the pointed spiniform projections on the shells of Mollusca; the spinous prominences on the test of Crustacea and insects; the fin spines as well as those on the opercula and scales of fishes; the generally movable processes of Echinodermata; the projecting rays and processes of Radiolaria, etc., etc. The vertebral column and also the processes from the separate vertebræ are known as spines, but as these are distinctly internal structures, they will not be considered in this connection.

In nearly all classes of organisms spines have been developed independently, and simply represent cases of parallel development of similar structures or morphological equivalents. They possess analogy of form without necessary homology of structure, and accordingly have no common phylogenetic connection. Therefore, if the relationships between the smooth and spinose forms belonging to any group of animals or plants can be traced, and the simplest and most primitive condition in each case, as well as the highest stage of progressive development, can be ascertained, their relative significance from an evolutionary standpoint may be confidently determined.

Growth of a Spine.

The growth of a spine is either direct an progressive, or indirect and regressive.   It is direct when it is developed by

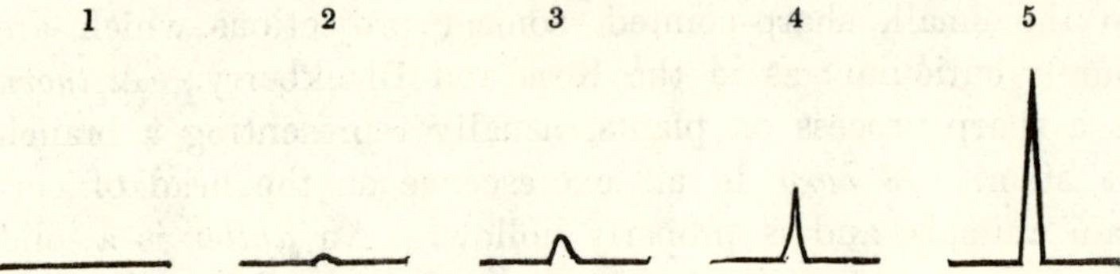

FIGURES 1–5. — Different stages in the growth of a spine.   1, plane surface; 2, slight elevation; 3, node; 4, short spine; 5, completed simple spine.

the addition of new tissue.   In this way growth is attained in the antlers of a Deer, the horns of a Cow, the ordinary spines of Brachiopoda, Mollusca, and Crustacea, and in other similar examples covering the majority of cases.   Growth is indirect, however, when the spine represents atrophy or suppression of an organ through the loss of its accessory parts; as in the thorns of the Locust and the Barberry, the spiniform termination of the stems of the Pear, or the spurs on the Python.

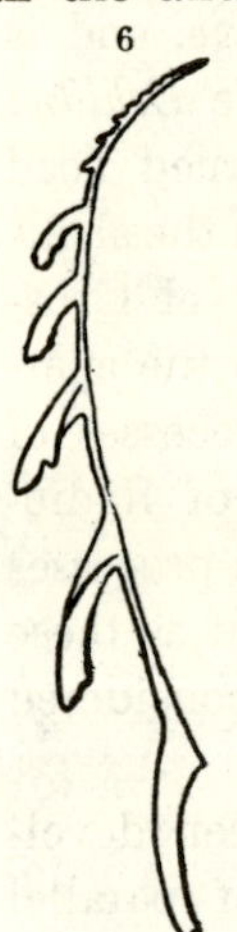

FIGURE 6. — A profile of a single radiating ridge of *Spondylus princeps;* showing the series of flattened spines.

The direct development of a spine is essentially the same process in all cases.   At a given point on the surface of an organism, there first appears a slight elevation, which becomes higher and higher, and is usually conical in form.   This cone represents the simplest type of spine; and among animals and plants most spines conform to this primitive pattern (figures 1–5).

Often there are various kinds of surface ornaments, which by growth and differentiation develop into spines.   By rhythmic, alternating areas of accelerated or retarded growth, the concentric laminæ on many molluscs may produce spines, as shown in figure 26.   In the same way the radiating ridges may be

diversified into a row of spines, as represented in figure 6.
Further, the surface may be reticulate, with longitudinal
and transverse lines, and at the points of intersection, nodes
and often spines are formed after the manner shown in
figures 7–12. The longitudinal or vertical lines may become
obsolete, leaving the spines to be borne on the transverse or

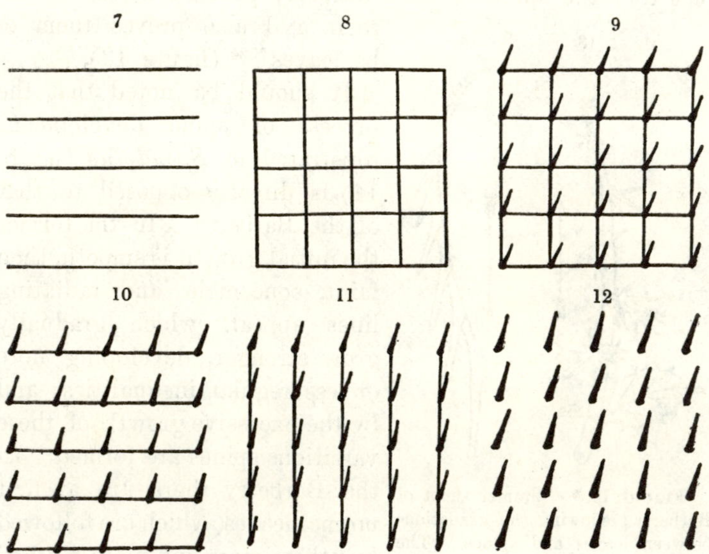

FIGURES 7–12. — Diagrams; showing growth and differentiation of ornament
into spines. 7, surface with parallel lines; 8, surface with regular reticulate lines;
9, same, with spines developed at the points of intersection; 10, same, with the
vertical lines obsolete, but still represented by the vertical rows of spines; 11,
same, with the horizontal lines obsolete, but still represented by the horizontal
arrangement of the spines; 12, same, with all lines obsolete, but both series
represented by the vertical and horizontal arrangement of the spines.

horizontal lines (figure 10). In other cases the horizontal
lines disappear, leaving the spines on the vertical lines
(figure 11). Finally, both horizontal and vertical lines become obsolete, and then only the spines remain, as shown in
figure 12.

The indirect production of spines is not always evident, for
if the ontogeny or phylogeny of the individual is unknown,

its direct or indirect development cannot be determined.   An excellent example of indirect, or regressive, growth of spines is afforded in the common Barberry (*Berberis vulgaris*), on the summer shoots of which are shown most of the gradations "between the ordinary leaves, with sharp bristly teeth, and leaves which are reduced to a branching spine or thorn.   The fact that the spines of the Barberry produce a leaf-bud in their axil also proves them to be leaves "[24] (figure 13).

It should be noted that the process of spine development illustrated in *Spondylus* (figure 14) is directly opposed to that of the Barberry.   In the former the initial growth is smooth, then faint, concentric, and radiating lines appear, which gradually grow stronger, developing more or less regular inequalities; and by the excessive growth of these variations spines are formed.   In the Barberry there are at first normal leaves, which are followed by others more and more toothed and bristly, until the leaf is represented by a branching spine, while finally spines only are formed.   The *Spondylus* represents a progressive increase in growth to produce the spines, while the Barberry exhibits a progressive decrease of growth, or an "ebbing vitality," as it has been termed by Geddes.[20]

FIGURE 13. — Summer shoot of Barberry; showing the gradations between leaves and spines.   The arrow indicates the direction of growth. (After Gray.)

FIGURE 14. — Profile of one of the primary rays of *Spondylus imperialis;* showing the series of spines.   The arrow indicates the direction of growth.

The spines are the final results of both the direct and indirect modes of production; the direct, through a process of building on new tissue, and the indirect, through a process of dwindling away to all but the axial elements.   These differences are graphically expressed in figures 13 and 14.

Attention should be called to the four kinds of spine production in different organisms. (1) In the Radiolaria, Echinoidea, the Giraffe, Cattle, and the Rhinoceros, the spines or horns are persistent, and grow by additions to the original structure. The new tissue may be superficial, sub-terficial, interstitial, or formed by synchronous resorption and growth. (2) In the Crustacea and Articulata generally, and in the Deer, Elk, etc., the spines are moulted, or shed, periodically. In their various stages, these types (1 and 2)

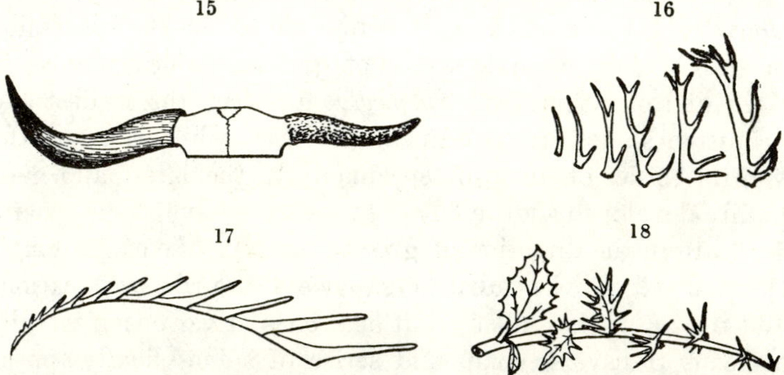

FIGURE 15. — Example of spine growth by simple increscence. Horn (left) and horn-core (right) of Ox. (After Owen.)

FIGURE 16. — Stages of spine growth by successive replacement. Antler series of Red Deer, at ages of 1, 2, 3, etc., years. (After Owen.)

FIGURE 17. — Stages of spine growth by serial repetition. Profile of a series of spines on one of the primary radii of *Spondylus imperialis*.

FIGURE 18. — Stages of spine growth by decrescence. Transformation of leaves into spines in *Berberis vulgaris*. (After Gray.)

can be studied only by means of separate specimens con-secutive in age, or by observing the metamorphoses in one individual. (3) In the shells of Brachiopoda and Mollusca, the stages of growth of the individual are generally retained throughout life, and the successive development of spines may be studied, therefore, in a single example. (4) Spines produced by suppression, as in the Barberry, express their origin through a series of gradations between separate parts; while in others suppression is brought about by the loss of structures.

The first type mentioned develops horns or spines by simple increscence (figure 15); for example, the Ox: the second, by successive replacement (figure 16); as in the Deer: the third, by serial repetition (figure 17); for example, *Spondylus*: the fourth, by decrescence (figure 18); for example, the Barberry.

Localized Stages of Growth. — By the multiplication of surface ornaments through the process of interpolation, many Mollusca present stages of spine development in two directions. (1) The normal series is represented by the succession of spines along a single sector of growth. For instance, in the radial plications of a *Spondylus* or *Lima*, the earliest and primitive spines are found near the beak, while those on the ventral border of an adult specimen are the latest and most highly developed (figure 30). These successive stages, therefore, are in the direction of growth, and may be called longitudinal. (2) By the radial divergence of the ribs or plications and the interpolation of additional ones at various intervals, as many transverse compound series of spines finally appear along the periphery as there are primary radii. Hence, in a given case, there may be two radii continuing to the beak, then by interpolation there are successively 5, 11, 23, etc., radii, the highest number being found at the periphery (figures 19, 20). Moreover, by taking the distal spines on these 23 rows, there result the same stages of spine development as shown in the longitudinal series along any primitive plication (figure 20). A pelecypod shell like *Spondylus* is here used to illustrate this process, but the application may also be made to the Brachiopoda as well as to the conical non-coiled Gastropoda. In a coiled form like a cephalopod or an ordinary gastropod, the longitudinal lines would follow the whorls spirally, and the transverse lines would correspond to the lines or increments of growth of the shell. Species in which the radii are all introduced at an early stage of growth (many species of *Cardium, Pecten, Lima*) or in which the radii multiply by regular dichotomy would show, of course, only the longitudinal series, for at the margin of the shell the radii would be of the same size and age, and the spines uniform.

The foregoing example illustrates an important principle

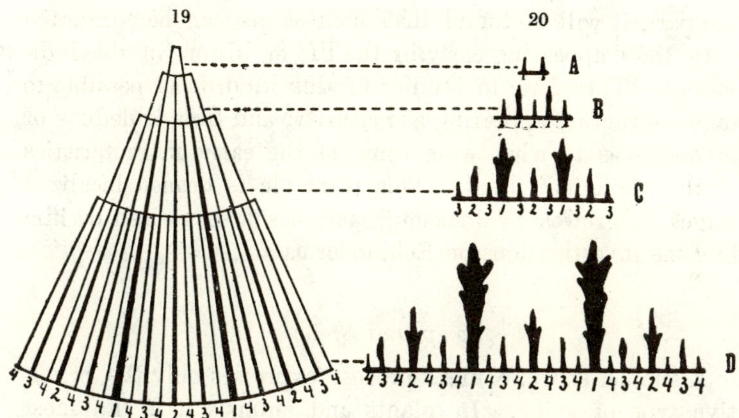

FIGURE 19. — Sector; showing in diagram the multiplication of radiating lines by interpolation. The two primary radii (1, 1) are the only ones continuing through the whole four zones. The first zone has 2 radii; the second, 5; the third, 11; and the fourth, 23.

FIGURE 20. — Profiles of the spines produced on the various radii at the four zones; as indicated in the preceding figure. A, the spines on the two primary radii of the first zone; B, the spines on the second zone, showing the growth of those on the two primary radii (1, 1), and the small spines on the newly interpolated radii (2, 2, etc.); C, the spines on the radii in the third zone; D, the spines at the bottom of the fourth zone. The two large compound spines are on the two primary radii. Their development may be traced by following them through A, B, C, to D. The next three longest spines (2, 2, 2) are tricuspid, and represent the stage of spine development attained by the spines on the radii which were interpolated on the second zone. The next six smaller spines (3, 3, 3, etc.) are on radii which were introduced on the third zone. The twelve small spines (4, 4, 4, etc.) are on the radii introduced on the fourth zone. Thus there are four stages of spine growth shown on the lower margin of the fourth zone, and these correspond to the four stages exhibited by the series of spines on one of the primary radii running through the four zones.

of ontogeny; namely, that in organisms which repeat various parts during their growth, these parts will develop or pass through a series of stages corresponding to the initial and subsequent stages of the parts repeated. In this way structures appearing late in the ontogeny of the individual will present primitive infantile and adolescent characters. Further development, if such takes place, will pass through a progressive series of ontogenetic changes, and if the stages of growth are by serial repetition and thus are retained in the part, it will be found that such stages can be correlated with those appearing early in the life or history of the individual. Therefore, in studies of this kind, it is possible to take a structure appearing at maturity, and from it deduce or predicate as to what were some of the early characteristics of the whole individual. This principle is termed localized stages of growth by Jackson,[37] and was first noticed by him in some investigations on Echinodermata.

Compound Spines.

A simple, sharp, conical process expresses only the primitive type of spine. In plants and animals it is the most

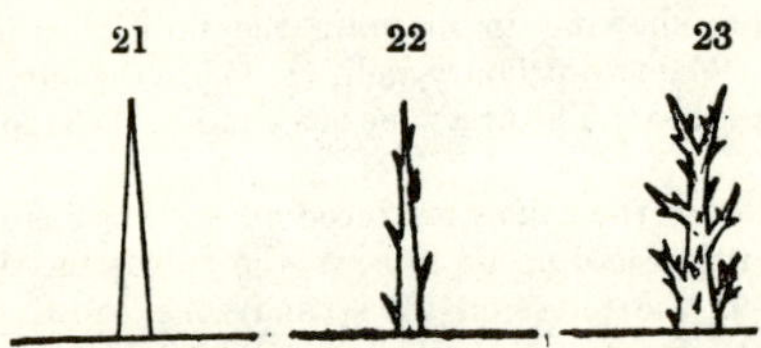

FIGURE 21. — Simple spine.
FIGURE 22. — Spine, with lateral spinules.
FIGURE 23. — Spine, with forked apex and lateral spinulose spinules.

common form found, and is the first stage of spine differentiation. From this type the myriad forms of spines known in the organic world are produced by almost insensible gradations. It is needless to attempt a detailed description of this infinite variety; but, as a single illustration, some of the leading forms of spine differentiation among the Radiolaria are here shown (Plate I). These figures are taken from Haeckel's "Report on the Radiolaria,"[26] and generally represent enlargements of from 100 to 400 diameters. Probably no other class of organisms presents greater variety, and many of the forms are repeated again and again, not only in various species of this group, but elsewhere both in the animal and vegetable kingdoms.

Whenever the development of a compound spine can be studied, it shows a gradual progress from the simple to the complex (figures 21–23).   The antlers of the Red Deer (*Cervus elaphus*) furnish a familiar example.   Fawns of the first year have antlers with only a single prong, a short front tine being added the second year; then "year by year as they are renewed they acquire a greater and still greater number of tines and branches, till they finally attain the complete stage, when their owner is termed a ' royal hart ' "[44] (figure 16).   Although somewhat conventionalized, the primary series of spines on the *Spondylus* shown in figure 20 exhibits the passage from simple to compound forms.   An inspection of many species of *Murex* will show the stages in series presenting a greater complexity.

After spine development has reached its maximum growth and differentiation, evidence of old age may be exhibited in two ways: (*a*) The spines may be reduced by resorption, decay, or abrasion, and finally become obsolescent; or what is of greater import (*b*), they may gradually cease to be developed, as is especially shown in organisms in which spine growth is by serial repetition.   Thus, in *Spondylus calcifer*, a young individual measuring about two inches across has marginal spines fully an inch in length.   Even longer spines are found when the shell reaches a width of four inches.   On attaining a maximum diameter of about six inches, spine growth gradually ceases, and the margin of the valves is entire and nearly smooth.   At this stage shell secretion is confined to excessive thickening of the valves.   These senile stages of spine growth will receive further consideration under the discussion of ontogeny and phylogeny of spinous species.

Application of Law of Morphogenesis. — The manner in which spines arise from plane surfaces, or from the growth or modification of superficial structures, and also through the decadence of organs, has now been noticed. The spine may thus be taken as a unit for comparison, and its various stages of growth, which were shown to have a definite sequence, may be used in correlation to determine relatively the degree of spine specialization attained by any organism. Furthermore, enough data have been already given to lead to the suspicion that spines may represent the limits of ornamental or superficial differentiation or variation. At this point in the discussion this statement must be considered as more suggestive than conclusive. The proof of its reality will be more clearly shown later on.

Ontogeny of a Spinose Individual.

With few exceptions the embryonic and larval stages of all organisms are devoid of specialized surface features. In other words they are without ornament and without weapons. The exceptions to this rule seem to be readily explained under the principles of larval adaptations and accelerated development. Cases of the latter kind, therefore, can hardly be considered as exceptions, since they represent, not real larval features, but former adult characters which have been pushed back or which develop earlier so as to appear eventually in the larval or later embryonic stages. In the very earliest stages of embryonic development, the truth of the first statement becomes obvious, and accordingly the protembryonic, mesembryonic, metembryonic, neoembryonic, and typembryonic stages are without surface ornaments or spines.

Among Mollusca, the protoconch, periconch, and prodissoconch, or the early larval shells, are smooth and without ornament. Even the prodissoconch of very highly spinose species, as in *Spondylus*, is as smooth as that of the plainest species of *Ostrea*, *Anomia*, *Avicula*, etc. Likewise, the protoconch of the most specialized or most retrograde cephalopod is perfectly plain. In the nepionic stages the spiny *Murex* is without spines. In the Brachiopoda the protegulum, or early larval shell, is always without sculpture; while the nauplius of Crustacea and the protaspis of Trilobita are generally spineless. The young of horned vertebrates are almost universally hornless, the Giraffe being the only mammal born with horns. The very young seedlings of plants are likewise spineless. In insects the embryonic stages generally have simple cuticles, but in the larval stages of this class and the Crustacea, a great variety of spines and ornamental characters is developed. Altogether, it may be asserted that spines do not appear during the embryonic stages of animals and plants, and that their initial development is commonly post-larval.

Examples illustrating the ontogeny of a spinose form could be multiplied indefinitely, and taken from nearly every class of organisms. In all cases practically the same sequence of events relating to the development of spines would be found. The organism would first be smooth, without sculpture or ornament, like the young of other organisms. At some stage of the ontogeny the beginnings of spines would appear, and develop first into simple, and later, according to the stage of differentiation attained, into compound spines. This progression would finally reach the maximum, spine growth would cease, and the surface of the organism would inversely revert to an early and more primitive type without spines. Normally these changes would represent the infantile, adolescent, mature, and early and late senile periods of the life of the organism. In some cases, however, the stages of spine growth, or acanthogeny, do not agree with the ontogeny of the entire individual in respect to time, and here acceleration and the phylogeny of the species will be found to offer the proper explanation of the divergence.

As simple examples of the ontogeny of spiniferous species, the Mollusca afford especial advantages, owing to the fact already noticed, that the stages of development are commonly preserved in a single individual. In figure 24 the larval shell, or prodissoconch, of Pelecypoda, or bivalve shells, is represented, and shows the usual type throughout a large portion of the class. The succeeding shell growth of the dissoconch is at first generally smooth, save for the fine concentric lines of growth (figure 25). In ornamented or spinose species, however, irregularities in the growth lines soon appear (figures 26, 27), and these shortly assume the characteristic surface sculpture of the normal adult. Thus the prodissoconch of *Avicula sterna* is represented at *p*, figure 25, and is followed by regular concentric growth during the nepionic

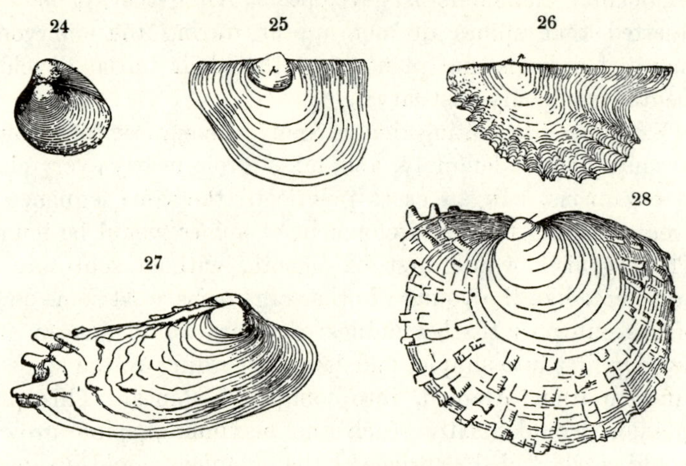

FIGURE 24. — Prodissoconch of *Ostrea virginiana*. × 43.
FIGURE 25. — Each stage of *Avicula sterna; p*, prodissoconch. × 19.
FIGURE 26. — Young *Avicula sterna;* showing the beginning of spine growth. × 3.
FIGURE 27. — Young *Saxicava arctica*. × 19.
FIGURE 28. — Young *Anomia aculeata;* prodissoconch succeeded by early smooth and later spinous dissoconch growth. × 30. (Figures 24–28 after Jackson.)

stages. In figure 26 the spiny characters of early adolescence are added to the previous stages, and in later stages the spines become more and more emphatic.

In *Spondylus* the prodissoconch is the same simple form, and is succeeded by a nearly smooth Pecten-like stage, during which the animal was free (figure 29). After fixation the growth is very irregular and ostræiform for a time, until the shell rises above the object of support, when all the most characteristic features of surface ornamentation become fully developed (figure 30). As the shells approach maximum growth, the spines gradually become shorter, and in old age none are developed, even those of early growth being removed by the action of boring animals and by solution (figure 31).

It seems unnecessary to increase the number of examples showing the ontogeny of spinose individuals. The Deer and

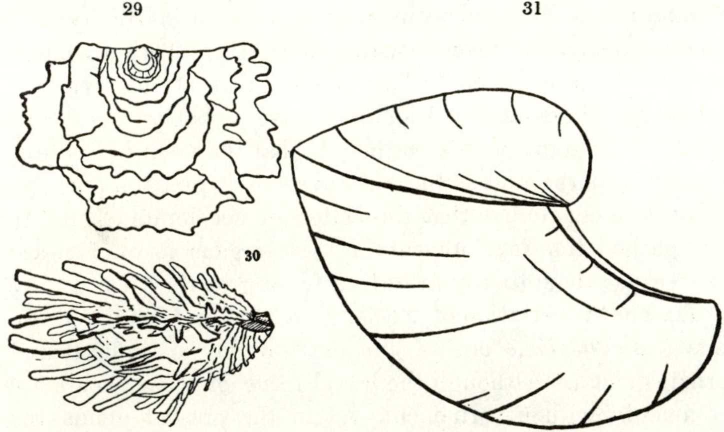

FIGURE 29. — Young *Spondylus princeps*. Right valve; showing pecteniform stage succeeded by ostræiform growth. Taken from apex of adult specimen; presented by R. T. Jackson. × 3.

FIGURE 30. — Side view of *Spondylus calcifer*, about one-third grown; showing the characteristic spinous growth. ½.

FIGURE 31. — Side view of *Spondylus calcifer*; showing the greatly thickened right valve and the entire absence of spines over the whole shell. ¼.

the Ox may be again cited in this connection. Both are born without horns, but during adolescence the antlers of the Deer become longer and more complicated with each renewal, while the horns of the Ox are longer and more twisted. In old age, when the Deer has passed his prime, the antlers are more obtuse, and exhibit a tendency toward decline and obliteration. Suppression of the antlers is accomplished by the removal of the cause of antler growth and specialization, so that the unsexing of the male results in small antlers, which are seldom branched, and become thickened by irregular deposits of bone (Owen [53]). Spines grow during the adolescence of the Horseshoe Crab, *Limulus polyphemus*, yet in old age they are obsolescent, being represented by rounded nodes.

As examples illustrating the accelerated development of spines in widely separated classes, the Giraffe among mammals and *Acidaspis* among Arthropoda may be selected. The Giraffe represents the continuance of a very primitive type of horn; namely, one covered with a hairy skin. They are never shed, and are common to both sexes. Out of this type all others found among the Mammalia have probably been developed. The point of interest here is that the young Giraffe is born with horns, and as these could serve no prenatal purpose, it must be concluded that the action of accelerated heredity has pushed the development of these organs so far forward as to cause them to appear during fœtal growth.

The next illustration of acceleration is taken from the Trilobita. *Acidaspis* is one of the most highly specialized and ornate genera. Although the larval forms of other genera are commonly without ornament, yet in the present genus the protaspis, or phylembryonic, stage partakes of this specialization in so far as to develop minute spines, which later become larger, more differentiated, and form a conspicuous feature of the adult. Other characters have been likewise shown to appear at an earlier period than in other genera, and the earlier inheritance of spines must be explained in the same manner.[8]

The facts, as stated, seem to warrant the conclusion, that in spinose organisms the very young are almost universally without spines. Acceleration may occasionally push their development into the embryonic and larval stages, but ordinarily they are not so subject to the action of this law as are some of the physiological and other structural characters. This will be explained as in part due to the lack of general plasticity, and because differentiated spine growth is the progressive limit of variation. Therefore there are no subsequent characters to displace them and crowd them forward in the ontogeny.

Phylogeny of Spinous Forms.

To interpret phylogeny in terms of ontogeny, according to the law of morphogenesis, or recapitulation, is perhaps easier than to trace a genetic sequence through a series of forms having a considerable geologic range. Taking the ontogenies of the animals already noticed, there is for the Pelecypoda the prodissoconch, which is correlated by Jackson[36] with *Nucula*, and a Lower Silurian nuculoid radicle is assumed for the Aviculidæ and allied forms. The first dissoconch growth produces a shell resembling *Rhombopteria*, a Lower and Upper Silurian type, and this is taken to represent the second stage in the phylogeny of *Avicula* (figure 25), *Anomia*, *Spondylus*, etc. Continuing the development of *Spondylus*, it .is found by Jackson that it passes successively through stages which may be correlated with *Pterinopecten* (Devonian), *Aviculopecten* (Devonian), *Pecten* (Carboniferous?), and *Hinnites* (Trias), while finally it assumes true spondyliform characters. These correlations agree with the geologic sequence of the genera, and are believed to indicate phylogenetic relationships. It may be further remarked that the early species of Spondyli are more truly pecteniform and hinnitiform than the later ones. The genus ranges from the Trias to the present. Zittel[73] remarks that "the oldest species are small, thin-shelled, and seldom much ornamented." Even in the Cretaceous, the majority of species are not far removed from *Pecten* and *Hinnites*. During the Tertiary the irregular, ostræiform, squamous, concentric, and spinous growth becomes more manifest, and at present most of the species show a great development and differentiation of the spines.

Thus, while *Spondylus* is normally considered as a spinose genus and the species are familiarly known as Spiny Oysters, yet, as it is traced back in geological history, the forms become less and less spinose, and their affinities and appearances are more and more in accord with non-spinose genera, until finally the prototype is a smooth, simple, delicate, unornamented shell.

'The simple antlers of the young Deer and Elk correspond in type with those of the adults of the Middle Tertiary Deer (Lydekker [44]), and it may be therefore assumed that the great number of branches and tines is a modern development. Further back in the Tertiary the ancestors of the Deer were without antlers, thus representing in phylogeny the new-born Deer of the living type. These correlations are made from comparisons of chronogenesis, or development in time, and ontogenesis, or development in the individual.

An example of a different kind will now be given to show more clearly a genetic sequence in forms. Among the Brachiopoda, *Atrypa hystrix* represents one of the terminal members or species of a line of varietal and specific differentiation, extending through the Silurian and Devonian. The type commonly known as *Atrypa reticularis* appears to have had its inception during the Ordovician; yet in the Silurian it is found as a conspicuous and fully developed form. Here, also, it has quite a wide range of variation, but there seems to be an insensible gradation between the extremes, which therefore cannot be considered as definite permanent varieties. There are, however, associated forms that have received distinctive specific names, which do not shade into each other. During the early and middle Devonian certain of these variations in the main stock of *A. reticularis* became more fixed, and at the time of the Hamilton sediments in New York, there are two forms known as *A. reticularis* and *A. aspera*, which apparently do not pass into each other. As time went on, these two types became more specialized and the divergence correspondingly increased, until in the Upper Devonian, in the Chemung sediments, there is a large many-plicated *A. reticularis*, as well as a form with very few plications and long marginal spines, *A. hystrix*. Hall and Clarke [31] thus summarize the stages leading to the formation of the spinose forms: " In the variant of *Atrypa reticularis*, occurring in the Niagara fauna at Waldron, Indiana, the free concentric lamellæ frequently show a tendency to fold inward at the summit of the principal plications. The infolded edges fail to unite, and this tendency to the formation of tubules is apparently carried no further at this period. More extreme results were attained by the *Atrypa aspera* of the Hamilton shales, or possibly by its migrated ancestor, during the period of time represented by the deposition of the Lower Helderberg, Oriskany, and Upper Helderberg sediments. At all events, the *Atrypa spinosa* of the Hamilton shales is but an *A. aspera* with the lamellæ enfolded into tubular spines. Intermediate stages connecting these different phases are not present in this fauna. . . . This spinose form is continued into the Chemung faunas (*A. hystrix*), with some modification of expression, the spines being few and long, and the plication of the surface very coarse and quite simple; the shell in its decline thus representing a decided return to the primitive type of structure." H. S. Williams [72] has classified the variations in the stock of *A. reticularis* as to whether differentiation in the number of plications is increased or retarded, and concludes that the extremes are most strongly expressed at the close of the life-period of the race. The numerously plicated type represents the accelerated phase of the multiplication of radii, while *A. hystrix*, with its few and coarse radii, represents the retardation or suppression of this tendency.

The only great group of animals receiving its name from its characteristically spinose surface is the Echinodermata, or the spiny-skinned animals; yet it is extremely doubtful whether this name would have been used had the first studies of the group been based upon the Paleozoic representatives, especially the pre-Devonian species. The early Sea-lilies (Crinoidea), Cystideans (Cystoidea), Blastoids (Blastoidea), and Star-fishes (Asteroidea) had smooth or nearly smooth integuments. In its early genera, even the most typically spiny class of the whole sub-kingdom, the Echinoidea (Sea-urchins), had very minute and insignificant spines. It is only in the late Devonian and in the Carboniferous that truly spiny forms of Crinoids, Star-fishes, and Sea-urchins are found.

Of equal significance is the fact that the Echinodermata together with the plants represent the most primitive type of structure, one in which there is a more or less circular arrangement of the parts or organs. The Echinodermata are the highest development in this line of growth among animals. They culminated in past geological ages, and from them no direct line of descent can be traced (Bailey[2] and Cope[11]).

The conclusion from the study of the phylogenies of spinose forms is parallel to the one drawn from the ontogenies; namely, that the ancestors of spinose as well as non-spinose organisms were simple and inornate.

Categories of Origin

As previously shown, spines are formed either by growth or by suppression, and therefore the processes determining their production are either constructive through concrescence or destructive through decrescence. Each of these is in turn determined by forces from without the organism (extrinsic) or by forces from within (intrinsic). In this connection it is of no especial moment whether or not the intrinsic forces are primary or are an immediate or subsequent reflex from the extrinsic. The main thing is the direction of the dominant force, whether centripetal or centrifugal. If in some cases it can be shown that spine development has been accomplished by intrinsic forces in the organism, then this development may be brought about independently of the environment and possibly at variance with it. Also, if in other cases the extrinsic forces or the influences of the environment have caused spine growth, it may in some instances illustrate the formation and transmission of an acquired character, or at least the operation of organic selection.

The point has now been reached where it is impracticable to make a rigid classification of the direct factors or an exact determination of primary and secondary causes. It was remarked at the beginning of this paper, that single causes were not sufficient in every case to account for spine growth, and while it is comparatively easy to formulate abstract expressions or terms covering all possible cases, it will be found difficult to construe properly certain factors to fit into any particular conception. In illustration of this, the foregoing statements may be taken. Thus spines are formed by the only means possible, either by growth of new tissue or by decrease in old. Again, the forces must act from the interior or from the exterior; in other words, they must be intrinsic or extrinsic. But in some specific instance, while considering food, forces of nutrition, external or internal demands, reactions, etc., a question may arise as to the proper disposition to make of a spine developing primarily by external stimuli and becoming a defence and secondarily a weapon; yet which by differentiation in time loses some of its protective and offensive qualities, and by selection may be confined to one sex.

Growth and decline are underlain by the processes taking place in individual cells as well as in aggregates of cells, for spine growth must be considered in unicellular as well as multicellular organisms.

Ryder[61] has very philosophically discussed the correlations of volumes and surfaces of organisms, and has reached the conclusion that " the physiological function of a cell is also a function of its figure, i. e., of its morphological character; that is to say, cells tend to elongate in the direction of the exercise of their function." Out of this may be deduced the correlative conclusion that aggregates of cells having a like function also tend to elongate in the direction of the exercise of this function; and, further, it may be asserted that parts or portions of cells will act in the same manner.

A familiar illustration of these principles as applied to a single cell may be taken from the rhizopod *Amœba proteus*. When disturbed by incident forces in all directions, it assumes a globular form. Under continuous motion of its own, it is elongated in the axis of motion, its larger pseudopodia being thrust out in more or less the same direction. The presence of a favorable exciting cause, like a particle of food, produces extension of the protoplasm to envelop it.

Furthermore, as is well known, continuous extra-pressure on any part of an organism produces atrophy and absorption, and intermittent or occasional pressure causes hypertrophy and growth. That the pressure should be intermittent seems a necessary condition for hypertrophy, in order that the parts affected may have normal intervals allowing the active exercise of nutrition.[55] This may be regarded as a parallel statement of the law of disuse and use; the former causing organs or parts to dwindle away and lose their function, and the latter producing increased nutrition and growth.

This ratio of exchange between nutrition and waste is on the side of full or excessive cell-nutrition, producing growth in the parts affected, while deficiency of nutrition produces decline or suppression. If the successive increment constituting growth is along definite progressive lines towards higher structures, and the decrement affects the decline of useless parts or permits of the replacement of a lower by a higher structure, then the sum of the changes is progressive evolution.*

Growth, as stated, seems to require normal intervals for the proper exercise of nutrition, which involves an intermittence of the exciting or stimulating forces. Rhythm has been shown by Spencer[66] to be a necessary characteristic of all motion, and therefore in considering either the intrinsic or extrinsic forces acting on the structures of an organism, they must be rhythmic or intermittent. In the environment the most apparent changes are those of light and darkness, heat and cold, moisture and dryness, and variations in amount of oxygen, all of which affect an organism directly, and also through the accompanying variations in the character and amount of the food supply, the number of enemies, etc. These and most of the mechanical forces of the environment are therefore intermittent, and their resultant must have a

---

* This is very near Cope's idea of progressive evolution.

definite tendency, so that the effects are not with each change successively positive and negative to the same degree; that is, the same structures or adjustments are not alternately made and unmade.

It is generally recognized that there is a necessity for a force or energy in living organisms, which is not the immediate and direct result of external agencies, but upon which these fall and produce reactions. It is considered as a phase or kind of vital force directing growth, and therefore a growth force, or the bathmic force of Cope.[10] The internal energy of growth, involving the capacity or effort of responding to external stimuli, is termed entergogenic energy by Hyatt.[34] Without this power an organism would be unable to move or respond to external stimuli. The effect of the action of this kind of energy must be the resultant between "the structures already existent in the organism and the external forces themselves."[34] Since the growth force is within the organism, or inborn, it is one of the principal characters transmitted through heredity, and if it is in excess of the external forces, the modifications will be principally congenital or phylogenic. If, on the other hand, the external forces predominate, the modifications will be principally adaptive, or ontogenic. In each case the resultant is the actual visible effect of the two. If both are toward the establishment of similar structures, their effect will be the sum of the two; but if they are opposed to each other, the effect will be their resultant, the nature of which, as seen above, will depend upon their relative power.

These conclusions can be correlated directly with the developmental variations occurring in the life history of any great group of organisms. Any one who has studied the chronological development or the phylogeny of a class of forms cannot fail to have been impressed with the fact that all types of life are physiologically more plastic or subject to greater changes near their point of origin. That is, the maximum of generic, family, and ordinal differentiation is found at an early period, while the greatest specific differentiation occurs at a later period. This shows that the results of variation at first affect the physiological and internal structures, and that later the changes are mainly physical and peripheral.

One explanation of this would be that the forces of the environment are at first freely transmitted and produce internal modifications, and that later these characters become stable, making the effects of the external stimuli apparent in the superficial differentiation of the organisms.

In any event the modifications in function and structure are followed by modifications in surface, showing that the more important physiological and structural variations are the first to be subjected to heredity and natural selection, which tend to fix or hold them in check. Features of less functional importance, as peripheral characters, are the last to be controlled, and therefore present the greatest diversity, while in this diversity spinosity is the limit of progress. In order to be hereditable, the modifications through the environment must have induced correlative internal adjustments and changed forces which can be transmitted to offspring, and they in turn reproduce the specific modifications.

For the purpose of illustrating these statements, the evolution of the Brachiopoda and Trilobita will be taken. The Brachiopoda are divided into four orders, all of which appear in the Lower Cambrian and continue to the present time. Schuchert[64] states that "of the 49 families and subfamilies constituting the class, 43 became differentiated in the Paleozoic, and of these 30 disappeared with it ;" also, "of the 327 genera now in use, 227 had their origin in Paleozoic seas, or nearly 70 per cent of the entire class." Throughout the Cambrian, "differentiation was mainly of family importance." "Differentiation is most rapid near the base of the older systems, and diminishes the force from the older to the younger geologic divisions." The most rapid increase was in the Ordovician, the culmination was in the Devonian, and the rapid decline came with the Carboniferous. About six thousand species are known, and of these probably not more than one hundred and fifty are living.

Similar data are derived from the Trilobita. This group is found all through the Paleozoic, at the close of which it became extinct. Two of the three orders are found in the Lower Cambrian. The remaining order appeared just after the close of the Cambrian in the early Ordovician, yet through the whole of the remaining sediments not a single new ordinal type was developed. When applied to a single order, the same truth comes out. The order Proparia is one whose entire history can be traced, extending from the Ordovician through the Silurian and Devonian. All the families appear in the Ordovician; in fact not a single family type in this or the other orders was produced during the whole Silurian, Devonian, and Carboniferous.[6]

As the classes, orders, and families are based upon the physiological and important functional structural characters or differences, it is evident that at or near the beginnings of their life history is found the demonstration of the domination of phylogenic over ontogenic characters.

Conditions or Forces affecting Growth. — Since spines are purely organic structures, their production must follow the general laws of organic change. The forces considered as of most consequence are two: (1) the external stimuli from the environment, and (2) the energy of growth force. These, with their opposites (1 a) the restraint of the environment, and (2 a) the deficiency of growth force, are believed to include the chief active and passive causes, not only of spine production, but of growth and decline in general. Correlating these four causes with their constructive and destructive agencies, together with their extrinsic and intrinsic modes of action, as previously explained, there result (A) the external stimuli of the environment as an extrinsic cause of concrescence; (B) energy of growth force as an intrinsic cause of concrescence; (C) external restraint as an extrinsic cause of decrescence; and (D) deficiency of energy of growth force as an intrinsic cause of decrescence. The remaining vital forces (nerve force, or neurism, and thought force, or phrenism) are not primary, and, although doubtless affecting growth in higher organisms, cannot be original causes applicable to all forms of life, both plant and animal.

In tabular form, the divisions and relationships of the factors of spine genesis may be expressed as follows:

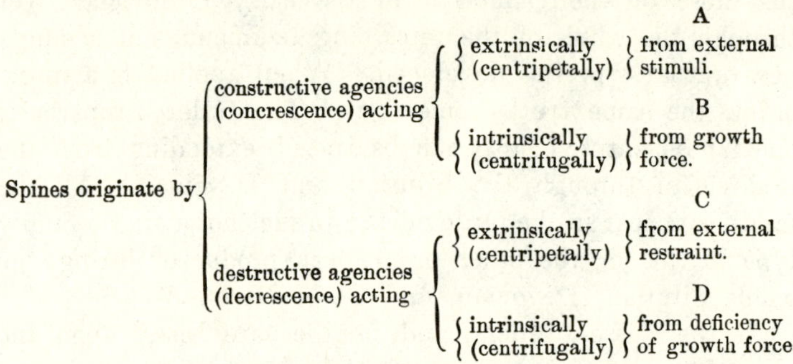

Under the last four divisions (A–D) it is proposed to discuss the origin of spines, and from the observations made, to derive certain conclusions regarding the significance of the spinose condition.

A. *External Stimuli.*

Under external stimuli are included all the forces of the environment (chemical, physical, organic, and inorganic) which, through their impact or influence on an organism, produce a consonant favorable change or disturbance. In general, it will be seen that the number of impressions and their power will depend largely upon the position and character of the surface upon which they impinge. The more exposed the position, the greater will be their strength and number, and if these stimuli or impressions are intermittent, and not so violent as to produce waste and rupture, growth will ensue. Under ordinary conditions, exposed parts will naturally be the first to receive sufficient stimulus to produce growth, and there will be normally a direct correlation between growth and stimulus. In a simple diagrammatic form, this would be expressed by a series of lines, the first representing a plane surface. Then, owing to the impossibility of maintaining a uniformly intermittent stimulus or a uniform response, some point or spot on this surface would grow in excess of the others. This difference would be augmented by the more favorable position of the spot to receive stimuli, further growth would take place, the growth force decreasing with the increase of distance, and the final action of these forces, stimulus and growth, would be to produce a pointed elevation. Such structures or outgrowths, especially when made of hard rigid tissue, would be termed spines under the general definition. The spine may be viewed as an attached organism, and its conical habit of growth would then conform to the law of radial symmetry, as determined by the physiological reaction from equal radial exposure to the environment. That all the irregularities of contour in all organisms have not developed into pointed processes or spines is not, therefore, the fault of the simple reciprocity between growth and external stimuli. This kind of development, however, requires a direct and immediate responsive external growth to the exciting force, which from various causes is frequently absent. Obviously, stimuli which result simply in motion or equivalent internal adjustments can have no effect toward spine production, so that only the results of such stimuli as bring about some accompaniment of superficial growth will be considered.

With the exception of perfectly spherical, freely moving forms, all organisms have certain parts which are more exposed to the forces of the environment than others, and from the principles already enunciated, such exposed parts under normal conditions will grow. This growth in the direction of function and stimulus, when acted upon by the hereditary functional and structural requirements of the organism, serves to produce the various external organs and appendages. But when the surface upon which the stimuli fall is not thus predetermined by heredity to grow into a certain organ or functional part, there results a normal responsive action between growth and stimulus, which, as already seen, tends to produce a conical or spiniform growth.

Under ordinary favorable conditions, simple external stimuli acting blindly through no agencies of selection would develop spines on all the most exposed parts, and tend to differentiate ornamental features. This has been the case with many organisms and colonial aggregates possessing no power of selection or not acted upon by any forces of determination, conscious or unconscious. In such cases spines may or may not serve for protection, and their function, if any, can be only determined separately for each case. If, however, the added function of offence is included, it is manifest that the spines must be located in special positions adapted to use for offensive purposes, as on the tails of some animals, and not necessarily over vulnerable parts. Here the selective agency of special adaptation is shown. Again, if while there is agreement in other essential characters, spines or horns are confined to either sex, it is evidently a case of sexual selection. Further, if they develop in harmony with the environment, or in a manner parallel to similar features of other organisms, it is through the operation of physical selection.

Altogether, under the general forces of external stimuli, there are five aspects in which to consider the production and growth of spines; namely,

A. *From External Stimuli.*

A 1. — In response to stimuli from the environment acting on the most exposed parts.

A 2. — As extreme results of progressive differentiation of ornaments.

A 3. — Secondarily as a means of defence and offence.

A 4. — Secondarily from sexual selection.

A 5. — Secondarily from mimetic influences.

B. *Growth Force.*

In unicellular organisms growth force, or bathmetic energy, must reside wholly in the germ cell, and therefore is concerned with reproduction as well as with cell differentiation.

In multicellular organisms the growth force is in both germ and soma cells, and its relative strength seems to depend upon its power to reproduce lost parts, often including germ cells as well as soma cells. In many of the lower classes the growth force is able to complete a structure or lost part without the stimulus of use, which in higher animals often seems to be part of the necessary requirements for growth.

Growth itself is the repetition of cells under nutrition and stimulus, and the latter may be hereditary or extra-individual. It is now recognized that since the division of a cell makes two unlike cells, each unlike the parent, such repetition will produce structures which present some degree of difference. The variation is therefore a necessary quality of growth, and its degree will change in response to the differentiation of the forces affecting growth.

When spines which have arisen from intrinsic growth force only are sought, it is apparent that they cannot be distinguished from those arising from external stimuli acting on and directing the growth force, unless in some instances they are found to be developed independently or even at variance with the environment. Because spines are sometimes useful to the organism, it is impossible to believe that they have originated from that cause, since their existence in some form must precede the capacity of making them useful. After they began to develop by either intrinsic or extrinsic forces, their being found useful would simply tend to their conservation and further development.

Variation which is not restricted by natural selection or a long line of hereditary tendencies is known as free variation. It is best exhibited in a stock which occupies for a considerable time a region favorable in respect to food, climate, and absence of dominating natural enemies. This relation has been called the period of "Zoic maxima" by Gratacap,[23] and has been further discussed by the same author, under the aspect of numerical intensity.[22] The most rapid rise of a stock is considered to be consequent to a favorable environment and high vitality.

In illustration of these points, the Achatinellæ of the Sandwich Islands afford a good example. The great number of species on these islands has probably been evolved since Tertiary times, and the process of specific delimitation is apparently still going on, for species are now to be found which did not exist fifty years ago (Verrill); also, a few species formerly common are now obsolescent or extinct. According to Hyatt, they all can be deducible from a single species which has differentiated in time through divergence, dispersion, and colonial isolation. In early times birds may have fed upon them, but the complete or partial extinction of the former by man has resulted in complete immunity for the arboreal Achatinellæ, and it is now common to find several of the most highly colored varieties feeding together on the same leaf. The modern importation of pigs, sheep, and mice on the islands has introduced an enemy to the terrestrial species, the effects of which are already being noticed. In specific differentiation and in individual variation, both Hyatt and Verrill regard the extraordinary development of this type as characteristic of free variation, under favorable conditions, in a plastic stock which has not yet reached its limits nor become fixed.

Among the Crustacea, the remarkable evolution of the genus *Gammarus* in Lake Baikal,[17] and of *Allorchestes* in Lake Titicaca,[19] seem to furnish parallel examples. *Allorchestes* ranges from Maine to Oregon and southward, through the United States, Mexico, and South America, to the Straits of Magellan. Before Lake Titicaca was explored, but one or two authentic freshwater species were known from both continents; yet from this lake basin alone, Faxon[19] has described seven distinct species, constituting the entire crustacean fauna with the exception of a species of *Cypris*. Several species are "remarkable among the Orchestidæ for their abnormally developed epimeral and tergal spines." These and the species of *Gammarus* from Lake Baikal will be referred to again later in this paper. It is simply desired here to indicate that these variations in *Achatinella* and

Allorchestes have arisen from a single parent stock, within a small geographic province. The natural interpretation seems to be (*a*) that the environment is favorable, as evinced from the great number of individuals; (*b*) that this has favored and increased the growth force; and (*c*) that, finally, the law of multiplication of effects, reproductive divergence,[67] the survival of the unlike, and the conservative forces of natural selection and heredity have directed the growth force, and produced the specific differentiation which is now found.

A factor of Evolution, called "Reproductive Divergence" by Vernon,[67] seems to be operative here, since it affords an explanation for a means of differentiation in a single stock under a common environment. As this factor has but recently been discussed, it may well be defined at this time, so as to enable a direct application to be made. Reproductive divergence assumes that in many species there will be greater fertility between individuals similar in color, form, or size, than between individuals not agreeing in these respects, and that in subsequent generations the divergence will become progressively greater in respect to the characteristic in question, so that finally the original stock will become separated into distinct varieties, sub-species, or species.

When, from any cause, the forces of nutrition are directed toward spine production, and when the direct results are accomplished in the reciprocal formation of one or more spines, there is often an apparent inductive influence or impulse given to growth toward the further production or repetition of spines. This may result in the formation of compound spines, or a group of spines, or even produce a generally spinous condition.

Naturally, spines arising through growth force may be useful for defence and offence, and the selective influences of sex and mimicry may also tend to greater development and elaboration. Furthermore, growth forces reacting on any external structures, as lines, lamellæ, ribs, nodes, etc., may tend to differentiate such ornaments into spines.

Therefore, under the general consideration of spines produced through growth force, the following factors are offered for consideration:

B. *From Growth Force.*

B 1. — Prolonged development under conditions favorable for multiplication.

B 2. — By repetition.

B 3. — Progressive differentiation of previous structures.

B 4. — Secondary development through the selective influences of defence, offence, sex, mimicry, and other external demands.

C. *External Restraint.*

Intermittent stimulus, as previously shown, produces growth in the direction of function. When the growth equals the waste, an equilibrium or static condition is reached, and no relative change occurs. The absence of either extrinsic or intrinsic stimulus will not be favorable to growth, and under such conditions an organ or structure may remain undeveloped, or, if already present in the organism, it may waste away and degenerate into a vestigial structure, or even disappear altogether.

On the other hand, it is well known that continuous pressure not only prevents growth, but in addition resorption takes place, and in this way the whole or a portion of a structure may be removed. These changes have frequently been studied in embryos, as well as in many internal structures, and are also familiar in the enlarged pedicle-openings of many Brachiopoda, caused by pressure of the pedicle, and in the similar opening for the byssal plug of *Anomia*. Packard [54] gives examples among the Crustacea and Insecta, which are clearly to the point. He says of the Crustacea, "It may here be noted that the results of the hypertrophy and overgrowth of the two consolidated tergites of the second antennal and mandibular segments of the Decapod Crustacea, by which the carapace has been produced, has resulted in a constant pressure on the dorsal arches of the succeeding five cephalic and five thoracic segments, until as a result we have an atrophy of the dorsal arches of as many as ten segments, these being covered by the carapace."

The restraint of the environment through unfavorable conditions is the antithesis of A, or the influence of constructive external stimuli, and is considered as the extrinsic operation of destructive agencies. It is evident that external unfavorable conditions will repress growth, with a resultant atrophy of the structures affected. In this way, also, the environment may cause the disuse of an organ, which by consequent suppression may dwindle away to a spine, as in the leaves and branches of desert plants, and the spurs of the Python [60] representing the hind limbs. It may likewise repress growth, as in the spines on the lower side of the poriferous coral *Michelinia favosa*,[40] representing aborted attempts at budding, the failure being due to the unfavorable position of the buds for securing food.

The restraint of the environment may also act in a mechanical manner to produce spines, as will be shown subsequently in some Brachiopoda and Trilobita. Furthermore, spines arising through any phase of external restraint, may secondarily come under the influences of natural selection, and be useful for protection and offence, or conform to other external demands.

Under the head of external restraint, therefore, are the following categories:

C. *From External Restraint.*

C 1. — Restraint of environment causing suppression of structures.

C 2. — Mechanical restraint.

C 3. — Disuse.

C 4. — Secondarily for protection, offence, etc.

D. *Deficiency of Growth Force.*

The growth force in organisms may be reduced in several ways, the most general and obvious modes being by an unfavorable environment, lack of physiological plasticity, too close interbreeding, pathologic influences, and parasitism. The first commonly implies a scarcity of food, or it may be that the temperature, moisture, light, elevation, or other conditions are unsuitable to the normal development. The lack of physiological plasticity affects growth force by its resistance to change, and is most strongly apparent in highly specialized forms. The effects of close interbreeding in reducing vitality are too well known to require further notice.

Only in exceptional instances can individual pathologic conditions have any effect on a stock. The retrogressive series of animals which are diseased in appearance, and are considered by Hyatt [34] as akin to pathologic distortions, are apparently types which have ceased to advance physiologically, and are therefore only adapted to special sets of conditions. In these, the pathologic or abnormal condition is racial instead of individual, and its cause seems to be a deficiency of vital power combined with great external differentiation, the final result being the assumption of characters belonging to second childhood and ending in extreme senility, with the loss of spines and other ornaments.

The life history of parasitic organisms shows their origin from higher normal types by a process of retrogression through loss of motion and disuse of parts. Their mode of living implies dependence upon the vitality of an immediate host, and altogether they may be deficient in the energy of growth force.

Any of the preceding factors, single or combined, acting upon an organism or group of organisms will produce suppression of structures or functions. Whether from external or internal causes, the waning and disappearance of characters are almost always inversely to the order of their development or appearance, either in the race or in the individual, and the most primitive or axial characters are therefore the most persistent and the last to disappear. In this way a leaf may be suppressed into a spine representing the midrib, a branch into a spiniform twig, a leg or digit into a spine, etc. As in other primary causes of spine genesis, there may also come secondary influences of protection, offence, etc., controlled by natural selection.

It will be convenient to consider spine production from lack of growth force under three heads:

D. *Deficiency of Growth Force.*

D 1. — Intrinsic suppression of structures and functions.
D 2. — Disuse.
D 3. — Secondarily for protection, etc.

Summary of Causes of Spine Genesis.

Before taking up in more detail the various causes of spine development, and illustrating them by means of examples drawn from a number of classes of organisms, it is well to restate the factors which are believed to induce spine growth. This is especially desirable from the fact that, through the operation of unlike forces, similar conditions may produce the same morphological results, as in the differentiation of ornamental lamellæ and ridges, which, either from external stimuli or dispersion of growth force, may develop into spines. In such cases it is difficult or impossible to distinguish the primary force, and the only satisfactory method is to discuss the subject under one head.

By carrying out this plan, and indicating the instances where the causes may replace or overlap each other, it may be shown how spines have originated, as follows: —

I. In response to stimuli from environment acting on most exposed parts. $(A_1.)$

II. As extreme results of progressive differentiation of previous structures. $(A_2, B_3.)$

III. Secondarily, as a means of protection and offence. $(A_3, B_4, C_4, D_3.)$

IV. Secondarily from sexual selection. $(A_4, B_4, C_4, D_3.)$

V. Secondarily from mimetic influences. $(A_5, B_4, C_4, D_3.)$

VI. Prolonged development under conditions favorable for multiplication. (B_1.)

VII. By repetition. (B_2.)

VIII. Restraint of environment causing suppression of structures. (C_1.)

IX. Mechanical restraint. (C_2.)

X. Disuse. (C_3, D_2.)

XI. Intrinsic suppression of structures and functions. (D_1.)

To illustrate the various causes of spine growth, representative examples which are believed to conform to the requirements will be selected from various groups of organisms. The number of spinose forms is so great that it will be impossible to give more than the briefest citation of a few of the leading types, especially those which have come under the notice of the writer; on this account the number of examples derived from the vegetable kingdom will be necessarily few.

I. *In response to stimuli from the environment acting on most exposed parts.* (A_1.)

The action of external stimuli falling on the most exposed parts of organisms is probably one of the most fundamental and fertile causes of spine production, since the relation between cause and effect is more direct and apparent here than by other modes of origin. In a general way it comprehends all the remaining causes coming under the head of external stimuli, but for present purposes it will be restricted by the elimination of secondary conditions, such as the indirect production of spines through differentiation of previous structures, and the action of external forces of selection.

The ruling forces in plants being so largely vegetative, or those of growth, and the cause of variation being principally physico-chemical and not molar, most of the modifications to produce spines will fall under other categories of origin (B, D) than the one now under discussion.

In the free swimming forms, however, as the desmids and

diatoms, the external relations are found to be very much like those of animals. The frustrule of the diatom *Attheya decora*[47] is quadrate in outline, and from the angles there extend sharp spinous processes, as represented in figure 32. The frustrule of the desmid *Staurastrum cuspidatum* is composed of two triangular halves, and the spines project from the vertices of the angles. Other species of *Staurastrum, Xanthidium* (*X. armatum*[47]), *Arthrodesmus* (*A. octocornis*[59]), etc., show similar spine growth from the most prominent portions of the frustrules. It is evident that in forms like these having angular outlines, any growth produced by external stimuli will naturally be greatest at the points of these angles, and in conformity with the previous analyses of these factors a spiniform extension of the tissues would result.

Among the freshwater Rhizopoda belonging to the Protoplasta (= *Amœbina*), the genus *Difflugia* affords good examples. *D. globulosa*[41] has a nearly spherical shell. In *D. pyriformis*[41] the shell is elongate pear-shaped, and generally round on the summit or fundus, though in rare instances a central spiniform elevation is developed. This tendency becomes fixed in *D. acuminata*[41] in which the shell in general form resembles the preceding species, but the fundus is commonly prolonged into a single acuminate process (figure 33), though occasionally two or three spines are found. In *D. corona*[41] there is a circlet of spines around the margin of the fundus besides the primary one in the centre. *Difflugia constricta*[41] is a variable form, with the top of the shell generally smooth, though sometimes it is acuminate, and occasionally it has two or even a cluster of spines (figures 34–36). *Euglypha mucronata*[41] has a terminal spine, as in *Difflugia acuminata.*[41] *Placocista spinosa*[41] is a flattened mitre-shaped form with a distinct edge, along which are numerous spines. It should be noted that no spines are developed on any portion of these freshwater Rhizopoda except those here mentioned.

The Nasellarian Radiolaria furnish many instances of a terminal spine from the summit of the silicious helmet or cup-shaped skeleton; as in *Eucyrtidium elegans*,[74] *Podocyrtis Schomburgki*,[74] *Tridictyopus conicus*,[26] *Cornutella hexagona*,[26] etc. Many of the primary, or axial, spines in other sub-orders probably originated according to I. In the Spumellarian forms especially, the principal spines project from the prominent portions; as in *Trigonactura triacantha*,[26] *Hymenactura copernici*,[26] *Rhopalastrum triceros*,[26] *R. hexaceros*,[26] etc. The existence of similar non-spinose species shows that the formation of spines is independent of the growth of the normal prominences; as in *Rhopalastrum malleus*,[26] *R. hexagonum*,[26] etc.

In the Foraminifera the configuration of certain forms is such that parts of the test are much more prominent than others, and in these exposed situations the spines are most frequently developed. Some of the triangular Textulariæ have spines at the two lateral angles on the oral side. Some of the individuals of *Textularia folium*[9] show that similar spines were developed at different stages of growth, so that, in a full-grown specimen, there may be two or three pairs of spines along the sides. Others, like *Verneuilina spinulosa*[9] and *Colivina pygmæa*,[9] develop spines from the points of each chamber. A number of species, also, show a single spine at the apex of the shell; as *Pleurostomella alternans*,[9] *Bolivina robusta*,[9] *Polymorphina sororia*, var. *cuspidata*,[9] etc. In the latter species the ordinary form is rounded or obtusely pointed at the fundus.

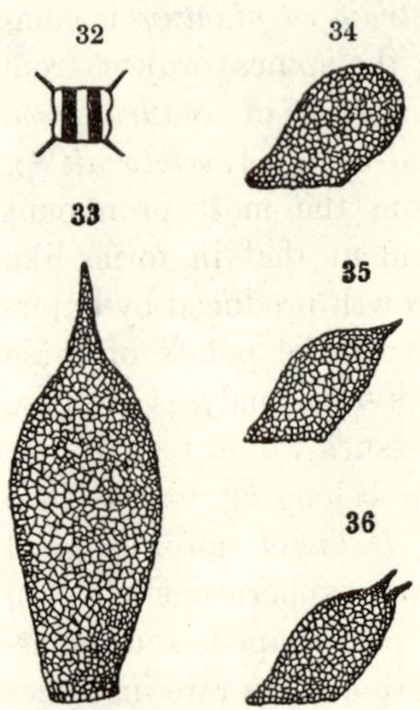

FIGURE 32. — *Attheya decora*, a diatom, with spines from the angles. — (From Mic. Dict.)

FIGURE 33. — *Difflugia acuminata*, a freshwater rhizopod; showing spiniform projection of the fundus. × 200. (After Leidy.)

FIGURE 34. — *Difflugia constricta*, a freshwater rhizopod, with rounded fundus. × 175. (After Leidy.)

FIGURE 35. — The same; showing a single spine on the fundus. × 175. (After Leidy.)

FIGURE 36. The same; showing two spines. × 175. (After Leidy.)

Some of the Infusoria have terminal spiniform processes, which, by analogy with other forms, have probably developed according to I; as *Ceratium tripos,*[9] *C. longicorne,*[9] *C. fusus.*[9]

The apertural spines on some of the graptolites are on the most exposed portions of the hydrotheca; as in *Monograptus spinigerus,*[52] *Dicranograptus Nicholsoni,*[52] *Retiograptus tentaculatus,* and *Graptolithus quadrimucronatus.* In many compound corals the corallites are polygonal from crowding, and the most exposed portions, the angles of the calices, often bear spines; as *Favosites spinigerus,*[30] *Callopora exsul,*[30] etc. The spines on the septa and costæ of corals probably originate by intrinsic forces (B), since they are internal growths not influenced directly by external stimuli.

The spines on the ventral sacs of Crinoidea are usually terminal, and in the most exposed situations; as in *Scytalocrinus validus,*[69] *Dorycrinus unicornis,*[69] *Aulocrinus Agassizi,*[69] etc.

The anterior and posterior pairs or rows of spines on the loricæ of some species of Rotatoria are in the most exposed places; as in *Anuræa squamula, Noteus quadricornis,* etc. The spinules on the tubes of *Spirorbis* are usually developed after it rises above the object of support so as to be exposed on all sides; as *Spirorbis spinuliferus.*[51]

The spinules at the corners of the angular cell apertures of many Bryozoa are in the most exposed situations, and probably arise through external stimuli; as in *Trematopora echinata,*[30] *T. spiculata,*[30] etc. The large marginal spines of the brachiopod *Atrypa hystrix*[31] probably owe their excessive development to external stimuli, though the phylogeny of the species shows that the spines first originated through the differentiation of the radiate and concentric ornaments.

In many pelecypods the siphonal region receives a great amount of stimulus, and the post umbonal slope is the part most exposed. Along this slope are found many of the spines, and generally the greatest differentiation of ornament. Examples of spines on post-umbonal slopes may be seen in *Callista sublamellosa* and young *Saxicava arctica* (figure 27).

Such spines represent periodic extensions of the mantle border, and in some cases the stimulus for this growth may come from internal causes. The spines on *Unio spinosus* and related species are believed by Mr. Charles T. Simpson to assist in anchoring the shell in the sand of swift running streams. In *Callista*, the young *Saxicava*, and the Unios mentioned, the spines occur on all individuals and at such an early period as to preclude any special sexual function.

In the Gastropoda the periodic extension of the shell over the posterior canal and the spiniform prominences formed on the labrum are situated in exposed places, or where the amount of stimulus is greatest; as in *Trophon magellanicus, Strombus pugilis, Fusus colus, Clavatula mitra, Melo diadema*, etc.

The spines on the larvæ of geometrid moths are usually on top of the loop, and are explained by Packard[54] as follows: "The humps or horns arise from the most prominent portions of the body, at the point where the body is most exposed to external stimuli."

When the origin and function of spines in a great many forms of animals, and especially among the higher classes, are examined, it seems almost impossible to decide whether a spine has been originated and perpetuated by free variation and heredity, or by the general action of external stimuli on the most exposed parts; and in the latter case, whether or not under the selective influences of use. Its origin in either instance may be through external stimuli, but in the latter, it falls under other captions than A_1; or, in other words, the external stimuli excite the growth force at certain points, and the growths so produced may be simply reciprocal without function or they may serve purposes of protection or offence. Thus the dorsal and rostral spines on the zoëa of the Decapoda are on the most exposed points, and seem to function as defensive structures. As soon as the legs become well developed or when the animal ceases to swim at the surface and hides among the stones, etc., at the bottom, these spines become reduced and are often succeeded by

others.   The spines of the adult are also usually efficient for
protection, but owing to the change in form of the animal
and change of habitat, the most exposed parts are different
from those of the larva, and the spines are frequently
developed where there were no larval spines; as in *Cancer
irroratus, Callinectes hastatus*, etc.   Again, the horned ungu-
lates show in their habits of sport, fighting, defence, and
procuration of food, that the exposed angles of the top of the
skull are subject to the greatest number of stimuli, and there
the horns are developed.   The connection between external
stimuli and growth is here most manifest, for it is impossible
to imagine the action of free variation or simple growth force
as resulting independently, in the evolution of horned out
of hornless species in several sub-orders of mammals, and
in every case determining the location of the horns on
the prominent angles of the skull, whether on the nasals,
maxillaries, frontals, or parietals.

It is well known that toads and frogs defend themselves
by using the head as a shield, and the cranial angles thus
receive the greatest amount of stimulus.   "There are natural
series of genera measured by the degree of ossification of the
superior cranial walls" (Cope[10]).   In the highest genera the
head is completely encased, and in some forms the projecting angles are developed into short horns.   The so-called
"Horned Toad" (*Phrynosoma*) has the same habit of defence,
and it is believed that this mode of protection or of receiving
impacts has given rise to the structure, by stimulating growth
at these points.

II.   *As extreme results of progressive differentiation of
previous structures.*   (A$_2$, B$_3$.)

The differentiation of existing ornamental structures into
spines has already been noticed in several instances in this
article.   It was shown that spines often arise by the elongation of nodes and tubercles or similar structures, by rhythmic
alternating areas of growth in lamellæ and ridges, and by

the growth of matter at the intersections of lines, lamellæ, ridges, etc.   Furthermore, it was indicated that this progressive differentiation could be produced either (*a*) by the direct action of external stimuli affecting the amount of nutrition brought to a certain structure, (*b*) by the stimulus and dispersion of growth force, or (*c*) by a combination of the two forces.   In this differentiation of the features which are generally called "ornamental," it will also be shown that the spine is the final result of progressive differentiation, and, as previously indicated, can be formed out of a variety of other structures.   The term "ornamental" is mainly one of human interpretation, and is used simply in apposition to "plain" or "simple;" for example, a clam cannot be imagined as consciously favoring a particular kind or arrangement of tubercles for ornamental purposes.

In a reticulate or cancellate surface formed by the crossing of raised lines, ridges, or lamellæ, it is evident that the causes or forces producing such structures will be increased at the points of intersection, and normally the amount of growth will here be greatest.   In this way it is possible to account for the very common presence of spines at the intersections of the radiating and concentric lines on many Mollusca and other organisms.

A few examples will now be given illustrating the differentiation of various structures into spines.

The points of intersections of the elements of the lattice in the Radiolaria are where spines are most frequently found; as in *Larnacalpes lentellipsis*, *Orosphœra Huxleyi*, *Carposphœra melitomma*, etc.[26]   In *Xiphosphœra pallas*, the ridges about the openings or meshes are granular, and the intersections are raised into spines.   Many of the discoid shells have their edges differentiated into spines; as *Heliodiscus asteriscus*, *H. cingillum*, *H. glyphodon*, *Sethastylus dentatus*, *Heliodrymus dendrocyclus*, etc.   When an edge becomes elevated and defined as a carina, this structure is also often spiniferous; as in *Tripocalpis triserrata* and *Astropilium elegans*.   The final differentiation of the radiate arrangement in the Radiolaria results in forms consisting only of a composite spine; as in the legion Acantharia.

In the Foraminifera there are many instances of the gradual differentiation of carinæ, ribs, costæ, etc., into spines. In *Bulimina aculeata*[9] the surface nodes and granules become developed into spines. In *Textularia carinata*[9] and *Cristellaria calcar*[9] the carinæ are spiniferous. The young of *Uvigerina aculeata*[9] is strongly costate, and later shell growth shows the costæ broken up into numerous spines. A related species (*U. asperula*[9]) has the whole test covered with spinules, which are sometimes arranged in lines, showing derivation from costæ. In *Truncatulina reticulata*[9] the carina is made up of confluent spines, often discrete along the edge, and sometimes entirely separated.

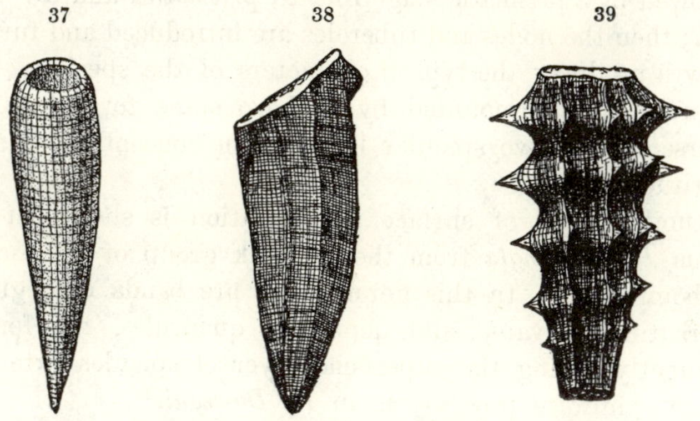

37   38   39

FIGURE 37. — *Cyathophycus reticulatus.* Ordovician. ½.
FIGURE 38. — *Dictyospongia Conradi.* Devonian. ½.
FIGURE 39. — *Hydnoceras tuberosum.* Devonian. ¼. (Figures 37, 38, 39, after Hall.)

To illustrate progressive chronogenetic and ontogenetic differentiation in a family of hexactinellid sponges.

The hexactinellid sponges belonging to the family Dictyospongidæ show some very clear instances of the progressive differentiation of ornament in time and in ontogeny. The Ordovician *Cyathophycus reticulatus*[28] is a turbinate form, with a rectangular mesh of longitudinal and transverse spicular rays (figure 37). At more or less regular intervals

4

some of the spicules are larger, thus dividing the surface into larger rectangular areas. In *Dictyospongia prismatica*[28] from the Devonian, the domination of eight of the longitudinal bundles of spicules has produced a prismatic form. *D. Conradi* is regularly an eight-sided pyramid or prism when young, but with the growth and elongation of the sponge it developed slight undulations, then nodes, and later prominent tubercles (figure 38). *Ceratodictya annulata* and *Hydnoceras nodosum*[28] show a further specialization in the formation of rings and nodes. Practically the limit to these specializations is attained in *Hydnoceras tuberosum*[28] (figure 39), *H. phymatodes*, and related forms. In *H. tuberosum* the apex representing the young stage or the initial growth is much like *Cyathophycus* or *Dictyospongia*. This is followed by a prismatic stage like *D. prismatica* and *D. Conradi*; then the nodes and tubercles are introduced and further growth produces the typical characters of the species. The tubercles are surmounted by a sharp spine formed at the intersection of two spicular laminæ, one concentric and one longitudinal.

Another type of surface specialization is shown in the genus *Physospongia* from the Keokuk group of the Lower Carboniferous. In this genus there are bands of regular, alternating, elevated, and depressed quadrules, the former frequently having the superficial layer of spicules extended into a spiniform process; as in *P. Dawsoni*.[28]

Among corals there is occasionally some evidence of the external differentiation of structures into spines. The epitheca of the Tetracorolla frequently shows, by means of low lines or low ridges, the number and direction of the septa, and in some of the later species these external septal lines are ornamented with rows of short spines or spinules; as in *Cyathaxonia cynodon*[18] and *Zaphrentis spinulosa*.[18]

Many Crinoidea and Asteroidea show the development of tubercles into spines, and the surface sculpture is often made up of ridges which bear strong spines at the points of intersection; as in *Gilbertsocrinus tuberosus*,[69] *Technocrinus spinulosus,*[69] *Actinocrinus lobatus,*[69] *A. pernodosus, Oreaster occidentalis, O. gigas, Retaster cribrosus,* etc.

The concentric laminæ of growth in the Brachiopoda are frequently differentiated into spinules; as in *Siphonotreta unguiculata,*[31] *Schizambon typicalis,*[31] *Spirifer fimbriatus,*[31] *S. pseudolineatus,*[31] *S. setigerus*[31] *Cliothyris Royssii,*[31] etc. Other species show the differentiation of the radii into spines; as *Acanthothyris spinosa*[74] and *A. Döderleini.*[15] In others the strong concentric laminæ passing over radii are often infolded into spines; as in *Atrypa spinosa.*[31]

Among the Mollusca innumerable examples could be cited showing clearly the differentiation of various ornamental features into spines. Some of these have already been discussed, but may be referred to again in this connection. Thus an illustration of the passage of concentric laminæ into spines is shown in *Avicula sterna*[36] and *Anomia aculeata*[36] (figures 26 and 28) and *Margaritiphora fimbriata,* etc. Many species of Gastropoda show the same types of differentiation. The differentiation of radiating lines or ridges into spines is equally common, and is well shown in *Spondylus* (figures 12, 14, 30), and in *Lima squamosus* (figure 40). In most of these cases the rib represents the progression of a fold in the edge of the mantle, while the spine is a process of a concentric lamina, and is usually more or less flat or tubular. Occasionally the rib becomes obsolescent, and is represented by a row of spines; as in some specimens of the gastropod *Crucibulum spinosum.* When the radiating and concentric ornaments are distinctly continuous, a reticulate or cancellate appearance is produced, and the points of intersection often bear spines; as in *Aviculopecten scabridus,*[29] *A. ornatus,*[29] *Actinopteria Boydi,*[29] *Pterinopecten spondylus,*[29] etc.

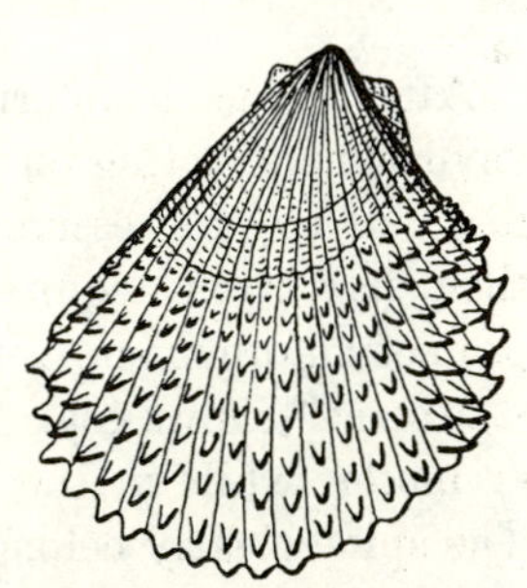

FIGURE 40. — *Lima squamosus.* Natural size. To show differentiation of radii into spines.

The raised lines or ridges on the legs and carapaces of Crustacea are frequently spiniferous; as *Gelasimus princeps*, *Gecarcinus ruricola*, etc. The radii on the shells of barnacles are sometimes differentiated into spines; as in *Balanus tintinnabulum* var. *spinosus*.[18]

In the higher animals the differentiation of ornamental features into spines is not common, especially as most of the forms are devoid of hard external parts. Among the fishes and reptiles certain lines and ridges on the head and body are often spiniferous, while in others the scales have spiniferous ribs.

III. *Secondarily as a means of protection and offence.*
(A₃, B₄.)

After spines have originated through the stimuli from the environment acting on the most exposed parts, or by growth force, or by progressive differentiation of previous structures, they may often acquire added qualities, one of which is to protect an organism from the attacks of many of its enemies.

Morris[49] shows that defence in animals is either mechanical or motor, while in the higher plants it is purely mechanical. The spine clearly belongs to the mechanical mode of defence, and in many animals may be efficient without motion. If motion is added, it then may serve not only for protection but for offence as well. Natural selection evidently could not originate a spine, but after one has appeared from any of the causes mentioned in the preceding paragraph, this agency could tend to preserve and allow the spine to develop along certain lines. The restrictions as a defensive structure would be those of efficiency, and therefore all the monstrous growths, vagaries, and ornamental spine features would arise independently of the action of protective selection, and would be accounted for by the operation of the forces of the environment, growth, and sexual selection. In this way the simple antlers of the Tertiary Deer may be imagined to have reached the highest degree of efficiency as weapons, by ordinary natural selection (figure 41). In most cases the subsequent increasing complexity of the antlers during more modern times cannot have improved their usefulness for protection or fighting (figures 42, 43), and probably arose through gradual specialization according to the law of multiplication of effects, acted on by the agency of sexual selection. In some species, as the Reindeer (*Rangifer tarandus*), the differentiation of the antlers has secondarily produced a useful structure. One of the brow tines in this species has become greatly enlarged and palmated, and serves to assist in removing the snow to uncover food. Evidently this has had something to do with the common retention of the antlers in both sexes.

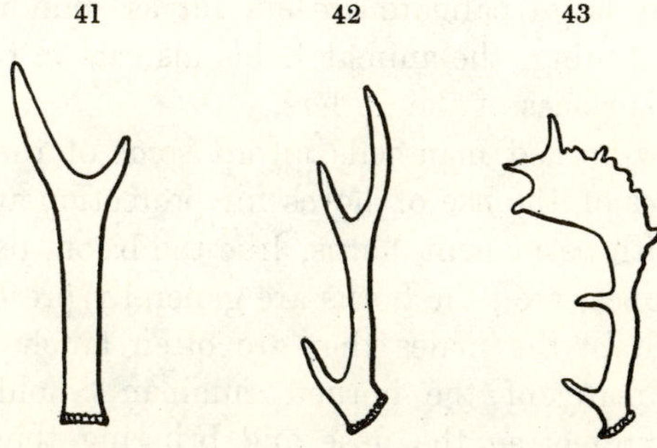

FIGURE 41. — Antler of *Cervulus* (?) *dicranoceros*. Pliocene.

FIGURE 42. — Antler of *Cervus pardinensis*. Pliocene.

FIGURE 43. — Antler of the Fallow Deer (*Cervus dama*). Reduced. (After Nicholson and Lydekker.)

Certain types of horns are common to particular regions, especially when the cattle are in a semi-wild state, as in the Western Plains of America. The Texas cattle have long, gently curved horns standing out from the head. Similar forms are prevalent in the cattle of southern Italy and in other warm temperate regions. Further north, the horns become more curved in a direction parallel with the head, and are therefore closer to the skull. The most northerly representative of the hollow-horned ruminants, the Musk-Ox (*Ovibos moschatus*), has the horns hanging down close to the skull and only curved outward in their distal portions.

Marsh suggests to the writer that these variations in the directions of the horns have been influenced by the climate. A warm climate permits the horns to stand out directly from the skull. Further north, or in a colder region, the frequent freezing of the horns and their consequent drooping has induced a natural drooping condition, and an Arctic climate has resulted in the production of horns closely appressed to the skull, in which position they cannot be affected by freezing temperatures.

Another possible service for antlers is also suggested by Marsh. As is well known, the male Moose is one of the most wary of the Cervidæ, and detects noises at great distances. The large palmate antlers act as sounding-boards, and, when listening, the animal holds his ears in the focus of the anterior surfaces of the antlers.

The hollow-horned mammals afford some of the most evident examples of the use of horns for protection and offence. In species with permanent horns, like the bison, oxen, goats, cattle, antelopes, etc., the horns are generally present in both sexes, though in the males they are often much the larger. In defence, many of the horned ruminants hold the head down, thus protecting the nose and bringing the top of the skull into prominence. In this position the horns are most effective. A similar posture is taken by the horned batrachians and lizards.

The Porcupine and Echidna rely largely on the protection afforded by their spines, and on this account they are sluggish in their movements, and make little effort to escape approaching enemies.

Many of the great horned Dinosauria of the Mesozoic are well provided with an armature of protective plates and spines on various parts of the body. In addition to an armature on the body *Triceratops* [46] had three large horns on the head, one median (nasal) and two lateral (supra-orbital). These were powerful offensive and defensive weapons. There were also other small nodes and spiniform ossicles around the posterior crest of the skull and on the jugals, forming a part of the general armor. In *Stegosaurus* [46] the efficient offensive and defensive weapons were the huge spines on the tail, and it is interesting to note, as a parallel to this condition, that the greatest nerve centres were in the sacrum, and therefore posterior also.

No group of vertebrates shows such a variety of protective and offensive characters as the fishes. Many of the older types were heavily plated, while in others the fin spines were greatly developed. Among modern forms the protective character of the spines is well shown in types like the Spiny Box-fish *Chilomycterus geometricus* and *Diodon maculatus*. A combination of mechanical and optical protection is afforded in the remarkable Australian Pipe-fish *Phyllopteryx eques* [25] (figure 49). This fish has numerous spines and ribbon-like branching filaments, the former giving it a mechanical defence, and the latter assisting in its concealment among sea-weeds, to which it bears a striking resemblance.

Spines for protection are extremely common among insects, even in larval forms. They have been so frequently noted as to require no elaboration here. Packard [54] has ably discussed the origin of nodes, tubercles, and spines among certain caterpillars. Among the forms which feed exclusively at or near the ground, he finds the body usually smooth, while those feeding on trees or on both trees and ground are often variously spined and tuberculated. These ornamental features arise from the modification of the piliferous warts common to all lepidopterous larvæ, and he concludes that the trees were more favorable for temperature, food, etc., than the ground, and that an increase of nutrition and growth force led to the hypertrophy of these warts into tubercles and spines. Having thus arisen, they immediately became useful for protection from birds and parasitic insects.

Among the Crustacea there are also numerous examples of protective spines. These may be confined to parts of the body and legs especially exposed, or the entire animal may partake of the spiny character, as in the crab *Echidnoceras setimanus*, where even the eye-stalks and antennæ are spiniferous. Others, like *Lithodes maia*, have the spines generally distributed over the carapace and legs. While serving for defensive purposes, this generally spinose character has probably reached its extreme development through the influence of repetition (B_2). The nauplius larva of *Lepas fascicularis* is very large, and has highly defensive spines which are explained by Balfour[3] as a secondary adaptation for protection. The larger spines on Trilobita, especially those from the genal angles and the axis, doubtless served protective purposes. The extremes of spinosity in

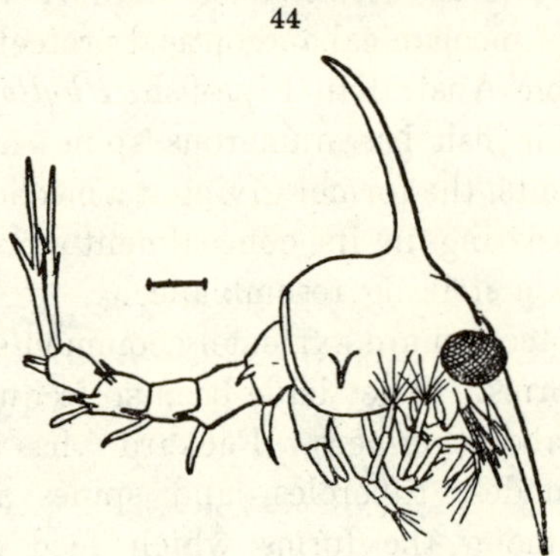

FIGURE 44. — Zoëa of the common crab (*Cancer irroratus*); lateral view. ×8. (After Verrill and Smith[68].)

this class are found in the various species and genera of the family Acidaspidæ, and also in many forms of *Arges, Terataspis, Hoplolichas*, etc.

Even among the star-fishes, which are so generally spinose, some forms have the spines so prominently developed on the most exposed portions of the animal that they evidently serve for protection; as *Acanthaster solaris, Echinaster spinosus*, etc.

The examples already given are sufficient to emphasize the fact that after spines are developed they may then often serve for protection and offence, and therefore be useful, their efficiency being controlled by natural selection resulting in the survival of the fittest.

Another process or kind of selection has been described by Verrill as " Cannibalistic Selection." He has shown that the young of carnivorous animals often prey upon each other, as in the larval forms of some Decapoda, or sometimes even before the escape of the young from the egg capsules, as in some of the Gastropoda. Here, of course, any natural variation in the newly hatched animals which would give an individual some advantage over its companions would tend to its preservation and to their destruction. In this way it may occur that the relative growth of spines in the zoëa of decapods has determined the survival of the well-armed individuals; as in the zoëa of *Cancer*[68] (figure 44), *Carcinus*, *Homarus*, etc.

IV. *Secondarily from sexual selection.* (A₄, B₄.)

The males and females of so many animals present differences in size, color, and ornament, that corresponding variations in the development of spines, horns, and antlers, might naturally be expected. That such differences actually occur in nature is evident. Every gradation can be found between horns or antlers common to both sexes and those confined to one sex. Probably the initial difference is as ancient as sex itself.

Sexual variations of horns are most familiar among the mammals. Some, as the Giraffe, Ox, Bison, and Reindeer, have them present in both sexes, though the antlers of the female Reindeer are smaller and more slender than in the male, and in the American variety are sometimes absent. Others, as in the Prong-horn Antelope, many sheep, goats, etc., have the horns usually quite small in the female, and well developed in the male. Lastly, the modern Deer, Elk, Moose, etc., have the antlers confined to the males alone, the female being entirely without them.

Some of the early deer (*Procervulus*) seem to have had antlers in both sexes, and in nearly all the families of the

Ruminata there are species without horns, other species with horns in both sexes, and still others with horns only in the male. In the wild state the presence or absence of horns and their character in any particular species seem to be well established, but in domesticated forms the greatest variety is found. Among domesticated cattle, presumably of one species originally, varieties are found without horns, and others with horns showing all degrees of twisting and length.

By protecting cattle from enemies, by forcing them into changed environment, and by varying amounts of nutrition, man has evidently brought the original stock into a condition of free variation. This state has been made use of in the production of endless varieties by selection and cross-breeding.

Darwin[14] accounts for the sexual selection affecting the growth of the antlers in the Deer as due to excess in the number of male individuals, and their struggles for supremacy in the possession of a mate. The antlers at the breeding season are strong and solid, and are therefore at their maximum of efficiency in each individual. They are shed at or before the time the young are born. Previous to the growth and maturity of the new antlers, the young are so far advanced as to be able to avoid being killed by the adult males. Furthermore, Darwin suggests that the excessive development of antlers into palmate and arborescent forms was probably an ornamental character attractive to the females. These complicated antlers not being the most efficient weapons, the fighting proclivities of the males would tend to favor the individuals with simple antlers, and to repress the more differentiated forms. Thus the two influences would be opposed to each other, though not necessarily equal. The law of the multiplication of effects may also have some force, since it may carry a structure beyond the bounds of efficiency. Even in one of the oldest horned mammals, the *Protoceras*[45] of the Miocene Tertiary, a great difference is seen in the horns of the two sexes. The female has little nodes or tubercles, which in the male rise to the height and prominence of the horns on the Giraffe, or are even relatively more pronounced.

The males of some other vertebrates have spiniform processes or spurs on their legs and wings serving particular functions. The spurs in birds are to be considered mainly as weapons which are used by the males in combats among themselves. They are developed on the metatarsal or metacarpal bones as bony processes ensheathed in horn. In the females the spurs are generally rudimentary. A kind of spur is also found on the hind limbs of the male *Echidna* and *Ornithorhynchus*, attached to the astragalus. It is perforated by a duct leading from a gland. The functions of the spur and of the secretion are unknown.

Many lizards, especially among the Chamæleontidæ, present striking differences between the sexes, and the males of some of them develop veritable horns like those in cattle, sheep, and other hollow-horned ruminants. Darwin[14] illustrates and describes a number of most interesting examples. One of them *Chamæleon Oweni* is here shown (figures 45, 46). The male has three horns, one on the snout and two on the forehead. They are supported by bony excrescences from the skull. From the peaceable nature of these animals, Darwin concludes that " we are driven to infer that these almost monstrous deviations of structure serve as masculine ornaments."

The males of the tropical American genus of fishes *Callichthys* " have the spines on the pectoral fins stronger and longer than those of the female, the spine increasing in size as the male reaches maturity " (Seeley[65]).

Among insects the males of many beetles belonging to the lamellicorns have long horns arising from various parts of the head and thorax. One of the best known forms is the Hercules beetle *Dynastes hercules*. Bateson[5] states that, in this and other genera, it is commonly found that the males are not all alike; but some are of about the size of the females and have little or no development of horns, while others are more than twice the size of the females and have enormous horns.   These two forms of male are called "low" and "high" males, respectively.   Among the males similar dimorphism in respect to size and length of horns occurs in *Xylotrupes gideon,* and in the stag beetle (*Lucanus cervus, L. titanus, L. dama*).

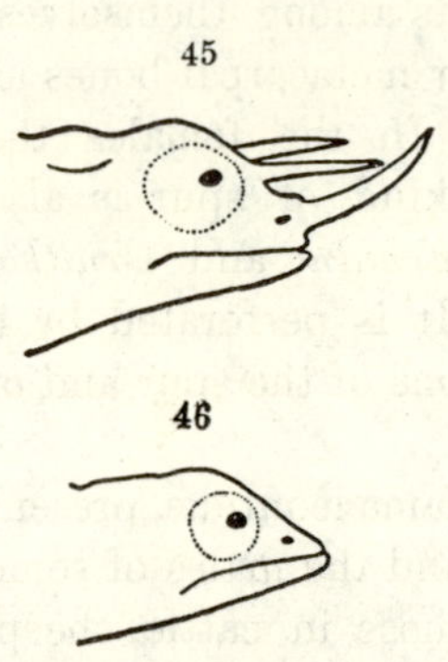

FIGURE 45. — Profile of head of *Chamæleon Oweni;* male. ½.

FIGURE 46. — Female of the same species. ½. (After Darwin.)

In many of these cases the horns are evidently protective, and not developed through the selective influences of the female.   In such cases the habits of the male are supposedly different from those of the female.   Thus Wallace [70] suggests that the horned males of the coleopterid families Copridæ and Dynastidæ fly about more, as is commonly the case with male insects, and that the horns are an efficient protection against insectivorous birds.   These interpretations clearly do not come under the definition of sexual selection as restricted to the choice of either sex.   Beauty, voice, or strength may influence the selection of a mate by the opposite sex, but when the habits of the sexes are different, and certain characters arise in response to this change, the explanation is then really found in the law of adaptation or physical selection.

V.   *Secondarily from mimetic influences.*   (A₅, B₄.)

Natural selection may aid in furthering and preserving a spinose organism after the spines have originated through any primary cause.   One aspect of this influence may be treated under the head of mimicry.   If, by their resemblance in form, color, or voice, any characters are similar to characters present in the surroundings of the animal, and afford a means of protection or are useful, they may be considered as mimetic in the broadest sense of the term.   Mimicry is usually restricted to a kind of special resemblance, and not to the cases of general resemblance afforded by an animal without significant colors in general harmony with its surroundings.

The influence of mimicry in the production of spines can only occur where the object mimicked is spiniform or spinose. Apparently this is rather infrequent and of little real importance as a factor of acanthogeny.

Insects and spiders have furnished the greatest number and variety of mimetic forms, both in their larval and adult conditions, and naturally would be expected to furnish examples of spines having mimetic significance. The object mimicked may be another species of insect or animal, in which case there is usually some offensive or defensive quality rendering the resemblance useful to the mimicker; or the whole or a portion of some plant or other object may be imitated, tending to the more or less complete concealment of the mimicking insect.

Satisfactory examples are not at hand, though doubtless many occur in nature, and some have been described, but not for the present purpose. A few will be cited here which seem to conform to the requirements.

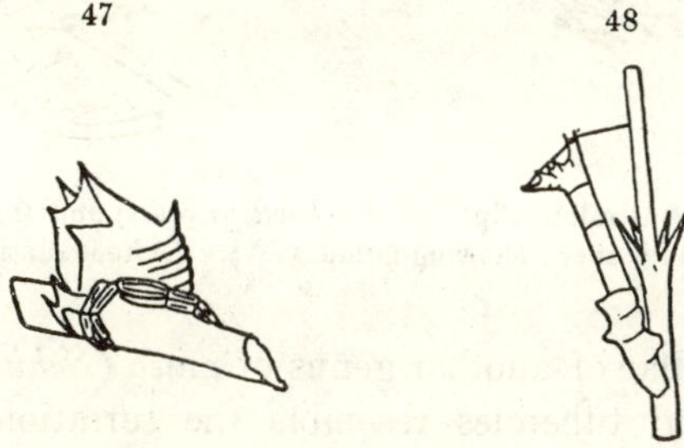

FIGURE 47. — Profile of a spider (*Cœrostris mitralis*) on a twig mimicking a spiny excrescence. (From Peckham, after Vinson.)

FIGURE 48. — The larva of the Early Thorn Moth (*Selenia illunaria*) resting on a twig; showing mimicry of stem and spiniform processes. ½. (After Poulton.)

A Madagascar spider (*Cœrostris mitralis*) is described by Elizabeth G. Peckham[56] as sitting motionless on a branch and resembling a woody excrescence with projections or spiniform processes (figure 47). Other spiny spiders of the Epeiridæ probably have similar protective mimetic features; as *Epeira spinea* and *Acrosoma arcuata*.

The larva of the Early Thorn Moth, as described and illustrated by Poulton,[53] bears a strong resemblance to the twig upon which it rests, even to the spiniform processes, axils, and buds (figure 48). Packard[54] cites a striking case of mimicry

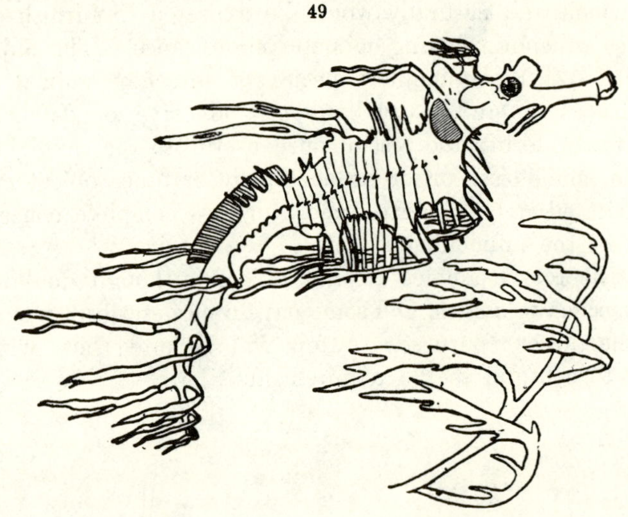

FIGURE 49. — Australian Pipe-fish (*Phyllopteryx eques*) and frond of sea-weed in lower right-hand corner; showing mimicry. ½. (After Günther.)

in the caterpillar of another genus of moth (*Schizura*), where the spines and tubercles resemble the serrations of a leaf "so that, when feeding on the edge of a leaf, the Schizuræ exactly imitate a portion of the fresh-green serrated edge of a leaf including a sere, brown, withered spot, the angular, serrate outline of the back corresponding to the serrate outline of the edge of the leaf."

The Australian Pipe-fish *Phyllopteryx*, previously mentioned under the head of spines for protection, shows the mimicry of a plant by an animal to a striking degree. This fish closely imitates a sea-weed (figure 49), and Günther[25] gives the following description of the spines and filaments on the species *Phyllopteryx eques:* "There is a pair of small spines behind the middle of the upper edge of the snout, a pair of minute barbels at the chin, and a pair of long appendages in the middle of the lower part of the head. The forehead bears a broad, erect, somewhat four-sided crest, behind which there is a single shorter spine. A horizontal spine extends above each orbit. There is a cluster of spines on the occiput, and from these narrow appendages are prolonged. On the nape of the neck is a long spine, dilated at the base into a crest, and carrying a long forked appendage. The back is arched, and on the under side are two deep indentations. The spines on the ridges of the shields are the strongest; they are compressed, are not flexible, and each terminates in a pair of short points. There is one pair of these spines in the middle of the back, and one on each of the three prominences of the abdominal outline; they terminate in flaps, which are long and forked. There are also very long compressed flexible spines without appendages, which extend in pairs along the uppermost part of the back, while a single series extends along the middle line of the belly. Small short conical spines run in a single series along the middle line of the sides, and along the lateral edges of the belly; and there is a pair of similar spines in front of the base of the pectoral fin. The tail, which is about as long as the body, carries the dorsal fin; it is quadrangular, and has sharp edges. It carries along its upper side five pairs of band-bearing spines, which terminate in branching filaments." *

The Horned Toad *Phrynosoma* bears considerable resemblance to the joints of the Prickly Pear, with which it is often associated, and it may be suggested that the likeness both in form and spinescence represents mimetic characters.

---

* The artist who copied Günther's figure for Leunis' "Synopsis der Thierkunde," 3d ed., by H. Ludwig (vol. i. p. 770, 1883), connected the fish with the adjacent fronds of sea-weed so as to form a single organism.

VI. *Prolonged development under conditions favorable
for multiplication.* (B$_1$.)

The prolonged development or existence of a stock under
favorable conditions for multiplication may be considered as
one of the primary influences favoring the production of
spines. This implies abundance of nutrition and comparatively few enemies outside of other individuals of the same or
closely related species. Under a proper amount of increased
nutrition the vitality and reproductiveness of a stock are
raised, and, other things being favorable, it is found that the
stock will give expression to what has already been described
as free variation. Hypertrophy is also very apt to be one
result of abundant nutrition, so that structures of little or
no use may be developed, and some of them comprise certain
features which are often called ornamental.

In the excessive multiplication of individuals it is evident
that there must be a great number of natural variations, and
that some of these will affect the pairing of the sexes in such
a manner as to accentuate and delimit certain variations.
Eventually there also comes a struggle for existence in
which favorable modifications have a decided advantage. In
this way it is believed that the great amount of differentiation found in some isolated stocks has been brought
about. Primarily, then, a favorable condition for nutrition
is assumed, which is followed by excessive numerical multiplication; while the natural variations are augmented and
governed by the action of reproductive divergence for which
such conditions are favorable. Secondarily, these variations
are subjected to the influences of cannibalistic selection,
defence, offence, sexual selection, and mimicry.

In illustration of the amount of differentiation attained by
a single stock under favorable conditions, the Amphipod
Crustaceans *Gammarus* and *Allorchestes*, found in lakes
Baikal and Titicaca, respectively, may again be noticed.

In respect to the number of species, *Gammarus* is very
sparsely distributed over the world, though in Lake Baikal

alone a hundred and seventeen species have been described by Dybowsky.[17] In contrast to this, it may be mentioned that but four freshwater species have been discovered in the whole of Norway. In Lake Baikal all the depths explored (to 1,373 metres) have furnished species. Those living near the surface are vividly colored, yet apparently make no attempts at concealment. Many of the species are also highly spinose, though not sufficiently armed to be protected from the fish. As these Crustaceans are voracious creatures, the spinose character has probably been favored by the agency of cannibalistic selection. The lake has a number of species of fish for which the Gammaridæ furnish excellent food, but the presence of a species of seal, predaceous fish, as well as the native fishermen, keep the fish below the danger point, thus allowing the Gammaridæ to become very abundant.

Similarly, in Lake Titicaca there is a wonderful specific development of a kindred Crustacean *Allorchestes*. One of the most spinose species (*A. armatus*) is also the commonest, and according to Faxon[19] occurs in countless numbers (figure 50).

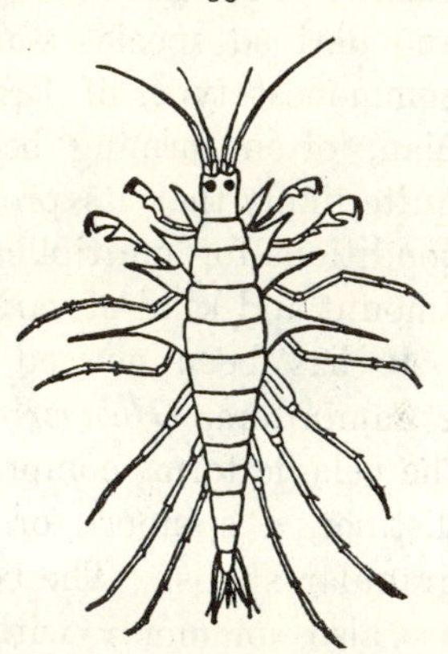

FIGURE 50.— *Allorchestes armatus*, a spiny amphipod from Lake Titicaca; female; dorsal view. Natural size = 9 mm. (After Faxon.)

Packard[54] shows that, among certain moths, the caterpillars as soon as they acquired arboreal habits met with favorable conditions in respect to food, temperature, etc., and that as spines and tubercles arose by normal variation, such features, being found useful for protection, were therefore preserved and augmented.

The differentiation of *Achatinella* has already been discussed (p. 36) as affording a striking instance of free variation among the Mollusca. The evolution of the Tertiary species of *Planorbis* at Steinheim, as described by Hyatt,[35] furnishes another example, though in neither case has the differentiation of structures proceeded far enough to result in spines. The costate form (*Planorbis costatus*) was tending toward that end, but did not attain it.

The series of Slavonian *Paludina* in the Lower Pliocene, as elucidated by Neumayr and Paul,[50] shows a somewhat further advancement. The species in the lowest beds (typus *Paludina Neumayri*) are smooth and unornamented. Higher in the strata they are angular and carinated, and at the top of the series the shells are carinated, nodose, and sub-spinose (typus *Paludina Hœrnesi*). The living American genus *Tulotoma* is closely related to the most differentiated species (*P. Hœrnesi*), and its approach to spinose features is more pronounced.

Under the phylogeny of spinose forms (pp. 23–25) an outline of the life history of the brachiopod *Atrypa reticularis* and derived species was presented. This being one of the commonest types of Brachiopoda in the Silurian and Devonian, often forming beds of considerable extent, it seems quite likely that its prolonged development under favorable conditions for multiplication must have had an effect on the amount and kind of variation.

It has been noticed by Brady[9] and others, that in the Foraminifera, *Globigerina bulloides*, *Orbulina universa*, etc., the pelagic forms comprise two varieties which are generally distinct, a spinous form and another with small minutely granular shells. The bottom specimens of the same species are also commonly without spines and often smaller. The interpretation seems to be that the large specimens indicate an abundance of nutrition which has also produced hypertrophy of the normal granules into spines. Some bottom specimens are large, but they are usually abnormal and of a monstrous or pathologic nature.

From the foregoing examples the conclusion to be drawn is that, with full nutrition, there comes a numerical maximum, and naturally with this a corresponding number of normal variations. Some of these modifications, as spines, have arisen by hypertrophy. After having thus originated by growth force, they may or may not be of use for offence, defence, or concealment, or in any way give their possessor a distinct advantage.

VII. *By repetition.*   (B₂.)

Under the consideration of spine production by repetition it is proposed to include local repetition or duplication of spines on or about a primary spine, the limit of this repetition resulting in a generally spinose condition.

It has been shown that intermittent stimulus produces growth, and furthermore that growth can take place only with proper nutrition. Under local stimulus the currents of the circulation or forces of nutrition are set up in an organism toward the centre of stimulation. The nutrient matter is brought to this point, and more or less of it is expended in building up a structure which is the reciprocal or direct resultant of the stimulus. Now, since all motion is primarily rhythmic,[66] and the repetition of parts an almost universal character among organisms,[5] it would appear that the foregoing conditions would be favorable to the repetition or reproduction of the structures. In this way it is easy to account for the growth of spines that cannot be explained as the direct result of external stimuli (A), or by any process of decrescence (C, D). The nature of the influence seems to be similar to induction in electrical physics, or to the force or stimulus of example in human conduct.

Stated as a concrete case, a simple spine produced by any primary cause may be taken, and it will be granted that the vital or physiological adjustments produced in its growth and maintenance have brought about or induced an harmonic condition in the adjacent tissues. Subsequent growth will most naturally repeat the previous structures, so that in addition to the primary spine there will be other smaller spines on or about it, together constituting either a compound spine or a group of spines.

Carrying this repetitionary process to a maximum, there would result a generally spinous condition. As a possible illustration of this, no class of organisms probably exhibits so many kinds and series of repetitions of all sorts of external structures as the Echinodermata, and it is significant that this is a typically spiniferous sub-kingdom.

Except in a few classes of organisms, compound spines are relatively rare as compared with simple spines. They are very common among the Radiolaria, which furnish the greatest complexity occurring anywhere in the organic world. (See Plate I.) They are also quite frequent among the Echinoidea, but more rare among the Asteroidea and Crinoidea.

Compound antlers are especially characteristic of the modern Deer family, though compound horns are but rarely found elsewhere among the mammals. The Prong-horn Antelope of America is the only living species of hollow-horned ruminant having this character. It of course is not intended that extra pairs of horns, which being separate, and often originating on different portions of the skull, should be considered as compound horns in the sense here employed. Likewise compound spines arising through suppression of organs or structures are not to be included here; as the compound thorns on the Honey-locust representing aborted branches.

The fin spines of fishes are often compound, and sometimes are made up of several elements; as in the spines of *Edestus* (*E. vorax*). Quite a number of Mollusca develop compound spines; as in many species of *Spondylus* and *Murex*. They are also not uncommon among the Crustacea and Insecta. Compound spines are infrequent in the Brachiopoda, being developed in but few species (*Spirifer hirtus* [31]). The Foraminifera also present but few examples (*Polymorphina Orbignii* [9]).

A number of generally or highly spinose types will now be noted to illustrate the limits of the repetition of spiny structures, the first spines having probably arisen through the operation of some primary cause, and the derived or secondary spines being produced, it is believed, by the law of repetition.

The Radiolaria have already been frequently mentioned, but as they are the most spiniferous of all classes of animals, and represent the highest degree of spine differentiation attained (figure 51 and Plate I), another brief notice will

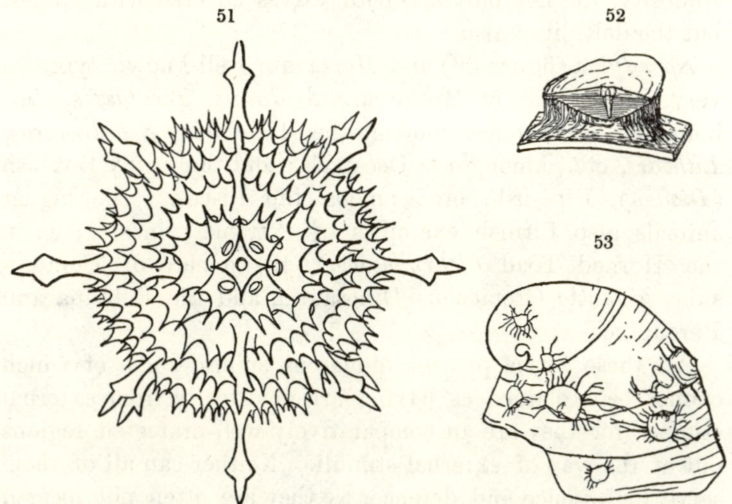

FIGURE 51. — *Acontaspis hastata*, a radiolarian; showing multiplication of spines by repetition. × 200. (After Haeckel.)

FIGURE 52. — *Strophalosia keokuk*, an attached brachiopod; showing the spines extending from the ventral valve to and along the surface of attachment. × 2.

FIGURE 53. — A gastropod shell (*Platyceras*) to which are attached a number of *Strophalosia keokuk*. Natural size.

be of interest. These spines furnish characters of high taxonomic value, although generally speaking they seldom have more than specific importance among other classes. The Echinoidea and Asteroidea must also be noticed in this connection, though from the nature and origin of their spines they do not conform to the mode of spine growth in other classes.

Productus, Productella, Strophalosia, Aulosteges, and *Sipho-*

notreta represent highly spinose genera among the Brachiopoda. *Strophalosia* is a form in which the ventral valve is cemented to some object. Whenever the valve rises well above the object of support, the spines are free like those frequently present on the dorsal valve; otherwise the spines extend root-like along the supporting surface (figures 52, 53).

Aulosteges presents a still further tendency to complete spinosity, for not only are both valves covered with spines, but the deltidium also.

Spondylus (figure 30) and *Murex* are well-known types of very spiny forms of Mollusca. *Acidaspis, Terataspis*, etc., hold the same place among the Trilobita; *Echidnoceras, Lithodes*, etc., among the Decapoda; and the Spiny Box-fish (*Diodon*), Pipe-fish, etc., among the Pisces. The higher animals also furnish examples of extreme spinosity; as in the Horned Toad (*Phrynosoma*), the genera of Ceratopsidæ, gigantic Cretaceous Dinosauria, and the Echidna and Porcupine.

All these forms present numerous spines, some of which cannot be explained as having arisen directly from external stimuli, for they are in comparatively well-protected regions out of the way of external stimuli. Neither can all of them serve for offence and defence, as they are often not located in the most advantageous positions; nor are they differentiated out of any previous ornaments or special structures. In fact, no factor of spine genesis except the one of repetition seems to be sufficient to account for their development.

VIII. *Restraint of environment causing suppression of structures.* (C₁.)

The previous categories of spine production (I–VII) have been brought about by some process of growth or concrescence through external and internal agencies. There still remains for discussion the formation of spines by processes of decrescence caused by extrinsic restraint (C) or by intrinsic deficiency of growth power (D). The lack of vitality or growth force generally stands so directly as the result of an unfavorable environment, that it is often difficult or impossible to distinguish between their action. Furthermore, as in the case of many parasites, it may be seen that the environment may be quite favorable as regards temperature, nutrition, etc.; but unfavorable in respect to motion and use of sensory and motive organs. From the almost universal degradation and retrogression of parasitic forms, it is necessary to consider these as intrinsically deficient, and therefore lacking in the qualities of growth force which normally favor a progressive evolution. Here, also, there are apparently two intimately associated causes. In an attached animal the absence of stimulus from disuse of an organ tends toward atrophy, and the retrogressive development serves to affect many organs in the same manner. The direct and indirect results of the restraint of the environment may therefore be expected to shade imperceptibly into each other, with only the extremes sufficiently distinct for separation.

The influence of an unfavorable environment as affecting the character and growth of plants and animals is well shown in desert or arid regions, and the flora has been made the subject of especial study by Henslow.[33] In such regions the first thing to impress the observer is the small size of the species. Next to diminutive size, the scantiness of life is a striking feature, for large areas are common in which life is almost wanting. An examination of these plants reveals a series of characters not usually present elsewhere, among which may be mentioned the development of a minimum amount of surface, constituting what is known as consolidated vegetation; next their uniform gray color, due either to excessive hairiness or a coating of wax; and lastly, their frequent spinescent characters.

The spines on desert plants are a feature of such general occurrence that it has led to the notion that vegetable spines are always associated with unfavorable conditions and are therefore suppressed structures. This is probably incorrect, for in plants, as in animals, spines may be developed by the progressive differentiation of previous structures; as in the angular edges of the leaf stems of many Palms becoming spiniferous, or, as will be shown, suppressed structures may arise from deficiency of growth force. In all cases spines may or may not serve for protection. Thus, while they are not always an indication of unfavorable environment, those occurring on desert plants may generally be so considered, for they are developed out of structures which are normally of vital physiological importance.

An animal or plant having spines and living in a favorable environment, involving freedom of motion for animals and abundance of nutrition without extremes of temperature or dryness for both animals and plants, will, it is believed, from the discussions and analyses of spine genesis in its various phases, develop these features in most instances without the sacrifice of organs and structures having important physiological and motor functions. Thus, ordinarily, among animals it is found that spines arise as excresceuces or outgrowths of exoskeletal or epidermal tissues, without seriously affecting the function of the organ or organs upon which they are located. Such cases may clearly belong to the most progressive series, and in fact usually occur there.

On the other hand, if it is found that a leg, a wing, a digit, or other organ is developed into a spine, this is always accomplished by a process of retrogression, resulting in the greater or lesser suppression of the part in question. It is also seen that this kind of spine occurs most frequently in retrogressive series or in others showing arrested development, and the necessary interpretation seems to be either that the environment is or has been unfavorable, at least so far as the particular organ or set of organs is concerned, or that the vital power has declined. Both influences are intimately associated, and the latter is often the direct result of the former.

The stunting effects of aridity and barren soil on common plants is familiar to all. Among the plants of the desert is found every evidence of similar stunting combined with adaptations to resist the unfavorable conditions of deficient water supply, excess of radiation, etc. The diminution in size applies not only to stature, but to the leaves and branches, especially the parenchymatous tissues or parts of the plant engaged in aërial assimilation. Consonant with these changes, the drought and other conditions produce a hardening of the mechanical tissues, which is of great aid in resisting the extreme heat and dryness of the desert. Sometimes a deposit of wax affords a similar protection.

The reduction of the leaves takes place in various ways. They may simply become smaller in every dimension and finally be reduced to mere scales, or an aphyllous condition may be established. They may grow narrower and narrower until only the hardened veins or midrib remain; or leaves may be developed only for a short time, and in the case of compound leaves after the shedding of the leaflets a spiniform leaf axis remains; as in *Astragalus Tragacantha*[38] (figures 55, 56). The suppression of branches tends toward the same end; namely, either to their complete disappearance or to their partial suppression into hard spiniform processes or thorns. Thus leaves, branches, and other parts of the plants may become reduced to their axial elements, bringing about what is commonly termed spinescence.

The spiny character of these plants is therefore one of the results of an arid environment, and it may or may not be of sufficient frequency to give an especial character to a particular desert flora. There is, moreover, a secondary influence which has an effect in determining the abundance of spinose plants in desert as well as in many other situations. This relates to the destruction of the edible unarmed species by herbivorous animals, and the comparative immunity of the spiny types. Thus, in old pastures, the prevailing flora is apt to be one that is offensive to grazing animals. This character is generally given by poisonous plants or those having a disagreeable flavor, or by those whose form or spiny structures afford protection.

This secondary influence by grazing animals may have had some effect in determining the particular abundance of spiny plants in certain desert regions, and their comparative infrequency in other similar regions. In either case the unfavorable environment brings about a suppression of structures, and one type of this action results in the production of spines. These represent the limits of retrogression before the part becomes entirely obsolete.

Wallace has criticised Henslow's views on the origin of xerophilous plants and their distribution. It is believed that the views here offered remove some of the objections, and bring the opinions of these authors into greater accord.

Under arid conditions bracts, stipules, leaves, and even branches may become spinescent. Some forms in which the spinose character has not as yet become permanently fixed by heredity, when transported or found living in moister and richer soils, develop normal leaves or branches, and lose their spinescence; others, like the Cactus, retain their spines under similar changes; while still others, as *Acanthosicyos horrida*,[33] cannot be artificially cultivated, and have become truly xerophilous types.

As examples of plants which lose their spines by cultivation, the Pear, species of Rose, Plum, etc. (Henslow), may be cited. According to Henslow,[33] others, as *Onomis spinosa*, have an especially spiny variety (*horrida*) living on sandy sea-shores, while in more favorable natural situations the same plant becomes much less spiny, and under cultivation loses its spines. M. Lothelier[42] also found that by growing the Barberry (*Berberis vulgaris*) in moist air, the spines disappeared, the parenchyma of the leaves being well formed between the ribs and veins. Dry atmosphere and intense light both favored the production of spines.

Henslow[33] cites the genus *Zilla* as a desert plant in which the branches are transformed into spines, *Echinops* for a similar modification of the foliage, *Fagonia* for spiniform stipules, and *Centaurea* for spinescent bracts. As further illustrations taken not only from desert plants but also from others commonly found in dry, rocky, or unfertile situations, the following examples may be taken, some of which are familiar cultivated species: The stunting of branches into spines is common among neglected Pear and Plum trees, and is a normal character in the Hawthorn, Honey-locust, *Cytisus* (figure 54), *Vella*, etc. Leaves transformed into spines are characteristic of the Cactaceæ of America, the columnar Euphorbiaceæ of Africa and southern Asia, and are also familiar in the half-shrubby Tragacanth bushes (figures 55, 56) so common in southern Europe, especially in the eastern portion, and in the ordinary Barberry (figure 13). Spiniform stipules are usually present in the species of *Robinia*, of which the Common Locust (*Robinia Pseudacacia*) furnishes a well-known illustration (figure 57). Spiniform bracts are best known among the Thistles (*Cirsium lanceolatum, C. horridulum*, etc.).

54   55   56   57

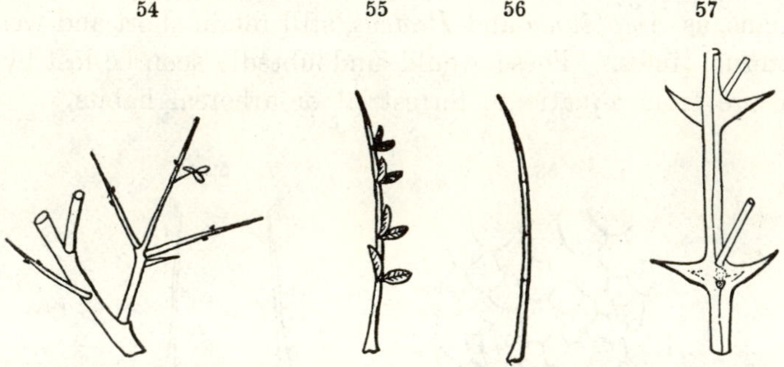

FIGURE 54. — The spiny *Cytisus* (*C. spinosus*) ; showing suppression of branches into spines. (After Kerner.)

FIGURE 55. — A single leaf of Tragacanth (*Astragalus Tragacantha*), from which the three upper leaflets have fallen. (After Kerner.)

FIGURE 56. — Leaf axis of the same, from which all the leaflets have fallen. (After Kerner.)

FIGURE 57. — Twig of Common Locust (*Robinia Pseudacacia*) ; showing spines representing stipules.

As the restraint of an environment acting on an animal so generally results in the disuse and atrophy of the organs affected, most cases will have to be considered under the head of disuse. Therefore, while the environment is the primary factor, its influences are mainly exhibited through secondary or resultant conditions. In some cases, however, it is possible to interpret a vestigial or suppressed structure directly into terms of an unfavorable environment. Thus, if the probable origin of the vestigial hind legs of a Python is considered, it leads to the belief that they represent legs which were of functional importance to some of the early ancestors of this snake. The gradual elongation of the body and the consequent change from a walking or direct crawling habit to a mode of progression chiefly by horizontal undulations, necessarily brought the legs into a relation with the environment which was unfavorable either for their function or growth. Their suppression is complete in most snakes, but in the Python the hind legs are represented by two spurs or spines (figures 58 and 59). On the interior of the body they are supported by vestiges of femora and ilia, showing their true affinities with hind limbs. Some snake-like batrachians, as *Amphiuma* and *Proteus*, still retain short and weak external limbs. These would undoubtedly soon be lost by a change from aquatic to terrestrial or arboreal habits.

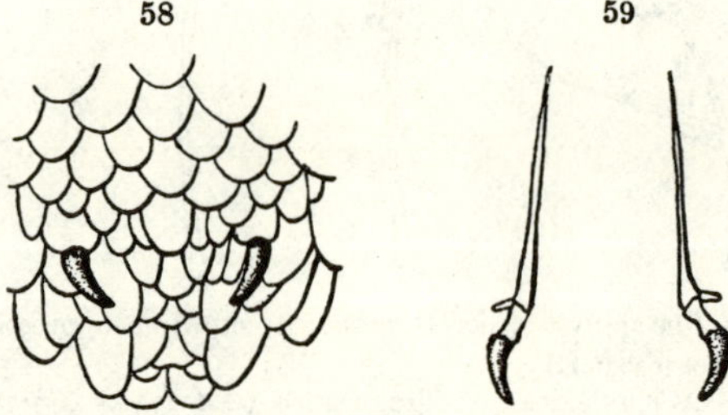

FIGURE 58. — Portion of skin of Python; showing the spurs which represent the suppressed or vestigial hind legs.　× ⅓.　(After Romanes.)

FIGURE 59. — Bones of suppressed legs of Python. All but the claw-like termination are internal.　× ⅓.　(After Romanes.)

In explanation of the nodes and spiniform processes on the epitheca of *Michelinia favosa*, it may be suggested that they represent aborted corallites or attempts at budding. This coral belongs to the order Porifera, which has been shown by the writer[7] to have very pronounced tendencies toward proliferation, and on the interior of the colony these attempts result in the production of mural pores. Most of the species of *Michelinia* are hemispherical or spherical. *M. favosa* is inclined to be pyriform in shape, rising above the object of support, and thus presenting a rather large epithecal surface. Manifestly the lower side of the corallum is unfavorably situated for the growth of corallites, and any efforts at proliferation on the part of the peripheral corallites is apt to result in stunted outgrowths. There is here a very close connection between restraint of environment and deficiency of growth force. If the whole corallum is taken into consideration, the restraint of the environment may be taken as preventing the growth of corallites on the lower side. If one of these single stunted corallites is considered, it may be said that the deficiency of growth force through lack of nutrition caused its suppression.

IX. *Mechanical restraint.* (C$_2$.)

Among the factors of spine genesis mechanical restraint is probably of the least importance. It can only rarely happen that an organism is forced to grow a spine contrary to the natural tendencies of normal development. Yet, as there are occasional types of spiniform structures which can be best explained as due to the mechanical restraint of the environment, it is necessary to notice them in order to make the categories of origin as complete as possible.

The illustrations will be taken chiefly from the Brachiopoda and Trilobita. The recent brachiopod *Mühlfeldtia truncata* is semi-elliptical in outline, and has a very short stout pedicle which holds the shell so closely to the object of support that the beak is truncated from abrasion and resorption. In specimens attached to a small branch of a coral, thus allowing the cardinal extremities of the shell to project beyond the object of support, the ends of the hinge are generally rounded. Specimens growing on a large flat surface have the cardinal extremities angular or sub-mucronate. Similar variations are to be observed in other living species of Brachiopoda (*Cistella*, some *Dallina*, etc.). Some of the extinct genera show more highly developed cardinal extremities which are often very characteristic of certain species, though considerable variation is found to exist. It is evident that these elongated hinge-lines have arisen from the mechanical necessities of a functional hinge, and their greater or less extent is also to a degree dependent upon the nature of the object of support, which furnishes a stimulus to the growing ends of the hinge. A marked example is shown in *Spirifer mucronatus,* with the cardinal angles extended into spiniform processes (figure 60). Similar features are presented by many other species of *Spirifer, Orthis, Leptæna, Stropheodonta,* etc.

FIGURE 60. — Dorsal view of *Spirifer mucronatus ;* Devonian ; showing spiniform cardinal angles. × ¾. (After Hall and Clarke.)

In the Trilobita the pygidium, or abdominal portion, consists of a number of consolidated segments, and the segments of the thorax are successively added in front of this tail piece. The first thoracic segment is therefore formed between the cephalon and pygidium, and its form is mechanically in agreement with the requirements of the animal for bending the body, and with the adjacent margins of the cephalon and pygidium. In a way it may be said that the segment is moulded by the adjacent parts, and may therefore take its form from the cephalon (figure 61), or from the pygidium, as in the examples following (figures 62–65).

During growth the new segments are added in front of the anal segment, so that after the number of abdominal segments is complete the thorax is increased by the successive addition of what in earlier moults were pygidial segments. By this means the pygidium generally controls or determines the character of the segments of the thorax. If the pleura of the pygidium are extended into spiniform processes, the pleural ends of the segments are also spiniform; as in *Lichas* (figure 64), *Ceraurus, Cheirurus* (figure 62), *Deiphon* (figure 63), *Acidaspis, Dindymene,* etc.

Likewise, if the pleura or their distal ends are directed posteriorly nearly parallel to the axis, the mechanical necessities of motion require that the portions of the free segments pointing backward should be free, thus making the ends of

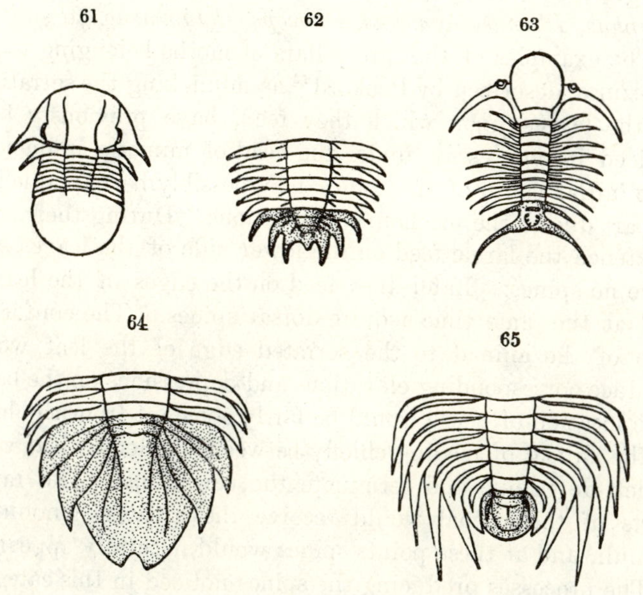

FIGURE 61. — *Illænus (Octillænus) Hisingeri*, Ordovician, Bohemia; a trilobite; showing spiniform pleural extremities of first thoracic segment, corresponding to the genal spines of the cephalon. × ¾. (After Barrande.[4])

FIGURE 62. — *Cheirurus insignis*, Silurian, Bohemia; pygidium and six thoracic segments. × ¾. (After Barrande.)

FIGURE 63. — *Deiphon Forbesi*, Silurian, Bohemia; entire specimen; showing spiniform pleura of segments corresponding in direction to those of the pygidium. (After Barrande.)

FIGURE 64. — *Lichas scabra*, Silurian, Bohemia; pygidium, with three thoracic segments; showing spiniform ends of pleura. × ¾. (After Barrande.)

FIGURE 65. — *Paradoxides spinosus*, Cambrian, Bohemia; pygidium and six free segments. × ¾. (After Barrande.)

the thoracic pleura generally appear as retrally curved spiniform extensions. Extreme examples of retrally directed pleura accompanied by small pygidia are shown in *Paradoxides* (figure 65), *Holmia, Olenellus, Elliptocephala*, etc.

Genera having the ends only of the pleura directed backward are generally less inclined to form spiniform terminations. In contrast with these it is found that all the Trilobita having the pleura directed outward, and with entire pygidial margins, do not ordinarily develop long pleural spines; as *Asaphus, Illænus, Agnostus, Phacops, Calymmene,* etc.

The examples of the caterpillars of moths belonging to the Schizuræ, described by Packard [54] as mimicking the serrations of the leaves upon which they feed, have previously been noticed in this essay, under the head of mimetic influences. The initial cause of the spines may possibly be explained as in part due to the mechanical conditions. During their early existence the larvæ feed on the lower side of the leaves, and have no spines. Later they feed on the edges of the leaves, and at the same time acquire dorsal spines. The conformation of the animal to the serrated edge of the leaf would produce corresponding elevations and depressions on the back. The location of these would be fairly constant from the habit of the animal of feeding chiefly between the denser leaf veins which determine and terminate the serrations. The raised parts of the animal would receive the greatest amount of stimuli, and at these points spines would naturally appear.

The processes producing the spines noticed in this category (IX) are classed with others under decrescence, for the reason that the growth is restrained or controlled by mechanical necessities. If the restraint were absent, it is probable that a more expansive growth would take place or that other structures would be correspondingly benefited.

X. *Disuse.* (C$_8$, D$_2$.)

In causing the reduction or atrophy of an organ, the effects of disuse have generally been recognized by most observers. In this way the origin of many of the so-called "rudimentary organs" has been satisfactorily explained by Darwin [14] and others. Two classes of structures are evidently comprised within the common definition of rudimentary organs; namely, nascent and vestigial organs.

Nascent structures indicate the beginnings or initial stages of organs, while vestigial structures are the remnants left after the functional suppression of organs. The suppression is usually caused by unfavorable conditions or by disuse, which produces either a retardation of growth or a retrogressive development. In both cases the results are similar. By retardation an organ is prevented or restrained from functional development and is therefore useless as a normal organ. By retrogression an organ gradually reverts to an initial type, loses its function, and becomes a vestigial structure. In most instances a change of food or habit or the substitution of a new and functionally higher structure causes the disuse of some organ which under previous conditions was of use to the animal.

Nascent structures, or the beginnings of organs, are generally made up of active tissues that only require stimulus and nutrition to perfect their function. On the other hand suppressed or vestigial structures are composed of comparatively inert tissue, and are in consequence largely made of the mechanical elements of secretion of the organism. It may therefore be considered that true rudimentary or nascent organs are potentially active, and suppressed structures are inert. It is with the latter class, the inert, that a study of spine genesis by atrophy is chiefly concerned.

The gradual loss of function through disuse, and the consequent loss of nutrition with the concomitant rapid decrescence of active tissues, bring about a change in the ratio of active and inert structures. The progression of this process naturally results in the production of a structure having a maximum of inert or mechanical tissues and a minimum of active constituents. Moreover, it has already been shown that the axial elements are the most persistent, and therefore the last to disappear; also that the peripheral appendages and outgrowths of any organ first show the action of decrescence. Evidently the conditions here described are favorable for the production of spines out of an organ primarily possessing distinct active functions. The axis of an organ gives the necessary form, and the hard tissue the structure, so that the whole will conform to the definition of a spine given early in this paper; namely, a stiff, sharp-pointed process.

The restraint of the environment was found to be one cause for decrescence of organs. Another, which is properly the subject matter of the present section, is disuse; and lastly, it will be seen how the deficiency of growth force may bring a similar suppression of structures.

There is considerable difficulty in selecting particular examples which will conform clearly to the strict requirements of these three categories. In a certain sense some of the examples of spines produced by decrescence may belong to more than one category. However, it does not prevent the acceptance of any one of the three as primary causes. Thus it may be urged that disuse has caused the atrophy of leaves into spines among many desert plants, or produced a similar reduction of the limbs in a Python. While this may be true from one point of view, yet the manifest unfavorableness of the environment in both seems to be a sufficient reason for making it the primary factor. On the other hand many parasites showing similar atrophies are not dependent upon a large number of active organs for their food and maintenance. After finding a host an abundance of food is at hand, and the environment may be considered a favorable one. All the organs, except those of nutrition and reproduction, then become more or less useless and dwindle away, leaving vestigial organs or disappearing altogether. Furthermore, a change of habit, as from climbing to flying, will necessarily cause the atrophy of some of the structures used for climbing and the hypertrophy of others for flying.

Most of the examples illustrating the production of a spine through the atrophy of an organ by disuse are to be found in the legs and digits of animals. The process bears considerable resemblance to the formation of spines on many plants by the suppression of leaves, branches, etc. They will be noticed here, although properly these vestigial structures among animals are more strictly of the nature of claws, or, at the most, spurs.

Many parasitic plants, especially among the Balanophoreæ, are reduced to a simple stem bearing the inflorescence. The leaves are represented by scales which are often spiniform, though seldom of sufficient stiffness to entitle them to be called spines. In desert plants, many of which have a similar type of growth, the hardening of the mechanical tissues by the effects of drought has converted similar leaf structures into spines, while the parasitic plants are not normally subjected to such continuous dryness and extreme heat, and therefore the mechanical tissues seldom become hardened.

Parasitic animals, especially among the Crustacea and insects, often show a reduction in the number of joints in the legs, and even in the number of limbs themselves. The terminal claws generally persist, and are sometimes longer than the rest of the leg; as in the Itch-mite *Sarcoptes Scabiei*, and in the female of the parasitic copepod *Lernæascus nematoxys* (figure 66).

Among many aquatic Crustacea and limuloids, the specialization and segregation of the ambulatory and swimming appendages toward the head or anterior regions of the body have produced a corresponding suppression of appendages on or near the extremity of the abdomen. This statement of fact is the basis of the principle of cephalization of Dana,[12] who applies it especially to the Crustacea, as follows: "There is in general, with the rising grade, an abbreviation relatively of the abdomen, an abbreviation also of the cephalothorax and of the antennæ and other cephalic organs, and a compacting of the structure before and behind; a change in the abdomen from an organ of great size and power and chief reliance in locomotion, to one of diminutive size and no locomotive power." Audouin's law that among the Articulata one part is developed at the expense of another may be also noticed here as affording a further explanation of the suppression of the posterior appendages correlative with the greater development of the parts anterior to them. In a Crustacean using its tail for propulsion, as the Lobster (*Homarus*), the telson is broad and flat, and the adjacent segment has a similar development of the appendages. In other forms, as the Horse-shoe Crab (*Limulus*) and the Phyllocarida, the tail is not used for propulsion, and at best serves chiefly as a rudder, while some of the legs on the anterior part of the abdomen or on the thorax are large and strong and are often provided with paddles. These groups, the limuloids and Phyllocarida, show a greater or less suppression of the last abdominal appendages, and in many genera the body terminates in a spiniform telson or tail spine. The process of suppression may or may not result in a spine. In the crabs the abbreviated abdomen is folded under the cephalothorax, and in *Lepidurus* and *Pterygotus* the telson is a scale or plate-like organ. For the most part, however, the abbreviation of the abdomen and the suppression of its appendages have reduced the telson to a spine; as in *Limulus* (figure 67), *Eurypterus*, *Stylonurus*, and *Prestwichia* among limuloids, and *Olenellus* among the Trilobita. In addition to a telson spine, the Phyllocarida have two lateral spiniform cercopods, the three spines together constituting the post-abdomen; as in *Ceratiocaris*, *Echinocaris* (figure 68), *Mesothyra*, etc.

Although the last abdominal segments of the Horse-shoe Crab have lost their appendages and show evidences of suppression, yet the tail spine is a large and useful organ, for it is of just the proper length to enable the animal to right itself after being overturned, which it is unable to do with its feet alone. The process of natural selection has doubtless in this way contributed to the development and retention of the long spine. This use cannot be ascribed to the tail spines of the Phyllocarida, though they evidently were important aids in directing movement, and also offered some degree of protection.

The terminal claws on the phalanges of the wings of some birds are nearly all that remain of the external fingers or digits. In the Hoactzin of South America (*Opisthocomus cristatus*) the young bird has a thumb and index finger, both provided with claws, and climbs about much like a quadruped, using its feet, fingered wings, and beak. According to Lucas,[43] a rapid change "takes place in the fore limb during the growth of the bird, by which the hand of the nestling, with its well-developed, well-clawed fingers, becomes the clawless wing of the old bird with its abortive outer finger." Similar claws or spurs occur on a number of other birds, some having functional wings, as in the example just described, and others having only vestiges of wings, as in the Wingless Bird of New Zealand (*Apteryx*, figure 69).

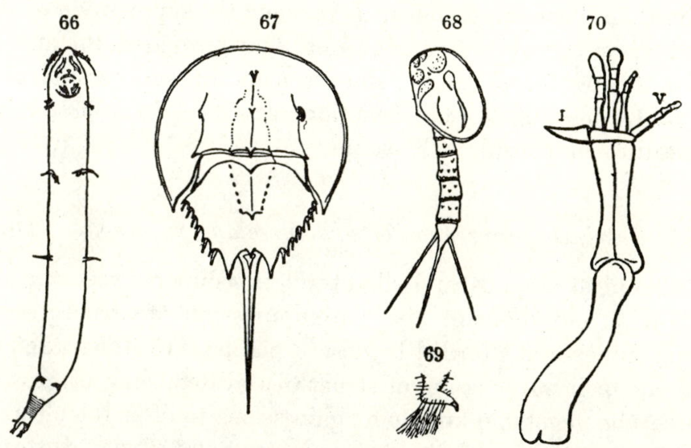

FIGURE 66. — Female of *Lernœascus nematoxys*, a parasitic copepod; showing suppression of limbs. Enlarged. (After Claus.)

FIGURE 67. — Horse-shoe Crab (*Limulus polyphemus*); showing telson spine and abbreviated abdomen. Reduced.

FIGURE 68. — A Devonian phyllocarid (*Echinocaris socialis*); showing spiniform telson and cercopods.

FIGURE 69. — Wing of *Apteryx australis*. $\times \frac{1}{8}$. (After Romanes.)

FIGURE 70. — Skeleton of right fore limb of the Jurassic Dinosaur *Iguanodon bernissartensis*; showing partially suppressed first digit. $\times \frac{1}{30}$. (After Dollo.[16])

Another example may be taken from the Dinosaurian Reptiles. The Jurassic genus *Iguanodon*, from England and Belgium, belongs to a group (Ornithopoda) in which the number of functional digits varies from three to five in the manus, and from three to four in the hind foot. In this genus the hind foot had three functional toes, representing the second, third, and fourth of a normal pentadactyl foot.

The first is represented by a slender tarsal bone alone, while the fifth is completely suppressed. The manus, or fore foot, of this animal shows the second, third, fourth, and fifth digits of functional importance as digits, while the first is shortened and atrophied to the condition of a stout spur, standing out at right angles to the axis of the leg, as shown in figure 70. The fore legs of *Iguanodon* and others of the same order were short, and apparently used more for prehension than locomotion, and in *Iguanodon* the suppression of the pollex, or thumb, into a spur doubtless provided the animal with a powerful weapon. Here is seen the suppression of a digit by loss of normal function, resulting in a protective structure of considerable value.

XI. *Intrinsic suppression of structures and functions.* (D₁.)

The most obvious and direct relationship between an unfavorable environment and the suppression of structures to form spines was afforded by desert plants. In illustration of the intrinsic suppression of structures by deficiency of growth force, the vegetable kingdom again seems to offer the clearest evidences of a like relation between cause and effect. Instead, however, of taking an unfavorable environment, in the present instance a favorable environment must be assumed, and then a type which expresses in various ways its deficiency of growth force must be sought.

In the desert plants it was found that no single family exclusively constituted the desert flora, but that a considerable variety of types was present, and that some of these belonged to perfectly normal families commonly living under ordinary favorable conditions. Moreover, it was evident that there were certain types of form and habits of growth which were especially characteristic of plants living in desert or similar unfavorable regions. Therefore, to illustrate clearly intrinsic restraint or suppression of structures it will be necessary to take an environment which, in most respects, may be considered as favorable, and also a type of plant life presenting evidences of a deficiency of growth force.

The great groups of plants commonly known as brambles and climbing plants appear to meet most of the requirements. They abound in regions where the greatest luxuriance of vegetation is found, and are therefore chiefly characteristic of the tropics. Kerner[38] estimates that there are two thousand species of the true climbing plants in the torrid zone, and about two hundred in temperate regions. Tropical America has the largest number of species, the flora of Brazil and the Antilles being especially rich. In the sombre depths of the tropical forest the climbing plants, or "lianes," are not so abundant as in the open glades and along the edge of the forest, where the amount of light is greater and the conditions of existence are more favorable. As far as richness of soil, amount of light, and degree of temperature are concerned, it must be admitted that their environment is as favorable as that of any of the associated plants having different habits of growth. The difference between the strong and erect plants and the comparatively weak and climbing forms is therefore not an extraneous one. It resides within the plant structures themselves, and is an intrinsic character or an expression of hereditary vital forces.

The law of recapitulation demands that each individual during its development shall pass through an epitome or recapitulation of its ancestral history. In view of the fact that the young seedlings of climbing plants and brambles have the erect form and proportions of normal erect foliage stems, it is safe to infer that they have been derived from erect forms. Further evidence is afforded from the absence of climbing plants in the earlier terrestrial floras. It is obvious, therefore, that they have been developed out of erect forms by a process of degradation.

The next striking feature to be noticed in climbing plants is their extreme slenderness, due to the general suppression of the plant body. They may attain lengths not reached by the highest trees, and yet the diameter of the trunk is but a minute fraction of the length. The Climbing Palm, or Ratan, has stems of great length and tenuity. It has been stated that stems two hundred metres long have been observed having a uniform thickness of only from two to four centimetres.[38] The diameter of such a stem would be only one or two ten-thousandths of its length. The length of the internodes is another conspicuous character in climbing plants, and both this and the slenderness of the stems suggest the results obtained by growing ordinary plants in the dark, where the conditions are adverse to increased vitality.

The transfer of function from one part of the plant to another, usually by a process of retrogression or degradation, is also very common. The first growth above the ground is a leafy stalk. Later, after the plant has attained a considerable height, the lower portion puts out quantities of rootlets and loses its foliage. The rootlets may be mere dry threads or points of support for the stem; or, if they happen to encounter a crevice containing soil, they develop into true absorbent organs. In others the ends of the growing stems or any point on the stems, upon reaching the earth, may put out vigorous roots. These facts seem to show a lack of positive differentiation throughout the plant, which admits of the substitution of a lower structure for a higher by the suppression of a higher function.

Lastly, the general spininess of climbing plants and brambles is a well-known and conspicuous character. Kerner[38] says that "most, if not all, plants which weave into the thicket of other plants are equipped with barbed spines, prickles and bristles." These spiniform processes seem to fall naturally into two classes: first, those produced by the suppression of stipules, leaves, petioles, branches, etc.; and second, those appearing as simple eruptions on the surface.

The suppression of normal plant organs into special structures, as tendrils and claspers, is extremely common, and, as already shown, this process, if carried far enough without complete suppression, will favor the production of a spiniform growth representing the axial elements of the organs.

The classes of organs thus affected are practically the same as those in desert plants, though varying somewhat in manner and degree. The consolidated type of plant body is naturally absent, for in this respect the diffuseness of climbing plants is quite antithetical. It does not seem necessary to give a long list of examples among the climbers, illustrating the suppression of organs into spines. Although

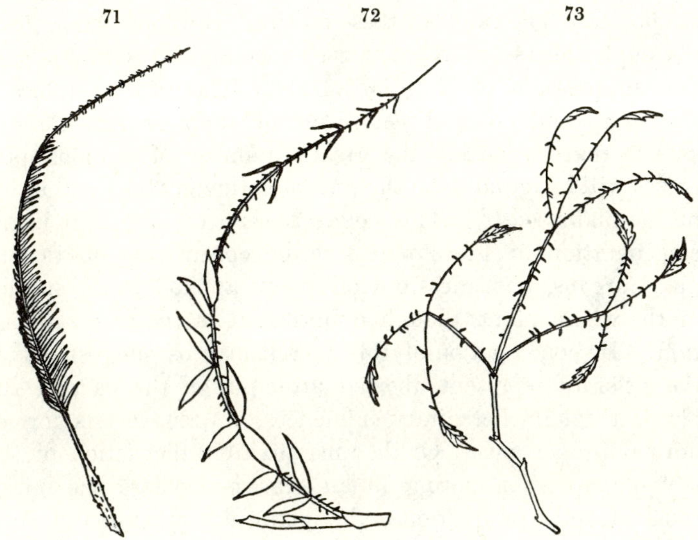

FIGURE 71. Leaf of Ratan (*Dæmonorops hygrophilus*). Reduced. (After Kerner.)

FIGURE 72. Leaf of Ratan (*Desmoncus polyacanthus*). Reduced. (After Kerner.)

FIGURE 73. Bramble (*Rubus squarrosus*). Reduced. (After Kerner.)

apparently not of rare occurrence, spines produced in this way are not as common as among desert plants. Two figures of the pinnate leaves of Ratan are introduced here to show the suppression of a number of the terminal leaflets into spines (figures 71, 72). In *Machærium* the stipules are converted into thorns.[62] A tropical *Bignonia* (*B. argyroviolacea*) has normal full-sized simple leaves, and suppressed leaves bearing two opposite leaflets on one stalk, and ending "in a structure which divides into three limbs, with pointed hooked claws, and which is not unlike the foot of a bird of prey." [38]

By far the greater number of spines on climbing plants are of the nature of prickles, and are not produced by the suppression of any particular organ or organs, but appear usually without any very definite order. They represent outgrowths of the superficial layers, and hypertrophied plant hairs, or trichomes. The cause of these cortical eruptions is not clear, although they seem to be intimately connected with the general suppression of the plant body. They are therefore a secondary and not a direct result of suppression. Bailey[2] asserts that "probably the greater number of spinous processes will be found to be the *residua* following the contraction of the plant body." This connection is very apparent in the consideration of the suppression or contraction of various plant organs, but is less obvious when applied to the surface of the whole plant, though doubtless it is the true explanation. In continuation of this idea it may be suggested that the prickles represent aborted attempts on the part of the plant, through hereditary influences, to recover its former normal proportions. Or they may exhibit the action of the law of repetition acting in an organism where the initial cause of spine production is the intrinsic suppression of such structures as leaves, petioles, stipules, etc. The subsequent repetition of spines on other parts of the organism results in a series of homoplastic spines which are not homologous with those first formed.

The prickles on climbing plants and brambles may often serve for purposes of protection (D_3), and enable the plant to cling to a support, but these utilitarian properties cannot be considered as an initial cause. Natural selection, also, probably has fostered the development of certain types of spiny climbers and the production of adaptive characters. Nevertheless, in studying these forms, it is necessary to revert to the original consideration of the localized suppression of normal plant structures, and to the general suppression of the plant body as affording a more primary conception of the causes and modes of spine growth among climbing plants.

In many cases of retrogressive series of animals there seems to be a close parallelism with some of the characters observed among the climbing plants. If the Ammonite family during the Cretaceous, or near the close of the Mesozoic, is taken as an example, it cannot be said that the environment of these old-age or pathologic series is unfavorable in respect to food, temperature, etc., for with them are associated many vigorous progressive series of other organisms. Neither can it be said that in many cases the animals perished on account of over-specialization, though this was evidently the cause of the extinction of a large number. The return to a condition of second childhood in old age cannot be called a progressive specialization, since it clearly points to a deficiency of growth force.

Old-age types, or phylogerontic forms, among animals may show the same attenuation or suppression of the body as do climbing plants. Thus *Baculites*, considered by Hyatt as a typical phylogerontic type, has a very attenuate shell, and some species, after attaining a certain diameter, cease to increase in any direction except length. On account of being a chambered shell, it is manifest that the growth of the animal must have practically ceased, while its secretive activities were continued and confined largely to lengthening the shell. Other related genera of Cephalopoda show a similar attenuation of the shell, evincing a stoppage of growth in the animal. Among the Mollusca it seems quite likely that attenuation of form often accompanies decreased growth power.

The pathologic varieties of the Steinheim *Planorbis*, as described by Hyatt,[35] or of the recent *Planorbis complanatus* described by Piré,[57] are further illustrations of this attenuation accompanying the uncoiling of the shell. The sedentary *Magilus*, immersed in its coral host, is also an example, for not only does the shell cease to increase in diameter, but the whole interior, except a small cavity at the end, is filled with a solid deposit of lime. Similar examples could be multiplied indefinitely. Since, however, but few of them are spiniferous, their consideration does not properly come within the scope of the present discussion, though, as is well known, some of the attenuate forms often enlarge and contract periodically, such enlargements frequently leaving prominent laminæ or nodes that are sometimes differentiated into spines. They suggest the observations on growth, senescence, and rejuvenation, made by Minot,[48] who showed that in guinea pigs from a very early age the increments of growth are in a steadily decreasing ratio to the increase of weight of the animal. This led to the general conclusion that the whole life of an individual is a process of senescence or growing old.

Spines arising by a real pathologic or diseased condition of the individual can have little or no effect in producing a normal spiniferous variety or species. However, some note should be taken of them, especially as they may be congenital, and thus appear through several generations. In the human species the peculiar skin disease known as ichthyosis sometimes produces spiniform excrescences, and the victims are commonly called "porcupine-men." The most celebrated instance was the Lambert family. Haeckel[27] gives the following account of this family: "Edward Lambert, born in 1707, was remarkable for a most unusual and monstrous formation of the skin. His whole body was covered with a horny substance, about an inch thick, which rose in the form of numerous thorn-shaped and scale-like processes, more than an inch long. This monstrous formation of the outer skin, or epidermis, was transmitted by Lambert to his sons and grandsons, but not to his granddaughters. The transmission in this instance remained in the male line, as is often the case." Other similar examples are cited by Gould and Pyle,[21] and the disease is described as "a morbid development of the papillæ and thickening of the epidermic lamellæ."

Categories of Interpretation

Having thus far examined the factors governing the origin of spines, and found that they could be grouped into a number of distinct categories, it is now desirable to interpret these results, and endeavor to arrive at the real significance of the spinose condition.

The two main generalizations which will be discussed are, first, that spinosity represents the limit of morphological variation, and, second, it indicates the decline or paracme of vitality.

Spinosity a Limit to Variation.

A number of data have already been given, leading to the belief that, on becoming spinose, organisms have reached a limit of morphological variation. They may continue to develop more and more differentiated and compound spines, but no new types evolve out of such a stock.

The subject may be treated in two ways, both leading to the same conclusion. First, the stages and processes involved in the growth of a spine itself may be studied, and next the development of spines in the ontogenies and phylogenies of animals and plants may be examined.

The growth of a spine has already been described, and it was shown that this type of growth may arise from specialization of other ornamental features, such as nodes, ridges, and lamellæ, and also from the decadence of leaves, legs, etc. These observations and numberless others which could be made, will be sufficient to show that almost any kind of superficial structure, as knobs, tubercles, ridges, laminæ, reticulations, etc., has by differential growth been changed into spines; also, that organs of various kinds, as legs, branches, leaves, etc., have by atrophy been reduced to spines. In each case the parts in their development pass through the various intermediate stages, and clearly show that the spine is a result and not a mean. Moreover, none of these structures or organs are developed through the contrary process; namely, that of beginning with spines and passing through stages corresponding to laminæ, ridges, tubercles, etc. The spine is the limit, and out of it no further structure is formed.

It is necessary to make some mention here of the movable spines of Echinodermata, which appear to form an exception to the foregoing statements. There seems to be no doubt that the fixed and movable spines, the pedicellariæ, the paxillæ, and the spheridia, are homologous structures, and that all begin as spiniform skeletal outgrowths, which by subsequent growth and modification produce the structures mentioned (Agassiz [1]). The echinoderm skeleton, including spines, etc., is deposited in the midst of living tissue, and in the case of the spines cannot be directly correlated with the spines of other classes of organisms, which are either very deficient in vitality or are dead structures as soon as completed. After the movable spines of echinoderms are fully developed, the living portion is often confined to the base, and the shaft becomes simply a dead structure upon which encrusting organisms may find lodgment, a condition seldom occurring in the living spines. These finished spines never develop into anything else, and are the structures which conform to the present discussion. The embryonic condition of the spines and pedicellariæ shows that they are really more internal than external structures, and therefore remain under the full control of the ordinary processes of growth, resorption, and modification by living tissues. Furthermore, the movable spines are of such functional importance that no close homologies can be made with ordinary spines found in other classes of organisms.

In tracing the ontogeny of a spinose form, it has been found (pp. 18–22) that each species at the beginning was plain and simple, and at some later period spines were gradually developed according to a definite sequence of stages. Usually after the maturity of the organism the spines reach their greatest perfection, and in old age there is first an over-production or extravagant differentiation, followed by a decline of spinous growth, and ending in extreme senility with their total absence.

There are abundant reasons for believing that the radicles of groups are undifferentiated and inornate, and whenever a class has had a long existence it has been by the continuance of such radical types or by the development of secondary or tertiary radicles, which, though differing in internal characters, still retain a primitive simplicity in superficial features. The early stages of ontogeny of any form should agree with the radical stock, and, as already noted, these stages are simple. Hyatt[34] says on this point: "the evidence is very strong that there is a limit to the progressive complications which may take place in any type, beyond which it can only proceed by reversing the process and retrograding. At the same time, however, the evidence is equally strong that there are such things as types which remain comparatively simple, or do not progress to the same degree as others of their own group. Among Nautiloidea and Ammonoidea these are the radicle or generator types. No case has yet been found of a highly complicated, specialized type, with a long line of descendants traceable to it as the radicle, except the progressive; and all our examples of radicles are taken from lower, simpler forms; and these radicle types are longer-lived, more persistent, and less changeable in time than their descendants."

A few examples will now be taken from the life histories of large groups. In the Brachiopoda the order Protremata, containing most of the spinose forms, has 4 genera and 22 species in the Cambrian of America, 20 genera and 173 species in the Ordovician, and 30 genera in the Silurian. "Then began a steady decline, with extinction in the Carboniferous of North America. In the Triassic of Europe this order is sparingly represented by small species, and is there essentially restricted to the family Thecidiidæ, which continues to have living representatives in the Mediterranean

Sea " (Schuchert[64]). The super-family Strophomenacea of this order is the longest lived, and excelled in amount of specific differentiation, there being 608 species in North America alone (Schuchert). In this super-family the early families and genera were without spines, it being only when *Chonetes* is reached that the first spines are found in the order. In this genus they are along the hinge and seem to make up for the weak and obsolescent pedicle. Greater spine growth occurs in the genera *Productella* and *Productus*, where, in extreme cases, the surfaces of both valves are thickly studded. During the Carboniferous the spiny Producti attained their maximum both in number, length of spines, and in individual size, for here occur the largest species of all Brachiopoda. This was the climax. The Permian genera are chiefly degenerate forms (*Aulosteges, Strophalosia*), and with the close of the Paleozoic the family Productidæ became extinct. The order Protremata, to which this family belongs, likewise underwent a rapid decline, and only two simple types continued on into the Mesozoic, while but one declining representative is living at the present time.

Among the Ammonoidea the chief spiny forms are those occurring just before the final extinction of the group and representing the beginning of the decline of the order (*Crioceras, Toxoceras, Ancyloceras, Hamites*, etc.). In the Dinosaurian Reptiles the great horned forms, *Triceratops, Torosaurus*,[39] etc., mark the extinction of the entire order. The great horned mammals of the Eocene, the Dinocerata, have left no descendants, and the giant Brontotheriidæ, after undergoing various horn modifications through the Miocene, continued no further.

It is not desirable, however, to convey the impression that the spines or horns are alone responsible for this wholesale extinction. It has been shown that they are undoubtedly often an expression of extreme specialization, and generally they represent the limits to which superficial structures may be differentiated. Although there may be other expressions for similar conditions, yet the presence of spines is one, if not the most evident, marker of the attainment of these limits. The presence of a spine on an organ or part indicates the limit of progression or regression of that part or organ. If the spinose condition is general, or if it dominates important functions, it then indicates the limit of progression and regression of the organism.

Spinosity the Paracme of Vitality.

The physiological interpretation of spinosity is a correlative of the morphological aspect of the same condition, and, as it was found that spinosity was a limit to morphological progress or regress, it will now be shown that it also indicates the paracme or decline of physiological progress. Both inferences are drawn from the individual or ontogenetic standpoint, as well as from the racial or phylogenetic.

In the spinose individual the decline of vitality has been studied by Geddes [20] in thorny plants. He concludes that they show a " gradual death from point backwards (i. e. *ebbing vitality*)." The requisite evidence is afforded in the experience of gardeners who generally consider spiny plants as " always given to die back," or, as otherwise expressed, they " often prune themselves." It is difficult to adduce the same kind of evidence among animals, though there may be some degree of semblance between this self-pruning of spiniferous plants and the growth, death, and shedding of the antlers of the modern Deer. Stronger evidence of the relations of spinosity to the organism is afforded in the consideration of spines as consisting wholly of the mechanical tissues. They are more or less dead structures and are usually without special physiological function. Hence, in so far as the whole or a part of an organism is spinose, it represents the ratio between the mechanical and active tissues, or between the inert and living structures.

Morris [49] correlates the mechanical and motor defences of animals and plants in a manner bearing upon this subject as follows: " If we examine the whole range of the animal kingdom, we find every phase of combination of mechanical and motor defence, the motion growing more sluggish as the defensive armor grows more efficient. But in the whole kingdom motion persists as one of the defensive agencies. No animal exists without some power of motion, by whose aid it withdraws or otherwise escapes from danger." He also notes that the plant kingdom, with the exception of the minute, swimming forms, possesses no defensive motion, and that mechanical defence alone exists. Under mechanical defence are included thorns, spines, etc., together with chemical appliances; as in plants with poisonous or disagreeable juices. These facts lead to the conclusion that, in proportion as animals are spinose or armored, they exhibit a vegetative type of structure, and have retrograded.

It has been shown elsewhere in this article, that the greatest development of spinose organisms occurs just after the culmination of a group, and, as this period clearly represents the beginning of the decline of the vitality of the group, the spines are to be taken as the visible evidence of this decadence. A similar observation has been made by Packard,[54] who after passing in review the geological development of the Trilobita, Brachiopoda, and Ammonoidea, states that " these types, as is well known, had their period of rise, culmination, and decline, or extinction, and the more spiny, highly ornamented, abnormal, bizarre forms appeared at or about the time when the vitality of the type was apparently declining."

Furthermore, it is now commonly agreed that all groups have been most plastic near their point of origin, or, in other words, that during their early history all the important or major types of structure have been developed. Their subsequent history reveals the amount of minor differentiation and specialization they have undergone. Apparently, most of the early impulses of growth, whether from the environment or from vital forces, resulted in physiological changes producing fundamental variations in function and structure. The later influences of environment and growth force are expressed in peripheral differentiation, and show that the racial or earlier characters had become fixed, and that the later or specific features were the chief variables. The stimuli which, during the early life history of a group, were expended in internal or physiological adjustments, later produce external differentiation, and in this differentiation spinosity is the limit. The presence of spines, therefore, indicates the fixity of the primary physiological characters, together with the consequent inability of the organism to change due to its decreasing vitality.

Conclusion.

Just as all the features of terrestrial topography are included between the limits of plains and mountains, and the mountains are considered as the limit of progressive accidentation, so the spines of animals or the monticules and pinnacles of their surface may be considered as the limits of progressive differentiation. The primitive base level, or peneplain, becomes elevated, and by erosion is cut up into tablelands, mesas, and buttes, with intersecting valleys. The valleys are gradually deepened, and the country becomes rougher until a maximum is reached. Then follows a reduction of the inequalities of the surface, and finally, in old age, the smooth, gently rounded outlines of geographic infancy again appear. So in organisms the smooth rounded embryo or larval form progressively acquires more and more pronounced and highly differentiated characters through youth and maturity. In old age it blossoms out with a galaxy of spines, and with further decadence produces extravagant vagaries of spines, but in extreme senility comes the second childhood, with its simple growth and the last feeble infantile exhibit of vital power.

The history of a group of animals is the same. The first species are small and unornamented. They increase in size, complexity, and diversity, until the culmination, when most of the spinose forms begin to appear. During the decline extravagant types are apt to develop, and if the end is not then reached, the group is continued in the small and unspecialized species which did not partake of the general tendency to spinous growth.

Lastly, it must be determined whether spines are really hereditable characters, and therefore can be used in studying the phylogenies of groups. No one has yet been able to show any type or set of characters which cannot be transmitted from parent to offspring. Hyatt[34] says: " Every-

74	Ontogeny stages.	Ontogeny condition.	Phylogeny stages.	Phylogeny condition.	Chronology.
	Old age or gerontic	Paraplasis	Phylogerontic	Paracme	5
	Adult or ephebic	Metaplasis	Phylephebic	Acme	4
	Immature or neanic	Anaplasis	Phyloneanic	Epacme	3
	Larval or nepionic	Anaplasis	Phylonepionic	Epacme	2
	Embryonic	Anaplasis	Phylembryonic	Epacme	1

FIGURE 74. — Diagram and table; showing correlation of stages and conditions of development in the spinose individual, in its ancestry, and in time.

thing is inherited or inheritable, so far as can be judged by the behavior of characteristics." Furthermore, in a review of animal life, extinct and living, no one can fail to be impressed with the fact that especially near the close of the life history of a group, or in a series of highly specialized forms, spinose characters are often considered as of supra-varietal value, and are rated of specific, generic, and sometimes of family rank, or even higher. They have therefore acquired a fixed importance in these special groups, and are recognized in the same categories with physiological and structural characters. The differences which appear at an early period in higher genera are the bases of distinction among lower genera. If the spines or other similar features do not make their appearance in an individual until a late adolescent stage, they are usually of negative value in a scheme of classification. This agrees with the general principle recently suggested by Harris,[32] that when the main features of the ornament (= spines, etc.) are foreshadowed in the larval and early adolescent stages, they are to be regarded as of taxonomic value.

The preceding diagram illustrates the previous statements, and shows the correlation between the stages and conditions of growth in the ontogeny of a spinose individual, with its phylogeny, and also the chronology of groups containing spinose forms. The numbers indicating chronology simply refer to successive periods of time. In particular cases they may be long geologic ages; as Cambrian, Ordovician, Silurian, Devonian, and Carboniferous, or in other instances they may represent much shorter periods.

From the study of the ontogenies of spinose forms, it has already been ascertained that they were simple and inornate during their young stages; and from the phylogenies of the same and similar forms, it was likewise learned that they were all derived from non-spinose ancestors. It has also been shown that spines represent an extreme of superficial differentiation which may become fixed in ontogeny, and the further conclusion, that spinosity represents a limit to morphological and physiological variation, has been reached. Finally, it is evident that, after attaining the limit of spine differentiation, spinose organisms leave no descendants, and also that out of spinose types no new types are developed.

References.

1. Agassiz, A., 1873. — The Homologies of Pedicellariæ. *American Naturalist*, vol. vii.
2. Bailey, L. H., 1896. — The Survival of the Unlike. A collection of evolution essays suggested by the study of domestic plants.
3. Balfour, F. M., 1885. — A Treatise on Comparative Embryology, memorial edition.
4. Barrande, J., 1872. — Système Silurien du centre de la Bohême. Pt. I, 1852; supplement.
5. Bateson, W., 1894. — Materials for the Study of Variation treated with especial regard to discontinuity in the origin of species.
6. Beecher, C. E., 1897. — Outline of a natural classification of the Trilobites. *Amer. Jour. Sci.* (4), vol. iii.
7. —— 1891. — The development of a Paleozoic poriferous coral. *Trans. Conn. Acad.*, vol. viii.
8. —— 1895. — The larval stages of Trilobites. *American Geologist*, vol. xvi.
9. Brady, H. B., 1884. — Report on Foraminifera. Voyage of H. M. S. Challenger. Zoölogy, vol. ix.
10. Cope, E. D., 1887. — The Origin of the Fittest. Essays on Evolution.
11. —— 1896. — The Primary Factors of Organic Evolution.
12. Dana, J. D., 1876. — On Cephalization. Pt. V. Cephalization a fundamental principle in the development of the system of animal life. *Amer. Jour. Sci.* (3), vol. xii.
13. Darwin, Charles, 1854. — A monograph of the sub-class Cirripedia, with figures of all the species. *Ray Soc.*
14. —— 1871. — The Descent of Man and Selection in relation to Sex.
15. Davidson, Thomas, 1886–88. — A monograph of Recent Brachiopoda. *Trans. Linn. Soc.*, London, vol. iv.
16. Dollo, M. L., 1882. — Première note sur les dinosauriens de Bernissart. *Bull. Mus. Roy. d'Hist. Nat. de Belgique.*
17. Dybowsky, B. N., 1874. — Beiträge zur näheren Kenntniss der im Baikal-See vorkommender niederen Krebse aus der Gruppe der Gammariden.
18. Edwards et Haime, 1851. — Monographie des Polypiers Fossiles des Terrains Palæozoïques.
19. Faxon, Walter, 1876. — Exploration of Lake Titicaca by Alexander Agassiz and S. N. Garman. IV. Crustacea. *Bull. Mus. Comp. Zoöl.*, vol. iii.
20. Geddes, P., 1891. — On the Origin of Thorny Plants. *Rept.*

Sixteenth Meeting British Association for the Advancement of Science, for 1890, Leeds.

21. Gould and Pyle, 1897. — Anomalies and Curiosities of Medicine.
22. Gratacap, L. P., 1894. — The Numerical Intensity of Faunas. *American Naturalist.*
23. —— 1886. — Zoic Maxima, or Periods of Numerical Variations in Animals. *American Naturalist.*
24. Gray, Asa, 1872. — Lessons in Botany and Vegetable Physiology.
25. Günther, A., 1865. — On the Pipe-fishes belonging to the genus Phyllopteryx. *Proc. Zoöl. Soc. London.*
26. Haeckel, E., 1887. — Report on the Radiolaria. Voyage of H. M. S. Challenger. Zoölogy, vol. xviii.
27. —— 1876. — The History of Creation, English edition.
28. Hall, James, 1884. — Descriptions of the species of fossil reticulate sponges, constituting the family Dictyospongidæ. *Thirty-fifth Rept. N. Y. State Museum.*
29. —— 1884. — Lamellibranchiata I. *Pal. N. Y.*, vol. v, Pt. 1.
30. —— 1879. — The Fauna of the Niagara Group, in Central Indiana. *Twenty-eighth Rept. N. Y. State Museum.*
31. —— and Clarke, 1894. — Brachiopoda. *Pal. N. Y.*, vol. viii, II.
32. Harris, G. F., 1897. — Catalogue of the Tertiary Mollusca in the Department of Geology, British Museum (Nat. Hist.). Pt. I. The Australasian Tertiary Mollusca.
33. Henslow, George, 1895. — The origin of Plant-Structures by Self-Adaptation to the Environment, exemplified by Desert or Xerophilus Plants. *Jour. Linn. Soc., Botany*, vol. xxx.
34. Hyatt, A., 1895. — Phylogeny of an Acquired Characteristic. *Proc. Amer. Philos. Soc.*, vol. xxxii.
35. —— 1880. — The Genesis of the Tertiary Species of Planorbis at Steinheim. *Anniversary Mem. Boston Soc. Nat. Hist.*
36. Jackson, R. T., 1890. — Phylogeny of the Pelecypoda, the Aviculidæ and their Allies. *Mem. Boston Soc. Nat. Hist.*, vol. iv.
37. —— 1896. — Studies of Palæechinoidea. *Bull. Geol. Soc. of America*, vol. vii.
38. Kerner von Marilann, Anton, 1895. — The Natural History of Plants, their forms, growth, reproduction, and distribution, English edition by F. W. Oliver.
39. Kingsley, J. S., 1884. — The Standard Natural History. Vol. II. Crustacea and Insects.
40. Koninck, L. de, 1842–44. — Description des Animaux Fossiles que se trouvent dans le terrain Carbonifère de Belgique.
41. Leidy, Joseph, 1879. — Fresh-water Rhizopods of North America. *Rept. U. S. Geol. Surv. Terr.*, vol. xii.

42. Lothelier, M., 1890. — Revue general de Botanique. *Compte Rendu*, t. cxxi.

43. Lucas, F. A., 1895. — The Weapons and Wings of Birds. *Rept. U. S. Nat. Mus.* for 1893.

44. Lydekker, R., 1892. — Phases of Animal Life, Past and Present.

45. Marsh, O. C., 1897. — Principal characters of the Protoceratidæ. *Amer. Jour. Sci.* (4), vol. iv.

46. —— 1896. — The Dinosaurs of North America. *Sixteenth Ann. Rept. U. S. Geol. Surv.*

47. Micrographic Dictionary, 1883. — J. W. Griffith and A. Henfrey, fourth edition, J. W. Griffith, M. J. Berkeley, and T. Rupert Jones.

48. Minot, C. S., 1891. — Senescence and Rejuvenation. *Journal of Physiology*, vol. xii.

49. Morris, Charles, 1886. — Methods of Defence in Organisms. *Proc. Acad. Nat. Sci. Phila.*, vol. xxxiii.

50. Neumayr, M., and Paul, K. M., 1875. — Die Congerien- und Paludinenschichten Slavoniens und deren Faunen. Ein Beitrag zur Decendenz-Theorie. *Wien Geol. Abhandl.*

51. Nicholson, H. A., 1875. — Palæontology of the Province of Ontario.

52. —— and Lydekker, 1889. — A Manual of Palæontology, third edition.

53. Owen, Richard, 1868. — On the Anatomy of Vertebrates. Vol. III. Mammals.

54. Packard, A. S., 1890. — Hints on the evolution of the bristles, spines, and tubercles of certain caterpillars, apparently resulting from a change from low-feeding to arboreal habits; illustrated by the life-histories of some Notodontians. *Proc. Boston Soc. Nat. Hist.*, vol. xxiv.

55. Paget, James. — Lectures on Surgical Pathology, I.

56. Peckham, Elizabeth G., 1889. — Occasional Papers of the Natural History Society of Wisconsin. Vol. I.

57. Piré, Louis, 1871. — Notice sur le Planorbis complanatus (Forme Scalare). *Ann. Soc. Malac. de Belgique*, t. vi.

58. Poulton, E. B., 1890. — The Colors of Animals, their meaning and use, especially considered in the case of insects. *Internat. Sci. Series.*

59. Pritchard, A., 1861. — A History of the Infusoria, including the Desmidiaceæ and Diatomaceæ, British and Foreign, fourth edition.

60. Romanes, G. J., 1892. — Darwin and after Darwin. An Exposition of the Darwinian Theory and a Discussion of Post-Darwinian Questions.

61. Ryder, J. A., 1893. — The Correlations of the Volumes and Surfaces of Organisms. Contributions from the Zoöl. Laboratory
of the Univ. of Penn., vol. i.

62. Schenck, H., 1892. — Beiträge zur Biologie und Anatomie der
Lianen, Jena.

63. Schmankewitsch, W. J., 1872. — Ueber den Einfluss der physi-
calisch-chemischen Bedingungen auf die Organization von Branchipus. *Zeitschr. Wissensch. Zool.*, xxii.

64. Schuchert, C., 1897. — A synopsis of American Fossil Brachiopoda,
including bibliography and synonymy. *Bull. U. S. Geol. Surv.*,
No. 87.

65. Seeley, H. G., 1884. — Pisces, in Cassell's Natural History. Edited
by P. M. Duncan. Vol. V.

66. Spencer, Herbert, 1876. — Synthetic Philosophy. Vol. I. First
Principles, second American edition.

67. Vernon, H. M., 1897. — Reproductive Divergence: an Additional
Factor in Evolution. *Natural Science.*

68. Verrill, A. E., and Smith, S. I., 1874. — Report on the Invertebrate Animals of Vineyard Sound and adjacent waters.

69. Wachsmuth and Springer, 1897. — The North American Crinoidea
Camerata. *Mem. Mus. Comp. Zoöl.*, vols. xx, xxi.

70. Wallace, A. R., 1891. — Natural Selection and Tropical Nature.
Essays on Descriptive and Theoretical Biology, new edition.

71. —— 1894. — The Rev. George Henslow on Natural Selection.
Natural Science.

72. Williams, H. S., 1895. — Geological Biology. An introduction to
the Geological History of Organisms.

73. Zittel, K. A. von, 1881–1885. — Handbuch der Palæontologie.
I Ab., II Bd.

74. —— 1896. — Text Book of Palæontology. Trans. and edited by
C. R. Eastman. Vol. I, Pt. I.

II

STRUCTURE AND DEVELOPMENT OF TRILOBITES

II

STRUCTURE AND DEVELOPMENT OF TRILOBITES

1. OUTLINE OF A NATURAL CLASSIFICATION OF THE TRILOBITES *

(PLATE II)

INTRODUCTION

WITH the possible exception of the barnacles, no group of arthropods has received more varied treatment by specialists than the trilobites. This taxonomic uncertainty has been due mainly to a lack of knowledge of the structure, and to certain real or fancied resemblances to *Limulus*.

The early references of trilobites to the mollusks, insects, and fishes need not be noticed, for since they have been made the subject of special study they have been commonly classed with the Crustacea and placed near the phyllopods by most observers. Quite a number of naturalists, however, still divorce the trilobites and limuloids from the Crustacea and ally them with the arachnids. It is not proposed at this time to discuss the homologies of *Limulus*, but the trilobites show the clearest evidence of primitive crustacean affinities, in their protonauplius larval form, their hypostoma and metastoma, the five pairs of cephalic appendages, the slender jointed antennules, the biramous character of all the other limbs, and their original phyllopodiform structure. They differ from *Limulus*, not only in most of these regards, but

---

* *Amer. Jour. Sci.* (4), III, 89–106, 181–207, pl. iii, 1897.

also in not having an operculum.   From this and all other arthropods they are distinguished by having compound eyes on free-cheek pieces which apparently represent the pleura of a head segment that is otherwise lost, except possibly in some forms of stalked eyes and in the cephalic neuromeres of later forms.   The most recent discussions as to the affinities of trilobites are to be found in the papers by Bernard,[7, 8, 9, 10] Kingsley,[28] Woodward,[34] and the writer,[5] where from the facts presented their intimate relationships with the Crustacea follow as a necessary corollary.

Previous Classifications.

The various schemes of classification that have been applied to the trilobites since that of Brongniart,[11] in 1822, have been enlarged and revised by various authors, until at the present time no particular arrangement of the families or genera can be said to be endorsed.   The one which is generally recognized as manifestly faulty, that of Barrande,[3] is the one most commonly found in text-books and special memoirs. Barrande's definitions and limitations of the generic and family groups were natural and accurate, showing a most complete knowledge of generic and specific values, but in the grouping and arrangement of the families he selected characters of secondary rank.

Of all the investigators who have attempted any classification of the families, J. W. Salter[32] seems to have had the clearest insight into the important value of certain characters, and to have approached nearest to a natural system.   In zoölogical research the study of ontogeny and the principles of morphogenesis were then scarcely recognized as having any direct application.   It is quite remarkable, therefore, that Salter, as early as 1864, should have singled out, as the basis of his sub-divisions, the characters which are the dominant variants in ontogeny.

It is not necessary in this place to discuss all the classifications which have been proposed.   Barrande[3] gives a complete résumé down to 1850, and shows in a very satisfactory manner the weak points of each, furnishing strong reasons as to why they are unnatural and therefore untenable. The underlying principles of these early attempts at a classification are here briefly summarized: (1) The first classification of trilobites was advanced by Brongniart,[11] in 1822, in which all the forms previously known as *Entomolithus paradoxus* were shown to belong to five distinct genera. (2) Dalman,[16] in 1826, made two groups based upon the presence or absence of eyes. (3) Quenstedt,[30] in 1837, recognized the number of thoracic segments and the structure of the eyes as of the greatest importance. (4) Milne-Edwards,[28] in 1840, considered the power of enrolment as of prime value. (5) Goldfuss,[20] in 1843, made three groups, depending on the presence or absence of eyes and their structure. (6) Burmeister,[12] in 1843, accepted the two divisions of Milne-Edwards, and laid stress on the nature of the pleura and the size of the pygidium. (7) Emmrich's first scheme,[17] in 1839, was founded on the shape of the pleura, the presence or absence of eyes and their structure. (8) The later classification of the same author,[18] published in 1844, was based on whether the abdomen was composed of fused or free segments, and the minor divisions depended chiefly on the structure of the eyes and the facial suture. (9) Corda,[15] in 1847, placed all trilobites in two groups, one having an entire pygidial margin, and the other with the pygidium lobed or denticulate. (10) McCoy,[25] in 1849, took the presence or absence of a facet on the pleura for a divisional character. As this is an indication of the power or the inability of enrolment, it does not differ materially from the schemes of Milne-Edwards and Burmeister.

Zittel,[35] in a historical review brought down to 1885, includes in addition the schemes of Barrande and Salter, and remarks that the basis of Barrande's general grouping, namely, the structure of the pleura, has neither a high physiological nor morphological meaning. Both Barrande and Salter recognize nearly the same families, with slight differences, and the latter adopts a division into two lines, based on the number of body rings and the size of the pygidium. These include and are themselves included in four groups, founded on the presence and form of the facial suture and the structure of the eyes.

Haeckel* has recently given the trilobites their full value in a classification of the articulates. Although he has not advanced a detailed classification, still it is desirable to review the ordinal groups which he proposes. He considers the Trilobita as a legion under the first class, Aspidonia, of the Crustacea, which is characterized as being without a nauplius larval form and as having a pair of pre-oral antennæ. In this class is also included the legion Merostomata, the Trilobita being especially distinguished by the number and character of the legs. The writer[5] believes that it is now satisfactorily demonstrated that the protaspis, or early larval form of the trilobite, is a protonauplius, and homologous with the nauplius of higher Crustacea. Therefore the Trilobita cannot remain in the Aspidonia, as here defined.

Haeckel further divides the Trilobita into two orders, the first, the Archiaspides (or Protrilobita), and the second, the Eutrilobita (or Pygidiata). The Archiaspides are represented by the families Olenida and Triarthrida, and are distinguished by the absence of a real pygidium, and by the complete homonomy of the numerous body segments and their phyllopodiform appendages. The families are themselves distinguished by the semi-circular or crescent-shaped cephalon and by the presence or absence of genal spines. The Eutrilobita are represented by the families Asaphida and Calymmenida, and are marked by the heteronomy of the body segments as expressed in the functional pygidium.

Salter, Burmeister, and Emmrich have, as previously noticed, attempted to use the comparative size and develop-

* Systematische Phylogenie der wirbellosen Thiere (Invertebrata). Zweiter Theil, 1896.

ment of the pygidium for dividing the trilobites into groups larger than families, and it seems evident from the present state of knowledge that it is impossible to make this character of more than family or even generic value. Many of the genera which must naturally be included in the Archiaspides have pygidia that cannot be said to be rudimentary, obsolete, or wanting in function. Even those genera having pygidia with few segments, as *Mesonacis, Holmia, Paradoxides, Selenopeltis, Dicranurus, Bronteus, Harpes*, etc., show in many other more important characters that they are highly differentiated and specialized forms, and that this feature is one expression of such development. The futility of the scheme is at once evident when a comparison is made between allied genera which present marked differences in the size and segmentation of the pygidium; as *Phacops* and *Dalmanites, Ceraurus* and *Encrinurus, Calymmene* and *Homalonotus, Harpes* and *Trinucleus, Mesonacis* and *Zacanthoides, Paradoxides* and *Dikelocephalus.*

The last classification to be noticed is that of E. J. Chapman,[13] in 1889, in which four sub-orders or primary groups are proposed, differing considerably from any previous arrangement, and based upon arbitrary features of general structure and configuration, especially the form of the glabella, whether wide, conical, or enlarged. Twenty-seven families are recognized. In this scheme *Trinucleus, Ampyx*, and *Æglina* form one section; *Paradoxides* and *Acidaspis*, together with *Phacops* and *Encrinurus*, another; all under one sub-order. Omitting the Agnostidæ, there are here considered in a single sub-order the most characteristic representatives of nearly all the types of trilobite structure. *Proëtus, Cyphaspis*, and *Arethusina* fall into three sections, under two sub-orders, although these genera, on account of their great similarity in essential points, are placed in a single family by most authors. A further analysis of this classification in its broader lines would be unprofitable. It is sufficient to state that the facts obtained from the study of the ontogeny of any species are completely in discordance with these classifications, and clearly demand other interpretations.

Rank of the Trilobites.

As to the rank of the trilobites in a classification of the Crustacea, there is also much diversity of opinion. They have long been regarded as an order, but any attempt to include them in this way under higher groups, such as the Entomostraca, Malacostraca, or Palæocarida, results in such broad generalities and looseness of definition as to render these divisions of little value. Even the Entomostraca, as restricted to the orders Phyllopoda, Ostracoda, Copepoda, and Cirripedia, seem heterogeneous and probably polyphyletic. Milne-Edwards,[27] Gegenbaur,[19] Walcott,[33] and others have considered the trilobites as belonging to a class of arthropods intermediate between the Crustacea and arachnids. Some recent authors, as Lang,[24] have attempted to overcome the difficulty by attaching them as an appendage ("Anhang") to the Crustacea. Kingsley,[23] on the other hand, has placed them as a sub-class of the Crustacea, leaving all the other Crustacea to come under a second sub-class, the Eucrustacea. The present state of knowledge of their structure and development is in favor of giving the trilobites the rank of a sub-class, but for purposes of comparison and correlation the fullest results can be brought out by recognizing the old and well-known sub-classes, — the Entomostraca and Malacostraca.

The following tabular view of the leading points of the comparative morphology of the three sub-classes is introduced to show, first, the claims of the Trilobita as an equivalent group, and, second, the progressive differentiation of characters. In nearly every particular the trilobite is very primitive, and closely agrees with the theoretical crustacean ancestor. Its affinities are with both the other sub-classes, especially their lower orders, but its position is not intermediate.

Comparative Morphology of Crustacea.

Sub-class I. Trilobita.	Sub-class II. Entomostraca.	Sub-class III. Malacostraca.
1. All marine. 2. Free.	Marine and freshwater. Free, parasitic, and attached.	Marine and freshwater. Free and parasitic.
3. Body longitudinally tri-regional. 4. Larva a protonauplius.	Various. Larva almost universally a nauplius.	Various. Larva generally a zoëa, a nauplius stage being often developed before hatching, except in *Euphausia* and *Peneus*.
5. Number of segments variable. 6. Cranidium of five fused segments.	Number of segments variable. Head of five fused segments to which, rarely, a thoracic segment is added.	Definite number of segments. Head of five fused segments to which one or more, or all of the thoracic segments may unite, forming a more or less complete cephalothorax.
7. Ocelli rarely present. 8. Paired compound sessile eyes on cheek pieces usually present.	Ocelli present throughout life. Paired compound eyes usually present ; stalked or sessile. Absent in adult Cirripedia and some Copepoda.	Ocelli absent in adult forms. Paired compound eyes usually present; stalked or sessile.
9. T h o r a x distinct ; number of segments variable, all free.	Thorax w i t h variable number of segments.	Thorax with eight segments, some of which are generally united with the head.
10. Abdomen distinct ; variable number of fused segments. 11. All segments of cranidium, thorax, and abdomen, except the anal segment, carry paired appendages.	Abdomen with variable n u m b e r of separate segments. Some segments without appendages.	Abdomen of seven generally free segments ; eight in Leptostraca. All segments usually carry appendages except the last one or two.
12. All appendages biramous except antennules.	Some appendages are modified and have lost biramous structure.	Some appendages have lost biramous structure.
13. Appendages typically phyllopodiform. Exopodite a swimming leg ; endopodite modified into a crawling leg.	Appendages g e n e r a l l y greatly changed in most orders ; phyllopodiform in young forms and throughout life in Phyllopoda.	Appendages typically phyllopodiform, b u t greatly modified in all but the lowest order (*Nebalia*).
14. All appendages of the head except antennules pediform.	Some appendages of the head modified into rowing organs, mandibles, or suckers.	Some appendages of the head modified into mandibles, or organs for seizing food.
15. Thoracic appendages ambulatory and swimming.	Thoracic appendages ambulatory, swimming, and seizing.	Thoracic appendages ambulatory, swimming, and seizing.

Sub-class I. Trilobita.	Sub-class II. Entomostraca.	Sub-class III. Malacostraca.
16. Abdominal limbs on all segments except the anal, phyllopodiform.	Abdominal limbs generally wanting.	Abdominal limbs often reduced except the last pair, which with telson frequently form a caudal fin. Chiefly branchial in some groups.
17. Coxal elements of all limbs forming gnathobases, which become manducatory organs on the head.	Coxal elements seldom forming gnathobases except on the head.	Coxal elements seldom forming gnathobases except on the head; never on the abdomen.
18. Respiration cuticular and by fringes on exopodites.	Respiration mainly cuticular and by the limbs and gill appendages.	Respiration cuticular and by the limbs and epipodites.

The more primitive characters of the trilobites as drawn from the foregoing table may be summarized as follows: (1) They are all free marine animals; (2) the animal has a definite configuration; (3) the larva is a protonauplius-like form; (4) the body and abdomen are richly segmented, and the number of segments is variable; (5) the head corresponds to the typical crustacean; (6) the thorax and abdomen are always distinct, the number of segments in each being variable; (7) all segments except the anal bear paired appendages; (8) all appendages are typically phyllopodiform; and (9) the coxal elements of all limbs form gnathobases, which become organs of manducation on the head.

It may be questioned by some whether the present state of knowledge of the ventral structure of trilobites warrants such general assertions as to details of organization. In the first place, it must be granted that there is a remarkable uniformity in the features of the dorsal crust, which naturally reflects to a degree the differentiation and variation of the organs and appendages of the ventral side. Furthermore, the actual appendages have been observed in such diverse and characteristic genera as *Trinucleus*, *Triarthrus*, *Asaphus*, *Ceraurus*, and *Calymmene*, and found to conform closely to a single type, so that it seems safe to assume a like agreement throughout.

Morphology of the Cephalon.

The structure of the trilobite head suggests homologies which should be noticed here, and if these correlations are based upon true structural likenesses, they serve not only to emphasize the primitive character of the trilobite, but also aid in interpreting certain organs and structures in the higher Crustacea.

The five fused somites of the crustacean head are generally believed to correspond to the third, fourth, fifth, sixth, and seventh neuromeres, leaving the first and second unrepresented either by distinct segments or appendages. These two neuromeres commonly constitute most of the cerebral mass above the œsophagus, and innervate the ocelli and paired eyes. In some the antennæ are innervated from supra-œsophageal ganglia, while in other forms their ganglia are infra-œsophageal. It was formerly supposed that the stalked eyes of the higher Crustacea represented appendages of a head segment, but this belief has been abandoned on account of the derivation of stalked out of sessile organs, as in *Peneus*, and also because the eyes do not always have a relatively fixed position, but may pertain to the first, second, or third head segments. Their structural position in the trilobites, however, is invariable, and it seems probable that in some families of higher Crustacea the eyes are in exact correlation, and may be similarly interpreted.

The writer[5] has previously discussed this question, and adduced reasons for considering the free-cheeks in trilobites as "the pleura of an oculiferous head segment." In many species (*Dalmanites, Æglina*, etc.) the free-cheeks are continuous, forming one piece extending around the front of the head, between the cranidium and the hypostoma, while in others there is a separate piece, the rostral plate, between the proximal ends of the cheek pieces holding a like position. These structures occupy the exact position of a true segment, and since, upon theoretical grounds, additional head segments are to be accounted for, the only satisfactory correlation is to consider them as such. Furthermore, the free-cheeks are distinctly separated from the cranidium by an open suture, and may be wholly converted into eyes, as in *Æglina armata* Barrande, or the unfaceted portion may be reduced to almost nothing, as in *Deiphon*. In such cases the parallelism is exact with true movable eyes. Bernard[7] concludes from his studies of *Apus* that the hypostoma is homologous with the annelid prostomium. This would make the hypostoma represent the first, and the free-cheeks the second of the obsolete segments. Thus the trilobite cephalon would fulfil the demand for additional evidences of primitive head segments, and account for the development of eyes separate from the cephalothorax as commonly restricted.

Supposed evidences of free-cheeks or of facial sutures have been recognized in *Limulus*, *Hemiaspis*, and *Bunodes*, but these seem really to correspond to the lines on the dorsal surface of the cephalon of *Harpes* and some *Trinucleus*, running from the glabella to the eye-spots and to the margin, and are not the sutures marking the limits of the free cephalic elements, as in *Asaphus* and *Proëtus*. *Limulus*, however, has a suture comparable to that in *Harpes* and *Trinucleus*, extending around the ventral border of the cephalothorax nearly to the posterior angles, and partly separating the ventral plate. In the process of moulting, this suture opens and enables the animal to free itself from its former test.

These interpretations may be employed to some advantage in correlating the segmentation of the trilobite cephalon. As previously stated, the recognized plan in the nervous system of a generalized crustacean requires that there should be a brain or supra-œsophageal ganglion innervating (*a*) the unpaired eye, (*b*) the frontal sensory organs and stalked eyes, and (*c*) the anterior antennæ; then a ventral nervous cord consisting of a succession of double ganglia innervating, respectively, the second pair of antennæ, the mandibles, the first pair of maxillæ, the second pair of maxillæ, and lastly each of the paired thoracic and abdominal appendages. Altogether, there are seven neuromeres pertaining to the head, and on the basis that each neuromere corresponds to an original segment, as on the post-cephalic region, there would need to be this number accounted for. The anterior segment, or number one in the trilobites, would be represented by the hypostoma; the second segment, by the paired eyes, free-cheeks, and rostral plate; the third, by the anterior lobe of the glabella and the first antennæ; the fourth, by the second lobe of the glabella and the second pair of antennæ; the fifth, by the third lobe of the glabella and the mandibles; the sixth, by the fourth lobe of the glabella and the first maxillæ; and the seventh, by the neck lobe, or occipital ring, and the second pair of maxillæ. The five annulations, or lobes, of the axis of the cranidium, since they primarily carry fulcra for the attachment of muscles supporting or moving the appendages, could thus be interpreted in terms of the ventral structure, making the first lobe the antennulary, the second the antennary, the third the mandibular, the fourth the first maxillary, and the fifth the second maxillary.

No other group of Crustacea furnishes such constant and well-developed structures representing the second theoretical head segment, which is obscure or obsolete in all the living groups, excepting probably the stalked eyes of some Crustacea and the movable ocular segment of the Stomatopoda. For this reason, in addition to the many other important differences previously noted in the table of comparative morphology, the trilobites are regarded as a sub-class, and the relative denomination and structural relations of this second segment, along with other characters, are considered as of sufficient physiological and morphological importance to determine the ordinal divisions.

Principles of a Natural Classification.

Most satisfactory and conclusive results have already come from the application of the law of morphogenesis, or the recapitulation theory, to various groups of animals, by means of which their natural classification and phylogenetic relations have been determined. Hyatt[21] says on this point

(1889): "We have endeavored to demonstrate that a natural classification may be made by means of a system of analysis in which the individual is the unit of comparison, because its life in all its phases, morphological and physiological, healthy or pathological, embryo, larva, adolescent, and old (ontogeny), correlates with the morphological and physiological history of the group to which it belongs (phylogeny)." It is also interesting to note that Agassiz[1] recognized in ontogeny a standard of classification. One of his strongest statements is as follows: "Embryology [= ontogeny] will in the end furnish us with the means of recognizing the true affinities among all animals, and of ascertaining their relative standing and normal position in their respective classes with the utmost degree of accuracy and precision."

These principles can be best applied in a group of animals which has a geological history more or less complete, and which is not wholly parasitic or greatly degenerated. It is of the greatest importance, also, to study the ontogeny of primitive and non-specialized species, because without very complete paleontological evidence the development of a much later derived form may be so involved with larval adaptations and accelerated characters as to be misleading.

The trilobites lend themselves to this treatment in fulfilling most of the necessary conditions. They have a known geological history stretching through the entire Paleozoic, from the beginning of the Cambrian to the Permian. Their structure is generalized and quite uniform, and no sessile, attached, parasitic, land, or freshwater species are known. The ontogeny of all the principal groups has been studied, including Cambrian, Ordovician, Silurian, and Devonian types.

The trilobites necessarily furnish little information of the stages of growth which may be classed as embryonic. The early embryonic stages are not preserved as fossils, and therefore may be omitted. In this category are the *protembryo*, or the ovum in its unsegmented and segmented stages (the so-called "eggs of trilobites" may of course represent any stage of embryonic development before the escape of the young); the *mesembryo*, or blastosphere; the *metembryo*, or gastrula; the *neoembryo*, or planula-like stage; and the *typembryo*, when the first distinctive features make their appearance. The first embryonic stage recognized in the trilobites can be referred to the *phylembryo*, as defined by Jackson,[22] when the animal may be clearly referred to its proper class. Since this period is distinctive for each class of animals and usually bears a separate name, it has been termed by the writer[5] the *protaspis* stage of trilobites. It closely approximates the protonauplius form, or the theoretical, primitive, ancestral larval form of the Crustacea. Like the homologous nauplius of modern higher Crustacea, it is the characteristic larval type common to the class. The nauplius is therefore considered as a derived larva modified by adaptation.

The post-embryonic stages of ontogeny have received the names *nepionic*, for the infantile or young; *neanic*, for the immature or adolescent; *ephebic*, for the mature or adult; and *gerontic*, for the senile or old. When especially applied to trilobites, the nepionic stages may include the animal when the cephalon and pygidium are distinct and the thorax incomplete. There would thus be as many nepionic stages as the number of thoracic segments. The neanic stages would be represented by the animal with all parts complete, but with the average growth incomplete. Final progressive growth and development of the individual would fall under the ephebic stage. Lastly, general evidences of senility would be interpreted as belonging to the gerontic stage.

Application of Principles for Ordinal Divisions.

In other classes of animals above the lower cœlenterates, the phylembryonic stage is the starting-point from which correlations are made, and out of which all the higher groups are developed by a series of changes along certain lines. The protoconch represents this period in the cephalopods and gastropods; the prodissoconch, in the pelecypods; the protegulum, in the brachiopods, and the protechinus, in the echinoids. In the trilobites the protaspis, as already stated, has the value of the phylembryo, and in its geological history and the metamorphoses it undergoes to produce the perfect trilobite accurate information can be gained as to what the primitive characters are, and the relative values of other features acquired during the long existence of the class.

The simple characters possessed by the protaspis are the following, as drawn from the study of this stage in all the principal groups of trilobites: Dorsal shield minute, not more than .4 to 1 mm. in length; circular or ovate in form; axis distinct, more or less strongly annulated, limited by longitudinal grooves; head portion predominating; axis of cranidium with five annulations; abdominal portion usually less than one-third the length of the shield; axis with from one to several annulations; pleural portion smooth or grooved; eyes, when present, anterior, marginal, or sub-marginal; free-cheeks, when visible, narrow and marginal. Examples, Plate II, figures 1, 5.

During this stage several moults took place before the complete separation of the pygidium or the introduction of thoracic segments. These brought about various changes; as the stronger annulation of the axis, the appearance of the free-cheeks on the dorsal side, and the growth of the pygidium by the introduction of new appendages and segments, as indicated by the additional grooves on the axis and limb. A full representation of the variety and succession of these early protaspis stages is presented in the writer's paper on the "Larval Stages of Trilobites." [5] Some of the conclusions and discussions in that paper are made use of here.

In the earliest or Cambrian genera the protaspis stage is by far the simplest expression of this period to be found. In the higher and later genera the process of acceleration or earlier inheritance has pushed forward certain characters until they appear in the protaspis, thus making it more and more complex.

Taking the early protaspis stages in *Solenopleura*, *Liostracus*, or *Ptychoparia*, it is found that they agree exactly with the foregoing diagnosis in its most elementary sense. Since they are the characters shared in common by all the larvæ at this stage, they are taken as primitive and accorded that value in dealing with adult forms possessing homologous features. Therefore any trilobite with a large elongate cephalon, eyes rudimentary or absent, free-cheeks ventral or marginal, and glabella long, cylindrical, and with five annulations, would naturally be placed near the beginning of any genetic series or as belonging to a very primitive stock.

Next must be considered the progressive addition of characters during the geological history of the protaspis, and in the ontogeny of the individual during its growth from the larval to the mature condition. It was shown in the paper already referred to, that there was an exact correlation to be made between the geological and zoölogical succession of first larval stages and adult forms, and therefore both may be reviewed together.

The first important structures not especially noticeable in all stages of the protaspis are the free-cheeks, which usually manifest themselves in the meta- or para-protaspis stages, though sometimes even later. Since they bear the visual areas of the eyes, when they are present, their appearance on the dorsal shield is practically simultaneous with these organs, and before the eyes have travelled over the margin the free-cheeks must be wholly ventral in position. When first discernible they are very narrow, and in *Ptychoparia* and *Sao* include the genal angles. In *Dalmanites* and *Cheirurus*, however, the genal angles are borne on the fixed-cheeks. If, as Bernard [7] concludes, the crustacean head has been formed by the bending under, to the ventral side, of the anterior segments of an ancestral carnivorous annelid, this furnishes a means of further determining and also of satisfactorily correlating the prime significance and importance of the free-cheeks.

Since the free-cheeks are ventral in the earliest larval

stages of all but the highest trilobites, and as this is an adult feature among a number of genera which on other grounds are very primitive, this is taken as generally indicative of a very low rank. It seems to mean that the second segment remains where it was mechanically placed, and retains its full somitic nature, though from the necessities of such a condition true ventral segments must soon disappear through modification into other structures or through disuse as segments. The genera *Harpes*, *Agnostus*, *Trinucleus*, and their allies agree in having well-developed, continuous, ventral free-cheeks, and constitute a natural group. As they possess one expression or type of the genesis of an important common character, based upon facts of development, it should stand as an ordinal character, and as such it is here taken. For this group the name Hypoparia is proposed. It is fully defined and its limitations established in the proper place in the classification.

The remaining genera of trilobites present two distinct types of head structure, dependent upon the extent and character of the free-cheeks. In both, the free-cheeks make up an essential part of the dorsal crust of the cephalon, being continued on the ventral side only as a doublure or infolding of the edge, similar to that of the free edge of the cranidium, the ends of the thoracic pleura, and the margin of the pygidium. They may be separated only by the cranidium, as in *Ptychoparia*, or by the cranidium and rostral plate, as in *Illænus* and *Homalonotus*, or they may be united and continuous in front, as in *Æglina* and *Dalmanites*. One type of structure is distinguished by having the free-cheeks include the genal angles, thus cutting off more or less of the pleura of the occipital segment. The genera belonging to this group constitute the second order, the Opisthoparia.

The third and last type of structure includes forms in which the pleura of the occipital segment extend the full width of the base of the cephalon, embracing the genal angles. The free-cheeks are therefore separated from the cranidium by sutures cutting the lateral margins of the cephalon in front of the genal angles. Genera having this structure are here placed in the order PROPARIA.

Several genera, as *Calymmene* and *Triarthrus*, have been described as having the facial sutures beginning at or cutting the apex of the genal angle, thus making it indeterminate whether they should be classed with the Opisthoparia or Proparia. It will be found, however, that some species of these genera leave no doubt as to the anterior or posterior position of the suture. The small genal spines of *Calymmene callicephala* Green are situated on the ends of the fixed-cheeks, while similar but larger spines in *Triarthrus spinosus* Billings are on the free-cheeks, making the former belong to the Proparia and the latter to the Opisthoparia.

Application of Principles for Arrangement of Families and Genera.

The remaining characters to be noticed have chiefly family and generic values, and naturally follow the preceding discussions. They are of great assistance both in determining the place of a family in an order, and the rank and genetic position of a genus in a family.

There is very satisfactory evidence that the eyes have migrated from the ventral side, first forward to the margin and then backward over the cephalon to their adult position. The most primitive larvæ should therefore present no evidence of eyes on the dorsal shield. Just such conditions are fulfilled in the youngest larva of *Ptychoparia*, *Solenopleura*, and *Liostracus*. The eye-line is present in the later larval and adolescent stages of these genera, and persists to the adult condition. In *Sao* it has been pushed forward to the earliest protaspis, and is also found in the two known larval stages of *Triarthrus*. *Sao* retains the eye-line throughout life, but in *Triarthrus* the adult has no trace of it. A study of the genera of trilobites shows that this is a very archaic feature, chiefly characteristic of Cambrian genera, and only appearing in the primitive genera of higher and later groups or as an evidence of degeneration. It first develops in the later larval stages of certain genera (*Ptychoparia*, etc.); next in the early larval stages (*Sao*); then disappears from the adult stages (*Triarthrus*); and finally is pushed out of the ontogeny (*Dalmanites*).

In *Ptychoparia, Solenopleura, Liostracus, Sao,* and *Triarthrus* the eyes are first visible on the margin of the dorsal shield after the protaspis stages have been passed through, and later than the appearance of the eye-lines; but in *Proëtus, Acidaspis, Arges,* and *Dalmanites*, through acceleration, they are present in all the protaspis stages, and persist to the mature or ephebic condition, moving in from the margin to near the sides of the glabella. Progression in these characters may be expressed, and in so far taken for general application among adult forms to indicate rank, as follows: (1) Absence of eyes; (2) eye-lines; (3) eye-lines and marginal eyes; (4) marginal eyes; (5) sub-marginal eyes; (6) eyes near the pleura of the neck segment.

The changes in the glabella are equally important and interesting. Throughout the larval stages the axis of the cranidium shows distinctly by the annulations that it is composed of five fused segments, indicating the presence of as many paired appendages on the ventral side. In its simplest and most primitive state it expands in front, joining and forming the anterior margin of the head (larval *Ptychoparia* and *Sao*). During later growth it becomes rounded in front and terminates within the margin. In higher genera, through acceleration, it is rounded and well defined in front, even in the earliest larval stages, and often ends within the margin (larval *Triarthrus* and *Acidaspis*). From these simple types of simple pentamerous glabellæ all the diverse forms among species of various genera have been derived, through changes affecting any or all the lobes. The modifications usually consist in the progressive obsolescence of the anterior annulations, finally producing a smooth glabella, as in *Illænus* and *Niobe*. The neck segment is the most persistent of all, and is rarely obscured. The third, or mandibular, segment is frequently marked by two entirely separate lateral lobes, as in *Acidaspis, Conolichas, Chasmops,* etc. Likewise, the fourth annulation carrying the first pair of maxillæ is often similarly modified in the same genera, also in all the Proëtidæ, and in *Cheirurus, Crotalocephalus, Sphærexochus, Ampyx, Harpes,* etc. Here again, among adult forms, the stages of progressive differentiation may be taken as indicating the relative rank of the genera.

The comparative areal growth of the free-cheeks is expressed by the gradual moving of the facial suture toward the axis. As the free-cheeks become larger the fixed-cheeks become smaller. In the most primitive protaspis stages and in *Agnostus, Harpes,* and *Trinucleus* the dorsal surface of the cephalon is wholly occupied by the axis and the fixed-cheeks, while in the higher genera the area of the fixed-cheeks becomes reduced until, as in *Stygina* and *Phillipsia,* they form a mere border to the glabella. Therefore the ratio between the fixed- and free-cheeks furnishes another means of assisting in the determination of rank.

The pleura from the segments of the glabella are occasionally visible, as in the young of *Elliptocephala,* but usually the pleura of the neck segments are the first and only ones to be distinguished on the cephalon, the others being so completely coalesced as to lose all traces of their individuality. The pleura of the pygidium appear soon after the earliest protaspis stage, and in some genera (*Sao, Dalmanites*) are even more strongly marked than in the adult state and much resemble separate segments. The growth of the pygidium is very considerable through the protaspis stage. At first it is less than one-third the length of the dorsal shield, but by successive addition of segments it soon becomes nearly one-half as long. In some genera it is completed before the appearance of the free thoracic segments, all of which are added during the nepionic stages. An interpretation of these facts to apply in valuing adult characters would indicate that a very few segments, both in thorax and pygidium, may be evidence of arrested development or degeneration.

On the other hand, the apparently unlimited multiplication of thoracic and especially of abdominal segments in some genera is also to be considered as a primitive character expressive of an annelidan style of growth. Genera like *Asaphus*, *Phacops*, etc., having a constant number of thoracic segments accompanied by other characters of a high order, undoubtedly represent the normal trilobite type.

These analyses and correlations clearly show that there are characters appearing in the adults of later and higher genera, which successively make their appearance in the protaspis stage, sometimes to the exclusion or modification of structures present in the most primitive larvæ. Thus the larvæ of *Dalmanites* or *Proëtus*, with their prominent eyes and glabella distinctly terminated and rounded in front, have characters which do not appear in the larval stages of ancient genera, but which may come in their adult stages. Evidently such modifications have been acquired by the action of the law of earlier inheritance, or tachygenesis.

In a classification of trilobites, for the purpose of illustrating the principles here enunciated, the ontogenies of *Sao* and *Dalmanites*, Plate II, figures 1–8, are selected. *Sao* belongs to the ancient family Olenidæ of the order Opisthoparia, and naturally may be expected to furnish very clear evidence as to the relations of many lower and older genera. *Dalmanites*, also, with its simple head structure, will give similar data regarding the Proparia.

The early protaspis stage of *Sao*, Plate II, figure 1, has no dorsal development of the free-cheeks, and with the elongate form of the cephalic portion may be compared with the cephala of *Agnostus* and *Microdiscus*, and therefore correlates with the Hypoparia. The cephalon, at a later period of development, when the animal has two free thoracic segments, Plate II, figure 2, shows the narrow marginal free-cheeks and distinct eye-lines. Here the resemblance to the cephala of *Atops* and *Conocoryphe*, Plate II, figures 14, 15, is very marked, and indicates that the Conocoryphidæ are genetically the first family of the Opisthoparia. When *Sao* has eight thoracic segments, Plate II, figure 3, the characters of the cephalon accord closely with *Ptychoparia* and *Olenus*, showing that these genera should precede it in arranging the genera of the family Olenidæ. Evidence is thus furnished for the proper position of the first two families of the order. Now, if the relative values of the differentiation of the glabella, the position of the eyes, and the size of the free-cheeks are considered in the light of the preceding analyses of these features, the remaining families of the order, as represented in their typical genera, naturally arrange themselves as indicated in Plate II, figures 18–23. There result (1) the Conocoryphidæ (represented by *Atops* and *Conocoryphe*, figures 14, 15); (2) the Olenidæ (*Ptychoparia* and *Olenus*, figures 16, 17); (3) the Asaphidæ (*Asaphus* and *Illænus*, figures 18, 19); (4) the Proëtidæ (*Proëtus*, figure 20); (5) the Bronteidæ (*Bronteus*, figure 21); (6) the Lichadidæ (*Lichas*, figure 22); and (7) the Acidaspidæ (*Acidaspis*, figure 23).

For the Proparia similar results are brought out by the study of the ontogeny of *Dalmanites* and by comparisons with the characters governing the sequence of families in the Opisthoparia. The narrow marginal free-cheeks place the Encrinuridæ and Calymmenidæ as primitive. The small or obsolete eyes and the larval form of the glabella in the former further show that this family should be placed at the beginning. The nepionic *Dalmanites*, with seven thoracic segments, has a head structure very similar to the adult *Cheirurus* (*Eccoptocheile*), figure 28, thus making the Cheiruridæ precede the Phacopidæ. The arrangement of families under the Proparia accordingly will be (1) the Encrinuridæ (*Placoparia* and *Encrinurus*, Plate II, figures 24, 25); (2) the Calymmenidæ (*Calymmene* and *Dipleura*, figures 26, 27); (3) the Cheiruridæ (*Cheirurus* (*Eccoptocheile*), figure 28); and (4) the Phacopidæ (*Dalmanites*, *Chasmops*, *Acaste*, *Phacops*, figures 29–33).

The sequence of families in the most primitive order, Hypoparia, may now be easily disposed of. The genera are so aberrant and offer such conspicuous differences from ordinary trilobites that it was considered better to delay their disposition until the variations in structure governing the arrangement of families in the higher orders were clearly shown. The degree of specialization of the glabella, of the form and character of the fixed-cheeks, and the great range in the number of segments in the thorax and pygidium are strong evidence that we are dealing with the terminal genera of the order, which must have attained its normal development in pre-Cambrian times. *Agnostus* and *Microdiscus* have so many protaspidian and larval characters that they must be considered more primitive than the other genera, although in some respects they show a high degree of specialization and even degeneration, as will be noticed under the family Agnostidæ. Moreover, *Harpes*, in its elongate cephalon, persistent ocelli, and many thoracic segments, is also quite primitive. *Trinucleus*, with ocelli present only in larval stages, a transverse cephalon, and genal spines belonging to the free-cheeks, is considerably higher and properly comes last in the order, thus making the arrangement of families as follows: (1) Agnostidæ (*Agnostus, Microdiscus,* Plate II, figures 9, 10); (2) Harpedidæ (*Harpes,* figure 11); and (3) Trinucleidæ (*Trinucleus, Ampyx,* figures 12, 13).

Diagnoses and Discussions.

Sub-class Trilobita.

Marine Crustacea, with a variable number of metameres; body covered with a hard dorsal shell or crust, longitudinally trilobate from the defined axis and pleura; cephalon, thorax, and abdomen distinct. Cephalon covered with acephalic shield composed of a primitively pentamerous middle piece, the cranidium, and two side pieces, or free-cheeks, which may be separate or united in front, and carry the compound sessile eyes, when present; cephalic appendages pediform, consisting of five pairs of limbs, all biramous, and functioning as ambulatory and oral organs, except the simple antennules, which are purely sensory. Upper lip forming a well-developed hypostoma; under lip present.

Somites of the thorax movable upon one another, varying in number from two to twenty-nine. Abdominal segments variable in number, and fused to form a caudal shield. All segments, thoracic and abdominal, carry a pair of jointed biramous limbs. All limbs have their coxal elements forming gnathobases, which become organs of manducation on the head. Respiration integumental and by branchial fringes on the exopodites. Development proceeding from a protonauplius form, by the progressive addition of segments at successive moults.

Heretofore it has been impossible to give an adequate diagnosis of the Trilobita, owing to the absence of information regarding certain important characters, and the obscurity of the information relating to some other features. It is believed that enough is now known to frame a definition of the class, which, in accuracy and completeness, will compare favorably with any based upon living groups. Such a definition brings out the fact that the differences between the trilobites and other large groups are clearly recognizable, and do not consist of a statement of anomalous characters whose real significance is unknown.

Arrangement of the Families of Trilobites.

Sub-class TRILOBITA.

Order A. Hypoparia.

Family 1. Agnostidæ.   Family 3. Trinucleidæ.
Family 2. Harpedidæ.

Order B. Opisthoparia.

Family 4. Conocoryphidæ.   Family 8. Bronteidæ.
Family 5. Olenidæ.   Family 9. Lichadidæ.
Family 6. Asaphidæ.   Family 10. Acidaspidæ.
Family 7. Proëtidæ.

Order C. Proparia.

Family 11. Encrinuridæ.   Family 13. Cheiruridæ.
Family 12. Calymmenidæ.   Family 14. Phacopidæ.

The order Opisthoparia, with nearly one hundred and fifty genera, has a much greater geological distribution than either of the others, and was by far the dominant group during the Cambrian and Ordovician, being represented by about eighty-five genera in the former age and forty-five in the latter. Nineteen genera of this order are found in the Silurian and ten in the Devonian, most of them having continued on from older ages. Four genera represent the order in the Carboniferous and one in the Permian, thus marking the extinction of the sub-class as well as the last genera of the Opisthoparia.

The comparative abundance and duration of the three orders are expressed in the table on page 133, from which it appears that the Hypoparia probably culminated in pre-Cambrian times, the Opisthoparia during the Cambrian, and the Proparia during the Ordovician.

In the following classification the families adopted by Salter[32] and Barrande[3] are in the main adhered to, and the number corresponds very closely with that in Zittel's "Handbuch der Palæontologie"[35] and also in the "Grundzüge"[36] of the same author. The order of arrangement, however, is very different. A great number of family divisions have been proposed, and undoubtedly many others will yet be made, but it is not within the province of this paper to determine the precise value and limitations of the families. This would require discussions of priority and synonymy, and otherwise detract from the direct purpose of the writer; namely, to establish a basis for a natural classification, and in this way to apply what is currently known and accepted regarding the trilobites. Nevertheless, some notice must be taken of several families and genera which for various reasons do not appear here. The family Aglaspidæ, including the genus *Aglaspis* Hall, proves to belong to the Merostomata and is therefore omitted. The family Bohemillidæ has been shown by the writer[6] to have no foundation, because the type of the genus *Bohemilla* Barrande was based upon a mutilated specimen of *Æglina*.

Several genera still commonly adopted are not here recognized in the lists under the families, since from the minute size of the individuals described and their immature characters they must be considered as the young of larger forms. Such are *Conophrys* Callaway, *Cyphoniscus* Salter, *Holometopus* Angelin, *Isocolus* Angelin, and *Shumardia* Billings. *Triopus* Barrande has been shown to be a chiton.

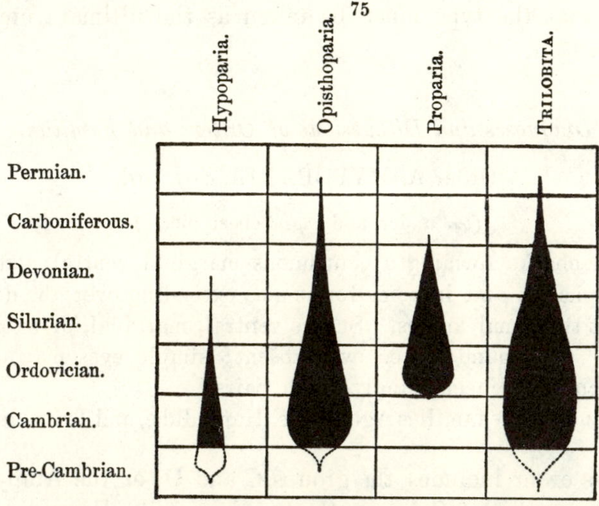

FIGURE 75. — Table of Geological Distribution of Trilobita.

Much could be said against some of the recognized genera, but, as with the families, the writer has preferred in almost every case to adopt, for the present, what has been commonly accepted, and thus to avoid the entanglement of dates and synonyms which would be out of place in any general discussions. The type species of every genus is here made the central idea. It is taken as representing the genus more closely than any fortuitous assemblage of diverse species, which the next investigator may show belong to another or to several genera. Our ideas of a genus are naturally based mainly upon the species with which we are most familiar. Until forced to make authoritative comparative statements, it does not occur to one that the type of the genus under consideration may be quite different. An American student's conception of *Homalonotus* will probably be formed largely upon the species commonly known as *H. delphino-cephalus* Green, from the Niagara, and *H. DeKayi* Green, from the Hamilton. The first time the type of the genus, *H. Knighti* Murchison, is seen he will be puzzled to place it. Similar examples could be multiplied indefinitely, and only show that the type must be taken as the ultimate unit of comparison.

Diagnoses and Discussions of Orders and Families.

Order A. HYPOPARIA, nov. ord.

(ὑπό under, and πἄρειά cheek piece.)

Free-cheeks forming a continuous marginal ventral plate of the cephalon, and in some forms also extending over the dorsal side at the genal angles. Suture ventral, marginal, or submarginal. Compound paired eyes absent; simple eyes may occur on each fixed-cheek, singly or in pairs.

Including the families Agnostidæ, Harpedidæ, and Trinucleidæ.

This order includes the groups C and D, or the Ampycini and Agnostini of Salter, and also the family Harpedidæ of that author, which he included in the Asaphini. The special recognition of characters, however, between Salter's groups and the order here proposed is different.

The presence of a part homologous with the free-cheeks of other trilobites has generally been more or less overlooked in the families of this order. In *Trinucleus, Dionide,* and *Harpes* the sutures have been correctly determined by Barrande.[3] Likewise, Angelin[2] gave the right structure in *Ampyx,* but in *Agnostus* this feature has escaped notice. The examination of extensive series of *Agnostus,* in the National Museum and in the Museum of Comparative Zoölogy,* has proved that under favorable conditions of preservation this genus shows a distinct plate, separated from the cranidium by a suture, and it can be compared only with the

---

* In the former, through the courtesy of C. D. Walcott and C. Schuchert, and in the latter, of A. Agassiz and R. T. Jackson.

free-cheeks in other trilobites, especially where they are continuous around the front of the cephalon, as in *Trinucleus* and *Ampyx*. The presence of a hypostoma in *Agnostus* was also determined. Even in the higher genera of this order the suture is frequently unnoticed in descriptions, but it can be seen in all well-preserved specimens. In *Trinucleus*[29] and *Harpes* it follows the edge of the cephalon, and separates the dorsal from the ventral plate of the pitted limb. Since eye-spots occur on the fixed-cheeks in the young *Trinucleus* and adult *Harpes*, it is probable that this character is a primitive one in this order, and has been lost in *Agnostus*, *Microdiscus*, *Ampyx*, and *Dionide*.

The ontogeny of *Sao, Ptychoparia, Triarthrus, Dalmanites,* etc., shows that the true eyes and free-cheeks are first developed ventrally, appearing later at the margin, and then on the dorsal side of the cephalon. Therefore the Agnostidæ, Trinucleidæ, and Harpedidæ have a very primitive head structure, characteristic of the early larval forms of higher families. Other secondary features show that this order, though the most primitive in many respects, is more specialized than either of the others, except in their highest genera. The characters referred to are the glabella and pygidium. Very few species show the primitive segmentation of the glabella, it being usually smooth and inflated, and resembling in its specialization such higher genera as *Proëtus, Asaphus,* and *Lichas.* The pygidium often fails to indicate its true number of segments. Some *Agnostus* and *Microdiscus* show no segments either on the axis or limb of the pygidium. *Trinucleus* and others may have a many-annulated axis and fewer grooves on the pleural portions. The number of appendages corresponds to the axial divisions, as determined by the writer.[4] The multiplication of segments in the pygidium and their consequent crowding makes them quite rudimentary.

Family I. AGNOSTIDÆ Dalman.

Small forms, having the cephalon and pygidium elongate, nearly equal, and similar in form and markings. Free-cheeks ventral, continuous; suture marginal or ventral. Eyes wanting. Thorax composed of from two to four segments, with grooved pleura. Cambrian and Ordovician.

Including the genera *Agnostus* Brongniart and *Microdiscus* Emmons.

The genera in this family are primitive in their form and structure, as shown by their ventral free-cheeks, marginal or ventral suture, elongate cephalon, and large pygidium. Some species have spines at the genal angles, corresponding to the interocular spines of *Holmia* and young *Elliptocephala*, and not to the spiniform projections of the free-cheeks. From their abbreviated thorax and progressive loss of annulations on the glabella and axis of the pygidium they must also be considered as degraded. *Microdiscus*, the earlier genus, has three or four free segments, and in some species (*M. speciosus* Ford) preserves the normal pentamerous glabella and annulated pygidial axis, while the later genus, *Agnostus*, has but two free segments, and has lost the annulations of both glabella and pygidium. Matthew[26] has described the protaspis stage of *Microdiscus*, which agrees with the similar stage of *Ptychoparia* and *Sao*.

Fully a dozen generic names have been proposed for forms of the general type of *Agnostus*, but none of them has ever come into current use. Nine were first published by Corda,[15] but as Barrande[3] subsequently showed that one was based on an *Orbicula*, another on a poor specimen of *Æglina*, and three others on a single species, this grouping soon fell into disuse. Moreover, Barrande was inclined to give no generic value to the form and lobation of the glabella, and therefore all the species were placed by him in the single genus *Agnostus*. At the present time more weight is given to the characters of the glabella and pygidium, as indicating generic differences in dorsal and ventral structure, so that further study may show the desirability of restoring such of Corda's names as were founded upon natural groups of this family.

Family II. HARPEDIDÆ Barrande.

Cephalon large, margined by a broad expansion or limb; glabella short and prominent. Free-cheeks ventral, continuous; suture marginal, following the outer edge of the limb. Paired simple eye-spots, or ocelli, single or double, at the distal ends of well-marked eye-lines on the fixed-cheeks, extending outward from the glabella. Thorax of from twenty-five to twenty-nine segments, with long grooved pleura. Pygidium (in *Harpes*) very small, composed of but three or four segments.

Cambrian to Devonian.

Including *Harpes* Goldfuss, *Harpina* Novák, and *Harpides ?* Beyrich.

The genus *Harpes* presents considerable variation in the lobes of the glabella. *H. ungula* Sternberg shows the full number of five lobes, but in some species, as *H. d'Orbignyianum* Barrande, the structure is like *Cyphaspis*, with separate basal lobes. *Arraphus* Angelin was apparently based upon a specimen of *Harpes* denuded of the pitted border. *Harpides* Beyrich is imperfectly known, but seems to belong here. The ocular ridges and tubercles on the fixed-cheeks, the broad limb, the glabella, and the narrow weak thoracic segments are all in accord with *Harpes*, though in other features it has affinities with the Conocoryphidæ.

In many respects *Harpes* is one of the most interesting genera of trilobites, since it is so unlike other forms. The broad hippocrepian pitted limb of the cephalon has its counterpart in *Trinucleus* and *Dionide*, although not so well developed in these genera. The cephalon is also comparatively longer and larger, both features being decidedly larval. It is the only family known in which functional visual spots, or ocelli, are situated on the fixed-cheeks. The young *Trinucleus* has similar eye-spots, or ocelli. The great number of free segments in the Harpedidæ is another primitive character, although the cephalon (in *Harpes*) still remains larger than the thorax and pygidium.

Family III.   Trinucleidæ Barrande.

Cephalon larger than the thorax or pygidium; genal angles produced into spines. Free-cheeks continuous, almost wholly ventral, carrying the genal spines; suture marginal or sub-marginal. Paired simple eyes, or ocelli, generally absent in adult forms; compound eyes wanting. Segments of thorax five or six in number, with grooved pleura. Pygidium triangular; margin entire; axis with a number of annulations; limb grooved.

Ordovician and Silurian.

Including the genera and subgenera *Trinucleus* Lhwyd, *Ampyx* Dalman, *Dionide* Barrande, *Endymionia?* Billings, *Lonchodomus* Angelin, *Raphiophorus* Angelin, and *Salteria?* W. Thompson.

The leading genera of this family form a tolerably homogeneous group, although each has sometimes been recognized as characterizing a separate family. *Trinucleus* and *Dionide* have a broad pitted border, but this hardly seems of sufficient importance to remove them far from *Ampyx*, since the three genera agree in nearly all important structural details, as the extent and character of the free cheeks, the glabella, the number of free segments, and the character of the pygidium. *Lonchodomus* and *Raphiophorus* of Angelin are commonly admitted as sub-genera of *Ampyx*.

Both *Salteria* W. Thompson and *Endymionia* Billings have been described as sub-genera of *Dionide* Barrande, though there is little positive evidence for this disposition of them. Until more perfect material representing these forms has been described, it will not be possible to decide satisfactorily upon their relationships or place in a classification. Therefore they are left with doubt in the present family.

Order B. OPISTHOPARIA, nov. ord.

(ὄπισθεν behind, and πάρειά cheek piece.)

Free-cheeks generally separate, always bearing the genal angles. Facial sutures extending forwards from the posterior part of the cephalon within the genal angles, and cutting the anterior margin separately, or rarely uniting in front of the glabella. Compound paired holochroal eyes on free-cheeks, and well developed in all but the most primitive families.

Including the families Conocoryphidæ, Olenidæ, Asaphidæ, Proëtidæ, Bronteidæ, Lichadidæ, and Acidaspidæ.

This order is nearly equivalent to group B, or the Asaphini of Salter, which included also the families Calymmenidæ and Harpedidæ, which belong elsewhere.

The families which are here placed under this order lend themselves quite readily to an arrangement based upon the characters successively appearing in the ontogeny of any of the higher forms. Thus *Sao*, *Ptychoparia*, and other genera of the Olenidæ have first a protaspis stage only comparable in the structure of the cephalon with the genera of the preceding order, the Hypoparia. Therefore this stage does not enter into consideration in an arrangement of the families of the Opisthoparia. In the later stages, however, there is a direct agreement of structure with the lower genera of this order. The nepionic *Sao*, with two thoracic segments (Plate II, figure 2), has a head structure agreeing in essential features with that in *Atops* or *Conocoryphe* (Plate II, figures 14, 15). A later nepionic stage, with eight thoracic segments (Plate II, figure 3), agrees closely with adult *Ptychoparia* or *Olenus* (figures 16, 17). These facts clearly indicate that the family Conocoryphidæ should be put at the base of this extensive order. Then, as *Ptychoparia* and *Olenus* are more primitive and simple genera than *Sao*, they, as typifying the family Olenidæ, should govern its position, which accordingly would be next after the Conocoryphidæ. In each case a family is considered as represented by its typical and most characteristic forms. It would be impossible to consider the advanced specialized genera of some families as representing their normal facies, for each one has undergone an independent evolution, and some characters have reached as great a degree of differentiation as will be found in much higher families.

It has been recognized that variations in the position of the eyes, the relative size of the free- and fixed-cheeks, and the degree of specialization of the glabella have a definite order in the ontogeny of any trilobite, and also that these characters have a greater taxonomic value than many others. Applying these principles in arranging the families which come under the Opisthoparia, we have the sequence as indicated above, beginning with the Conocoryphidæ and followed by the Olenidæ, Asaphidæ, Proëtidæ, Bronteidæ, Lichadidæ, and Acidaspidæ, in regular progression. See Plate II. figures 14–23.

Family IV. Conocoryphidæ Angelin.

Free-cheeks very narrow, forming the lateral margins of the cephalon, and bearing the genal spines. Fixed-cheeks large, usually traversed by an eye-line extending from near the anterior end of the glabella. Facial sutures running from just within the genal angles, curving forward, and cutting the anterior lateral margins of the cephalon. Eyes rudimentary or absent. Thorax with from fourteen to seventeen segments. Pygidium small and of few segments.   Cambrian.

Including the genera and sub-genera *Conocoryphe* Corda (= *Conocephalites* Barrande), *Aneucanthus* Angelin, *Atops* Emmons, *Avalonia* Walcott, *Bailiella* Matthew (= *Salteria* Walcott and *Erinnys* Salter), *Bathynotus* Hall, *Carausia* Hicks, *Carmon* Barrande, *Ctenocephalus* Corda, *Dictyocephalites* Bergeron, *Eryx* Angelin, *Harttia* Walcott, and *Toxotis* Wallerius.

The genera coming under this family present a number of very primitive characters such as are shown only in the larval stages of higher forms. The free-cheeks are narrow and marginal, and can be compared with those in the nepionic stages of *Sao* and *Ptychoparia*. The eyes have not been detected, but the presence of an eye-line suggests their possible existence. The variations of the glabella are very marked, and are as great as those which in higher forms attain some importance as family characteristics. In *Toxotis, Carausia,* and *Aneucanthus* the glabella expands in front, joining and forming part of the anterior margin, as in the glabella of the larval stages of *Solenopleura, Liostracus, Ptychoparia,* and *Sao.* *Ctenocephalus* and *Eryx* are slightly more advanced, as the glabella no longer marks the edge of the cephalon. In *Atops,** *Avalonia, Bathynotus,* and *Carmon* the glabella is cylindrical, distinctly defined, and limited within the margin, and in *Conocoryphe, Harttia,* and *Bailiella* it narrows anteriorly, and only extends about two-thirds the length of the cephalon. Generally in this family the glabella displays its primitive pentamerous origin. In *Bailiella* and *Carausia* two basal lobes are marked off from the fourth segment by oblique furrows, as in *Proëtus* and *Cyphaspis.*

From a phylogenetic standpoint the family Conocoryphidæ is at the base of this extensive order. As far as known, all the larval forms in the other families of the Opisthoparia agree in having the narrow marginal free-cheeks, bearing the genal angles. The eye-line is present in most of the adult Olenidæ, and in the early stages of all as far as known, so that the general average of the characters in the Conocoryphidæ represents the main larval features throughout the other families. They show, too, that although primitive in essential structure, differentiation through time has developed secondary features belonging to genera in higher families; as, for example, the basal glabellar lobes in *Bailiella.*

Family V.   OLENIDÆ Salter.

Cephalon larger than the pygidium, usually wider than long; genal angles commonly produced into spines. Free-cheeks separate. Facial sutures extending forward from the posterior margin of the cephalon along the eye-lobes, and either cutting the anterior margin separately or meeting on the median line. Eyes crescentic, reniform, or semi-circular, situated at the ends of eye-lines in all but highest genera. Trunk long, composed of from

---

* *Atops* (type *A. trilineatus* Emmons) seems to be a valid genus, and differs from *Conocoryphe* (type *C. Sulzeri* Schlotheim) in its glabellar characters, greater number of thoracic segments, and much smaller pygidium with fewer segments.

eight (?) to twenty-six free segments; rarely capable of rolling up. Pygidium frequently small; margin entire or spinose.

Principally Cambrian, but extending into the Ordovician.

Including the genus *Olenus* Dalman as the type, and the following genera and sub-genera, which should doubtless fall into several sub-family or even family groups: *Acerocare* Angelin, *Acrocephalites* Wallerius, *Agraulus* Corda, *Angelina* Salter, *Anomocare* Angelin, *Anopolenus* Salter, *Asaphelina* Bergeron, *Bavarilla* Barrande, *Bergeronia* Matthew, *Boeckia* Brögger, *Ceratopyge* Corda, *Chariocephalus* Hall, *Corynexochus* Angelin, *Crepicephalus* Owen, *Ctenopyge* Linnarsson, *Cyclognathus* Linnarsson, *Dikelocephalus* Owen (*Centropleura* Angelin), *Dorypyge* Dames, *Ellipsocephalus* Zenker, *Elliptocephala* Emmons, *Euloma* Angelin, *Eurycare* Angelin, *Holmia* Matthew, *Hydrocephalus* Barrande (= young *Paradoxides*), *Leptoplastus* Angelin, *Liostracus* Angelin, *Loganellus* Devine, *Menocephalus* Owen, *Mesonacis* Walcott, *Micmacca* Matthew, *Neseuretus* Hicks, *Olenelloides* Peach, *Olenellus* Hall, *Olenoides* Meek, *Oryctocephalus* Walcott, *Palæopyge* Salter, *Parabolina* Salter, *Parabolinella* Brögger, *Paradoxides* Brongniart, *Peltura* Angelin, *Plutonides* Hicks, *Proceratopyge* Wallerius, *Protagraulus* Matthew, *Protolenus* Matthew, *Protopeltura* Brögger, *Protypus* Walcott, *Pterocephalia* Roemer, *Ptychaspis* Hall, *Ptychoparia* Corda, *Remopleurides* Portlock, *Sao* Barrande, *Schmidtia* Marcou, *Solenopleura* Angelin, *Sphæropthalmus* Angelin, *Telephus* Barrande, *Triarthrella* Hall, *Triarthrus* Green, and *Zacanthoides* Walcott.

A complete study of this extensive family of trilobites would contribute much in the way of generic synonymy, and bring out the characters necessary for family determination and subdivision. This important work must be left for future investigation. So many genera have been described from separate cranidia or even pygidia as to make it impossible to deal with all of them in a systematic manner. The zeal to make the most out of the earliest known faunas has led many investigators to describe and recognize imperfect and poorly preserved material, and to establish genera upon very tenuous characters. Therefore, without a most intimate knowledge of all the forms, any grouping of the majority of the Cambrian genera into families or the limitations of the genera themselves must, as in the present instance, be taken tentatively and as necessarily incomplete.

A number of genera have been already made the types of family divisions; as *Paradoxides, Olenellus, Remopleurides, Ellipsocephalus, Ptychoparia,* etc. Some of them may be shown ultimately to possess characters of sufficient weight to be entitled to family distinction. A preliminary grouping of the best-known genera may be of some value here, and for the sake of convenience these divisions may be defined as sub-families. Four groups will be recognized, of which *Paradoxides, Oryctocephalus, Olenus,* and *Dikelocephalus* are taken as representative genera.

I. Paradoxinæ. — Including *Olenellus, Holmia, Mesonacis, Elliptocephala, Schmidtia, Olenelloides, Paradoxides, Zacanthoides,* and *Remopleurides.* Most of the genera are distinguished by their long narrow eyes, often extending more than half the length of the glabella, but more especially by the rudimentary character of the pygidium. In *Olenellus* the pygidium is a long telson-like spine. In *Holmia, Mesonacis, Elliptocephala,* and *Schmidtia* it is reduced to a small plate without distinct segmental divisions. In *Paradoxides, Zacanthoides,* and *Remopleurides* the axis may show from one to five annulations, while the limb may carry two or three pairs of spines or may be entire. In *Olenellus* and *Holmia* true facial sutures have been denied by some authors, but in their place false sutures are recognized. They are, however, evidently real sutures in a condition of symphysis, which often occurs in *Phacops, Proëtus, Phillipsia,* etc. Otherwise these genera would violate the first principle of trilobite structure, in not having the compound eyes on the free-cheek pieces. *Olenelloides* is a very striking form, but its pygidium is unknown, and the head structure is obscure. The elongate cephalon is a decidedly larval feature, and the genal and interocular (?) spines strongly suggest its immature condition, and point to the possibility of its being the young of *Olenellus* or a related form.

There has been much discussion as to the synonymy and value of most of the names proposed as genera or sub-genera in this group. *Paradoxides*, *Remopleurides*, and *Zacanthoides* are about the only ones that have escaped severe criticism in recent years. Taking the type of each of the others, it is found that *Elliptocephala* (1844) was based on the species *E. asaphoides* Emmons, *Olenellus* (1862) on *O. Thompsoni* Hall, *Mesonacis* (1885) on *M. vermontana* Hall sp., *Holmia* (1890) on *H. Kjerulfi* Linnarsson sp., *Schmidtia* (1890) on *S. Mickwitzi* Schmidt sp., and *Olenelloides* (1894) on *O. armatus* Peach. Some of these names are generally recognized as sub-genera of *Olenellus* (*Mesonacis*, *Holmia*, *Olenelloides*), while others are considered as synonyms (*Elliptocephala*, *Schmidtia*). The early genera were described from very incomplete material, and therefore lacked sufficient diagnostic characters to define them clearly. At the present time nearly or quite entire specimens representing the type species are known, and it is possible to compare all the essential features with some degree of accuracy. The main characters offering the greatest variation are (1) the number of thoracic segments and (2) their specialization into groups, (3) the relative development of the third free segment, (4) the number and position of the spine-bearing segments, (5) the form of the pygidium, (6) the presence or absence of interocular spines, and (7) the form of the cephalon. A simple variation in any one of these would not necessarily imply more than a specific difference, but the genera here mentioned exhibit marked changes in all or nearly all of these characters, and in any family should receive recognition. *Olenellus*, *Mesonacis*, and *Elliptocephala* are more closely related than the other forms, and probably have only a sub-generic value under *Elliptocephala*. In the first form with fourteen thoracic segments, the third is greatly enlarged and the fifteenth is the spiniform telson-like pygidium. In *Mesonacis* with twenty-six thoracic segments, the third is somewhat enlarged, and behind the narrow spine-bearing fifteenth segment there are eleven others without spines, followed by the small plate-like pygidium. In *Elliptocephala* with eighteen thoracic segments, the cephalon is broader, the third segment is not enlarged except in the young, and the fourteenth to eighteenth segments are narrower and spine-bearing.

II. Oryctocephalinæ. — Including *Oryctocephalus, Cteno-pyge, Olenoides,* and *Parabolina,* with large pygidia and all but the last one or two pleural elements continued into spines; also *Eurycare, Angelina, Peltura,* and *Protopeltura,* with smaller and shorter pygidia and denticulations of the margins corresponding to the pleural divisions.

III. Oleninæ. — Including *Olenus, Agraulus, Liostracus, Acerocare, Ptychoparia, Solenopleura, Ptychaspis, Leptoplas-tus, Loganellus, Sphæropthalmus, Parabolinella, Boeckia, Pro-ceratopyge, Ceratopyge, Protypus, Ellipsocephalus, Sao,* and *Triarthrus.* All these genera have small or medium-sized pygidia, with from two to eight annulations in the axis. Eyes medium to small, at the ends of distinct eye-lines in all but the latest genera, which preserve this character only during the young stages. Thoracic segments from eleven to eighteen.

IV. Dikelocephalinæ. — Including *Dikelocephalus, Asaphe-lina,* and *Crepicephalus.* Eight or nine thoracic segments. Pygidium wide, with the posterior lateral portion often pro-duced into broad spine-like extensions. *Dikelocephalus* is in many ways related to *Ogygia* and *Asaphus.*

Family VI. Asaphidæ Emmrich.

Cephalon and pygidium well developed; glabella often ob-scurely limited. Free-cheeks usually separate. Facial sutures extending forward from the posterior edge of the cephalon within the genal angles, and cutting the lateral or anterior margins, occasionally uniting in front of the glabella. Eyes smooth, well developed, sometimes of very large size, even occupying the entire surface of the free-cheeks. Thorax gener-ally composed of eight or ten segments, but varying from five to ten; capable of enrolment. Pygidium large, often with wide doublure. Cambrian, Ordovician, and Silurian.

The long list of genera in this family may be easily divided into two sections, which are often recognized as of family rank.

I. ASAPHIDÆ. — Including the genera and sub-genera *Asaphus* Brongniart (= *Cryptonymus* Eichwald), *Asaphellus* Callaway, *Asaphiscus* Meek, *Barrandia* McCoy, *Basilicus* Salter, *Bathyurellus* Billings, *Bathyuriscus* Meek, *Bathyurus* Billings, *Bolbocephalus* Whitfield, *Brachyaspis* Salter, *Bronteopsis* W. Thompson, *Dolichometopus* Angelin, *Gerasaphes* Clarke, *Holasaphus* Matthew, *Homalopecten* Salter, *Isotelus* DeKay, *Megalaspides* Brögger, *Megalaspis* Angelin, *Niobe* Angelin, *Ogygia* Brongniart, *Ogygiopsis* Walcott, *Phillipsinella* Novák, *Platypeltis* Callaway, *Ptychopyge* Angelin, and *Stygina* Salter.

This is a tolerably homogeneous group, although some of the Cambrian forms have a sufficiently archaic expression to make them seem a little out of place with genera of so pronounced a family type as *Asaphus, Niobe, Ptychopyge, Megalaspis,* and *Isotelus.*

The elements of the glabella are generally quite obscure, and even its limits cannot be clearly made out in late genera, as *Stygina* and *Asaphus.* The segmental nature of the glabella is clearly shown in *Ogygia, Ogygiopsis, Homalopecten, Asaphellus, Bronteopsis,* and *Bathyuriscus.*

The elements of the pygidium are obscurely marked in *Brachyaspis* and *Isotelus. Phillipsinella* is a very small form, and probably the young of an *Asaphus. Barrandia, Homalopecten,* and *Stygina* serve as transition genera to the Illænidæ.

II. ILLÆNIDÆ. — Including the genera and sub-genera *Illænus* Dalman, *Æglina* Barrande (= *Cyclopyge* Corda), *Bumastus* Murchison, *Dysplanus* Burmeister, *Ectillænus* Salter, *Holocephalina* Salter, *Hydrolenus* Salter, *Illænopsis* Salter, *Illænurus* Hall, *Nileus* Dalman, *Octillænus* Salter, *Panderia* Volborth, *Psilocephalus* Salter, *Symphysurus* Goldfuss, and *Thaleops* Conrad.

The Illænidæ form a much more compact group than the

preceding, characterized by having a rostral plate and by the very tumid form of the large cephalon and the obscure or obsolete boundaries of the glabella and occipital lobe. The pygidium often closely resembles the cephalon in size and form, and the axis is frequently scarcely defined.

Considerable variation is shown in the size, position, and direction of the visual surfaces. There is also a ratio between the size of the fixed-cheeks and the eyes. In proportion as the fixed-cheeks are large, the eyes are small, and as the area of the fixed-cheeks diminishes from a widening of the axis of the animal, the eyes become larger. Thus, in *Holocephalina*, with extremely large fixed-cheeks and narrow axis, the eyes are quite small. In *Illœnopsis, Dysplanus, Panderia*, and *Octillœnus* they are progressively larger, and in *Illœnus, Bumastus*, and *Nileus*, where the axis is wide and the fixed-cheeks are reduced, the eyes are relatively large. This variation reaches its limit in the species of *Æglina*, where the axis is very wide and the fixed-cheeks are reduced to almost nothing, so that the glabella and eyes make up the entire dorsal surface of the cephalon. In *Æglina princeps* Barrande the eyes extend about half the length of the cephalon. The eyes of *Æ. rediviva* Barrande bound the whole length of the sides of the head, and in *Æ. armata* Barrande the coalesced free-cheek pieces are almost wholly converted into a visual area, so that there is a continuous eye around the sides and front of the cephalon.

Variations in the position of the eyes are to be noted in nearly all the genera. In *Ectillœnus* and *Psilocephalus* they are in front of the middle of the length of the cephalon, and in *Dysplanus, Illœnopsis*, and *Holocephalina* they are near the posterior angles of the cranidium. *Panderia* has the eyes directed obliquely backward, and in *Thaleops* they are carried on conical extensions pointing outward.

Family VII.   Proëtidæ Barrande.

Cephalon about one-third of the whole animal; genal angles generally produced into spines; glabella tumid, with two lateral basal lobes defined by oblique furrows in front of the neck segment. Free-cheeks large, separate. Sutures extending from the posterior margin inward to the eyes, and then forward, cutting the anterior margins separately. Eyes prominent, often large. Thorax of from eight to twenty-two free segments, with grooved pleura. Pygidium usually of many segments; pleural and axial portions strongly grooved; margin entire or dentate.

Ordovician to Permian.

Including the genera and sub-genera *Proëtus* Steininger, *Arethusina* Barrande, *Brachymetopus* McCoy, *Celmus* Angelin, *Cordania* Clarke, *Crotalurus* Volborth, *Cyphaspis* Burmeister, *Dechenella* Kayser, *Griffithides* Portlock, *Phaëtonella* Novák, *Phillipsia* Portlock, *Prionopeltis* Corda, *Pseudophillipsia* Gemmellaro, *Schmidtella* Tschernyschew, *Tropidocoryphe* Novák, and *Xiphogomium* Corda.

The genera of this family readily fall into a series expressing more or less closely the development and specialization of various characters. *Arethusina* is the only genus retaining the archaic eye-lines, and both on this account and for the comparatively forward position of the eyes (itself a nepionic character), together with the large number of thoracic segments, it stands near the base of the series.

The eyes gradually approach the axis, and move backward through the genera *Tropidocoryphe*, *Cyphaspis*, *Proëtus*, *Prionopeltis*, *Phillipsia*, and *Griffithides*. Concurrent with this variation, there is a reduction of the fixed-cheeks and extension of the glabella. In *Arethusina*, *Tropidocoryphe*, *Cordania*, and *Cyphaspis* the fixed-cheeks are about the size of the free-cheeks, and occupy a large portion of the cranidium. They are more reduced in *Proëtus* and *Prionopeltis*, and in *Phillipsia* and *Griffithides* they form only a narrow border to the glabella. The lobation of the glabella varies greatly, and few species retain evidences of its original segmental nature. Some *Proëtus* and *Dechenella* show this feature, but in many *Phillipsia* and *Griffithides* the elements cannot be made out. In *Proëtus* there is often a small accessory lobe developed at the ends of the neck ring, which is only of interest as being homologous with similar lobes in many of the Lichadidæ and Acidaspidæ, where they often become very conspicuous. In all the Proëtidæ the oblique lobes of the fourth annulus of the glabella are also important in this connection, as here again is marked the inception of side axial lobes, which become prominent features in higher genera, indicating greater specialization of the organs and appendages of the head.

Family VIII. BRONTEIDÆ Barrande.

Dorsal shield broadly elliptical. Cephalon less than one-third the entire length; glabella rapidly expanding in front, with faint indications of lobes. Free-cheeks larger than the fixed-cheeks. Facial sutures extending from the posterior margin just behind the eyes abruptly inward around the palpebral lobes, and then diverging and cutting the antero-lateral margins separately. Eyes crescentic. Thorax of ten segments, with ridged pleura. Pygidium longer than cephalon or thorax; axis very short, with radiating furrows extending from it across the broad limb toward the margin; doublure very wide; margin generally entire. Ordovician to Devonian.

Including the single genus *Bronteus* Goldfuss (= *Goldius* de Koninck).

Many of the species of *Bronteus* (as *B. angusticeps* Barrande, *B. palifer* Beyrich) show a breaking up of the glabella into symmetrically disposed separate lobes, as in *Conolichas* and *Acidaspis*. The frontal lobe is transverse and much larger than the others. Back of it may be simple grooves marking the elements (*B. campanifer* Beyrich), or there may be one or two circular or elliptical swellings on each side of the axis (*B. angusticeps* Barrande), or, in addition, the axial portion may consist of several lobes. The reduction of the axis of the pygidium and the expansion of the limb meet with their greatest expression in this genus. *Lichas* shows the decline of these characters, the pygidial limb becoming more or less deeply lobed, and finally the lobes are represented by spines (*Arges*, *Terataspis*).   Further progression of these changes is shown in *Acidaspis*.

Family IX.   LICHADIDÆ Barrande.

Dorsal shield generally large and flat, with granulated test. Cephalon small, not more than one-fourth the entire length; genal angles spiniform.   Free-cheeks separate; sutures extending from the posterior margin obliquely inward to the eyes, and then almost directly forward, cutting the margin separately. Glabella broad, with a large, often tumid, central lobe and from one to three side lobes.   Eyes not large.   Thorax with nine or ten segments and grooved and falcate pleura.   Pygidium large, flat, commonly with toothed or notched margin corresponding to the pleural grooves; doublure very broad.

Ordovician to Devonian.

Including the genera and sub-genera *Lichas* Dalman, *Arctinurus* Castelnau, *Arges* Goldfuss, *Ceratolichas* Hall and Clarke, *Conolichas* Dames, *Dicranogmus* Corda, *Homolichas* Schmidt, *Hoplolichas* Dames, *Leiolichas* Schmidt, *Metopias* Eichwald, *Oncholichas* Schmidt, *Platymetopus* Angelin, *Terataspis* Hall, *Trochurus* Beyrich, and *Uralichas* Delgado.

Most of the forms of this family are above the average size of trilobites, and several species are among the largest of the class.   They are all thin-shelled, and were loosely articulated, so that entire specimens are extremely rare.

A great diversity is shown in the form and lobation of the glabella.   In *Lichas* (sens. str.), *Platymetopus*, and *Leiolichas* the anterior lobe dominates and is continuous with the axis. In *Hoplolichas* and *Homolichas* the lateral lobes are strongly defined, and each is nearly equal in size to the central lobe. *Dicranogmus*, *Oncholichas*, *Conolichas*, *Metopias*, *Arctinurus*, and *Arges* show the lateral lobes divided transversely into two or three smaller ones.   Lastly in *Ceratolichas*, and more especially in *Terataspis*, the central lobe becomes a prominent ovoid or globular extension.   These variations evidently indicate differences in the relative development of the appendages and organs of the head, and therefore are of considerable morphological importance.

The pygidium is composed of few distinct segments. The annulated portion of the axis is generally short, and the dentations on the border of the limb, corresponding to the pleural grooves, range from two to four on each side. *Leiolichas* is the only form which has an entire pygidial margin.

Family X.   ACIDASPIDÆ Barrande.

Dorsal shield spinose. Cephalon transversely semi-elliptical, quadrate, or trapezoidal; genal angles spiniform. Glabella with one large median axial lobe and two or three lateral lobes. Free-cheeks large, separate. Sutures extending from within the genal angles abruptly inward to the eyes, and then forward, cutting the anterior margin each side of the glabella. Eyes small, often prominent. Thorax of eight to ten segments, with ridged pleura extended into hollow spines. Pygidium usually small, with spinous margin.   Ordovician to Devonian.

Including the genera and sub-genera *Acidaspis* Murchison, *Ancyropyge* Clarke, *Ceratocephala* Warder, *Dicranurus* Conrad, *Odontopleura* Emmrich, and *Selenopeltis* Corda.

In this family and the Lichadidæ is shown the highest expression of differentiation and specialization of the Opisthoparia. The primitive pentamerous lobation of the axis of the cranidium is entirely obscured, and is only clearly seen in the protaspis and early nepionic stages. These two families are very closely related, the chief differences being noticed in the size and character of the pygidium, and the ribbed or grooved pleura. The Lichades are generally much larger and flatter, but the smaller and highly spinose forms of *Arges*, *Ceratolichas*, and *Hoplolichas* approach quite near some of the Acidaspidæ.

It has been customary of late years to regard all the species of this family as belonging to the single genus *Acidaspis*, and to consider the various sub-divisions bearing separate names as of the value of sub-genera. Clarke [14] has shown that, on the basis of priority, *Ceratocephala* is the first distinctive name ever applied to the group, and is therefore entitled to full generic recognition. He further recognizes *Odonto-pleura*, *Acidaspis*, *Dicranurus*, *Selenopeltis*, and *Ancyropyge* in the sub-ordinate position of sub-genera under *Ceratocephala*.

Order C. PROPARIA, nov. ord.

(πρό before, and πἄρειά cheek piece.)

Free-cheeks not bearing the genal angles. Facial sutures extending from the lateral margins of the cephalon in front of the genal angles, inward and forward, cutting the anterior margin separately or uniting in front of the glabella. Compound paired eyes scarcely developed or sometimes absent in the most primitive family, well-developed and schizochroal in last family.

Including the families Encrinuridæ, Calymmenidæ, Cheiruridæ, and Phacopidæ.

Salter's first division, Phacopini, included the two families Phacopidæ and Cheiruridæ. The Calymmenidæ were placed in his second division, the Asaphini.

This is the only order of trilobites which apparently begins within the known Paleozoic, and, unlike the other orders, it had no pre-Cambrian existence. The earliest forms of the Proparia came at the close of the Cambrian, in the lower Ordovician. Its greatest generic differentiation was attained early. There is a rapid decline in the Silurian and Devonian, and only one or two genera extend to the beginning of the Carboniferous.

In the Opisthoparia it was demonstrated that the Conocoryphidæ formed the natural base or most primitive family in the order, and was distinguished by the narrow marginal free-cheeks and the absence of well-developed eyes. It is of much interest and importance to be able to recognize, in the Proparia, a similar primitive family having characters in common with the former, but still clearly belonging to the higher order. *Placoparia*, *Areia*, and *Dindymene*, of the Encrinuridæ, constitute a group of apparently blind trilobites, with narrow marginal free-cheeks, presenting in general the appearance of *Atops, Conocoryphe, Ctenocephalus*, etc., of the Conocoryphidæ. The somewhat higher genera *Cybele* and *Encrinurus* have intermediate or transitional characters leading to the other families. The Cheiruridæ show a greater amount of differentiation and progressive and regressive evolution than any other in this order. *Crotalocephalus* and *Sphærexochus* seem to express the highest development, and *Deiphon* and *Onycopyge* show the effects of over-specialization, resulting in degeneration. The Calymmenidæ, in their small eyes and narrow free-cheeks, have decided affinities with the lower genera. The same may be said of *Trimerocephalus* of the Phacopidæ, though the other genera of this family possess large eyes, situated well back and close to the glabella. For these and other reasons, the family is placed at the end of the order, as expressing its highest development.

Family XI.　Encrinuridæ Linnarsson.

Cephalon narrow, transverse. Fixed-cheeks very large. Free-cheeks long, narrow, separate, sometimes with a free plate between the anterior extremities. Sutures extending from in front of the genal angles obliquely forward, and cutting the anterior margin in front of the glabella. Eyes very small or absent. Thorax of from nine to twelve segments, with ridged pleura. Pygidium generally composed of many segments; limb with strong ribs usually less in number than the annulations of the axis.　　　　　　　　　　　　　　Ordovician and Silurian.

Including the genera *Encrinurus* Emmrich (*Cromus* Barrande), *Areia* Barrande, *Cybele* Lovén, *Dindymene* Corda, *Placoparia* Corda, and *Prosopiscus* Salter.

The Conocoryphidæ were shown to be the radical of the order Opisthoparia, and for similar reasons the Encrinuridæ may now be taken as the primitive family of the Proparia.

The cephala of *Areia* and *Placoparia* have many resemblances to *Conocoryphe*, but the fixed-cheeks bear the genal angles and spines, while in the latter genus they are on the free-cheeks. In both families the free-cheeks are narrow and marginal, and the eyes are absent or rudimentary. Both these characters are decidedly larval. Other primitive and larval features belonging to the Encrinuridæ are the eye-lines in *Cybele* and *Encrinurus*, the undefined and expanded termination of the glabella in *Dindymene* and *Encrinurus*, and the pentamerous head axis in all but *Dindymene*, in which the four anterior lobes or annulations are obsolete. In *Encrinurus* the eye-line in meeting and joining the anterior lobe of the glabella sometimes gives the appearance of an extra lobe, as in *Ogygia* and *Paradoxides*.

Family XII.   CALYMMENIDÆ Brongniart.

Cephalon somewhat wider than long. Fixed-cheeks large; genal angles rounded or produced into spines. Glabella narrowing anteriorly. Free-cheeks long, separate, usually with a free rostral plate between the anterior extremities. Sutures extending from just in front of the genal angles, converging anteriorly, and cutting the margins separately. Eyes small; visual surface seldom preserved. Thorax of thirteen segments, with grooved pleura. Pygidium of from six to fourteen segments; axis tapering.   Ordovician to Devonian.

Including the genera and sub-genera *Calymmene* Brongniart, *Brongniartia* Salter, *Burmeisteria* Salter, *Calymmenella* Bergeron, *Calymmenopsis* Munier-Chalmas and Bergeron, *Dipleura* Green, *Homalonotus* Koenig, *Koenigia* (= *Homalonotus*) Salter, *Pharostoma* Corda, *Plœsiacomia* Corda, *Ptychometopus* Schmidt, and *Trimerus* Green.

The genera of this family naturally cluster around the two leading ones, *Calymmene* and *Homalonotus*. Closely related to the first are *Ptychometopus*, *Pharostoma*, *Calymmenopsis*, and *Calymmenella*, all agreeing in having the glabella well defined and marked by furrows or indentations at the sides, corresponding to its segmental nature.

The second group, including *Brongniartia, Trimerus, Homalonotus* (sens. str.), *Plæsiacomia, Dipleura,* and *Burmeisteria,* agree in having a low, not sharply defined, quadrate glabella, without distinct furrows or lobes. In general, the axis of the thorax and pygidium is much wider than in the first group, and the pygidium is more elongate and often pointed.

Family XIII. CHEIRURIDÆ Salter.

Glabella well defined. Free-cheeks small, sometimes much reduced. Sutures extending from in front of the genal angles inward to the eyes, and then obliquely forward, cutting the anterior margin in front and each side of the glabella. Eyes usually small. Thorax composed of from nine to eighteen segments, generally eleven; pleura often extended into hollow spines. Pygidium small, with from three to five segments; pleural elements commonly produced into spines.

Principally Ordovician and Silurian, but extending into the Devonian.

Including the genera and sub-genera *Cheirurus* Beyrich, *Actinopeltis* Corda, *Amphion* Pander, *Anacheirurus* Reed, *Ceraurus* Green, *Crotalocephalus* Salter, *Cyrtometopus* Angelin, *Deiphon* Barrande, *Diaphanometopus* Schmidt, *Eccoptocheile* Corda, *Hemisphærocoryphe* Reed, *Nieszkowskia* Schmidt, *Onycopyge* Woodward, *Pseudosphærexochus* Schmidt, *Sphærexochus* Beyrich, *Sphærocoryphe* Angelin, *Staurocephalus* Barrande, and *Youngia* Lindström.

As in other families, the most primitive genera are those in which the regular pentamerous lobation of the glabella is retained, with the eyes well forward, the free-cheeks narrow, and the fixed-cheeks ample. *Diaphanometopus, Anacheirurus, Eccoptocheile,* and *Cyrtometopus* agree in these respects, and therefore belong at the beginning of a phylogenetic list. *Ceraurus* and *Nieszkowskia* appear to branch off here, being characterized by the narrow transverse form of the cephalon and the great development of the two anterior pygidial pleura into hollow spines directed outward and backward. These features are simulated in *Deiphon,* in which, however, the prominent glabella is without distinct lobes, and the large pleural extensions of the pygidium do not belong to the anterior segment.   Its natural place is at the end of the series. F. Cowper Reed [31] has shown (in his memoir on the evolution of *Cheirurus* and its sub-genera, not including the other genera of the family) that the direct line from *Cyrtometopus* passes through *Cheirurus* to *Crotalocephalus*.   The genera *Pseudosphœrexochus* and *Amphion* also have relations with these genera and should be placed here.   There is next a group of forms with prominent globular glabellæ, leading from *Cheirurus* to *Sphœrocoryphe*, and including *Actinopeltis*, *Youngia*, and *Hemisphœrocoryphe*.   *Staurocephalus* should immediately follow these.   *Sphœrexochus* seems to be related to *Cheirurus* and *Actinopeltis*.   Like them it has two side lobes at the base of the glabella, and the anterior furrows are obsolescent, as in *Actinopeltis* and *Youngia*.   Lastly come *Onycopyge* and *Deiphon*, with their globular glabellæ without furrows, the spiniform fixed-cheeks, the thoracic and pygidial pleura, and the free-cheeks reduced to almost nothing, forming a small part of the doublure of the cephalon. The former genus has four spiniform pygidial pleura, two on each side, but in the latter two are reduced and the remaining pair is greatly enlarged.

Family XIV.   PHACOPIDÆ Salter.

Glabella tumid, widest in front.   Free-cheeks continuous, united anteriorly.   Suture extending from in front of the genal angles inward to the eyes, and then forward around the glabella. Eyes generally large, and always with distinct facets, schizochroal.   Thorax with eleven segments, with grooved pleura. Pygidium usually large and of many segments; limb ribbed; margin entire or dentate.   Ordovician to Devonian.

Including the genera and sub-genera *Phacops* Emmrich, *Acaste* Goldfuss, *Chasmops* McCoy, *Coronura* Hall, *Corycephalus* Hall and Clarke, *Cryphœus* Green, *Dalmanites* Emmrich (*Hausmannia* Hall and Clarke), *Homalops* Remele and Dames, *Monorachos* Schmidt, *Odontocephalus* Conrad, *Pterygometpous* Schmidt, *Symphoria* Clarke, and *Trimerocephalus* McCoy.

The last family of trilobites comprises forms which are commonly believed to be the most highly organized of the class, and certain it is that a high degree of organization is manifested. Some of the characters may be considered as progressive, while others are larval or possessed chiefly by the most primitive families, and are therefore to be looked upon as regressive. Schizochroal eyes occur in no other family, and this feature is apparently indeterminate. The complete union of the free-cheeks, carrying the doublure of the sides and front of the cephalon, can be best homologized with similar structures in some of the lowest genera, and is a retention of the complete ocular segment. The glabella, though considerably enlarged anteriorly, does not attain the degree of specialization shown in *Lichas* and *Acidaspis*. Only *Chasmops* and related forms (*Monorachos, Homalops, Symphoria,* and *Coronura*) have separate or accessory lobes. The margin of the cephalon shows even greater diversity than in any other family. It may be plain (*Phacops, Cryphæus*), notched (*Corycephalus*), denticulated (*Odontocephalus*), or extended in front as a spinose, spatulate, or dentate process (*Dalmanites nasutus* Conrad, *D. tridens* Hall, etc.). The pygidium has a range almost as great, though in this respect it is equalled in the Lichadidæ, Acidaspidæ, and some of the Olenidæ. In America the section typified by *Dalmanites* culminated during the lower Devonian. Not only are the largest forms found here (*Coronura diurus* Green, *C. myrmecophorus* Green, *D. tridens* Hall, etc.), but also the most ornate and specialized; as *Corycephalus, Odontocephalus,* and *Coronura*.

References.

1. Agassiz, L., 1873. — Methods of Study in Natural History, eighth edition.
2. Angelin, N. P., 1854. — Palæontologia Scandinavica. Pt. I. Crustacea formationis transitionis.
3. Barrande, J., 1852, 1872. — Système Silurien du centre de la Bohême. Part I. 1852 ; supplement, 1872.

4. Beecher, C. E., 1895. — Structure and Appendages of Trinucleus. *Amer. Jour. Sci.* (3), vol. xlix.

5. —— 1895. — The Larval Stages of Trilobites. *American Geologist*, vol. xvi.

6. —— 1896. — On the validity of the family Bohemillidæ, Barrande. *American Geologist*, vol. xviii.

7. Bernard, H. M., 1892. — The Apodidæ. A Morphological Study. *Nature Series.*

8. —— 1894. — The Systematic Position of the Trilobites. *Quar. Jour. Geol. Soc. London*, vol. l.

9. —— 1895. — Supplementary notes on the Systematic Position of the Trilobites. *Quar. Jour. Geol. Soc. London*, vol. li.

10. —— 1895. — The Zoological Position of the Trilobites. *Science Progress*, vol. iv.

11. Brongniart, A., 1822. — Histoire Naturelle des Crustacés fossiles. Trilobites.

12. Burmeister, H., 1843. — Die Organisation der Trilobiten.

13. Chapman, E. J., 1889. — Some remarks on the classification of the Trilobites as influenced by stratigraphic relations ; with outlines of a new grouping of these forms. *Trans. Roy. Soc. Canada*, vol. vii.

14. Clarke, J. M., 1892. — Notes on the Genus Acidaspis. Report of the State Geologist, *N. Y. State Mus.*, 44*th Ann. Rept.*

15. Corda, A. J. C. [and J. Hawle], 1847. — Prodrom einer Monographie der böhmischen Trilobiten. *Abhandl. böhm. Gesell. Wiss.*, Prag, vol. v.

16. Dalman, J. W., 1826. — Om Palæaderna eller de sa kallade Trilobiterna.

17. Emmrich, H. F., 1839. — De Trilobitis. Dissertation.

18. —— 1844. — Zur Naturgeschichte der Trilobiten.

19. Gegenbaur, C., 1878. — Elements of Comparative Anatomy, English edition (Bell and Lankester).

20. Goldfuss, A., 1843. — Systematische Uebersicht der Trilobiten und Beschreibung einiger neuen Arten derselben. *Neues Jahrbuch für Mineralogie*, etc.

21. Hyatt, A., 1889. — Genesis of the Arietidæ. *Mem. Mus. Comp. Zoöl.*, vol. xvi.

22. Jackson, R. T., 1890. — Phylogeny of the Pelecypoda. The Aviculidæ and their Allies. *Mem. Boston Soc. Nat. Hist.*, vol. iv.

23. Kingsley, J. S., 1894. — The Classification of the Arthropoda. *American Naturalist*, vol. xxviii.

24. Lang, A., 1891. — Text-book of Comparative Anatomy. English translation by H. M. and M. Bernard.

25. McCoy, F., 1849. — On the classification of some British fossil Crustacea, with notices of new forms in the university collection at Cambridge. *Ann. Mag. Nat. Hist.* (2), vol. iv.

26. Matthew, G. F., 1896. — Faunas of the Paradoxides Beds in Eastern North America. No. I. *Trans. N. Y. Acad. Sci.*, vol. xv.

27. Milne-Edwards, A., 1873. — Recherches anatomiques sur les Limules. *Ann. Sci. Nat.*, t. xvii.

28. —— H., 1834–40. — Histoire naturelle des Crustacés.

29. Œhlert, D.-P., 1895. — Sur les Trinucleus de l'ouest de la France. *Bull. Soc. Géol. France* (3), t. xxiii.

30. Quenstedt, F. A., 1837. — Beiträge zur Kenntniss der Trilobiten. *Archiv für Naturgesch.*, Bd. I.

31. Reed, F. R. Cowper, 1896. — Notes on the Evolution of the Genus Cheirurus. *Geological Magazine*, vol. iii.

32. Salter, J. W., 1864. — A Monograph of British Trilobites. Pt. I. *Pal. Soc.*, London, vol. xvi.

33. Walcott, C. D., 1881. — The Trilobite ; New and Old Evidence relating to its Organization. *Bull. Mus. Comp. Zoöl.*, vol. viii.

34. Woodward, Henry, 1895. — Some Points in the Life-history of the Crustacea in Early Palæozoic Times. Anniversary Address of the President. *Quar. Jour. Geol. Soc. London*, vol. li.

35. Zittel, K. A., 1881–1885. — Handbuch der Palæontologie, Bd. II.

36. —— 1895. — Grundzüge der Palæontologie.

List of Genera.

2. THE SYSTEMATIC POSITION OF THE TRILOBITES *

As a preface to these remarks, it may be stated that there is
no intention of indulging in a controversy regarding trilobite
affinities. Professor Kingsley, as a biologist and authority on
living arthropods, naturally approaches the subject from a
standpoint nearly opposite to that of a trilobite investigator
or paleontologist. The differences of opinion or interpretation held by each are generally more apparent than real, and,
as stated, depend mainly upon the point of view. Further,
it cannot be expected that students of Lang, Claus, and
Lankester will agree as to the value and significance of a
number of important characters, or upon certain theories
which have been the natural outcome of such differences.

In the study of trilobite morphology and classification I
have made homologies and correlations from theories, opinions, and observations which seemed most current and in
general favor in standard text-books. The chief purpose of
the investigation was to work out the structure and development of the trilobite, and to apply the information to a classification of the members of the group itself. The results have
been recently published in the *American Journal of Science*
(February and March, 1897). No attempt was made to revise
the classification of the animal kingdom from the trilobite
standpoint, nor even to determine the branches of arthropod
phylogeny. The discussion of the systematic position of
Limulus was carefully avoided, though this is usually considered the chief end of any trilobite theorizing. The affinities

---

* This paper was written to follow one by J. H. Kingsley, on "The Systematic Position of the Trilobites," published in the *American Geologist*, XX, 38-
40, 1897.

of the trilobites were manifestly closer to the Entomostraca and Malacostraca than to other arthropods, and therefore comparisons were drawn with these sub-classes of the Crustacea.

In the following remarks only the main points of difference between the views held by Professor Kingsley and myself are dwelt upon : —

If the trilobites are true crustaceans, as conceded, it is then fair to expect a more or less close agreement between the larval forms of both. In my paper on "The Larval Stages of Trilobites" (*American Geologist*, September, 1895) I endeavored to show this close agreement, and concluded that the protaspis stage of trilobites could be homologized with the nauplius larva of higher Crustacea. Professor Kingsley notes the following differences: (1) The differentiated median and pleural regions; (2) the segmented cephalic region; (3) the absence of a median eye: and (4) paired eyes.

As to the first, I do not think the differentiation is much greater than in the nauplii of *Apus*, *Cyclops*, *Lucifer*, and others in which there are side regions. The pleural regions cannot be considered as highly specialized characters, since they are common to many groups, and each segment is considered as primarily consisting of tergum, pleura, and sternum.

(2) The segmentation of the protaspis is very feeble in the earliest stages, and is evidently emphasized from the fact that the fossils are viewed as opaque objects and exhibit strongly any inequalities of surface features, while living nauplii are studied as translucent objects. Furthermore, any such difference cannot be real, since the nauplius shows its true segmented nature in its paired appendages.

(3) The apparent absence of a median eye in the trilobite protaspis could be taken as of some value were it not that the fossils are not more than one millimetre in length, and even under the most favorable conditions could hardly be expected to show such small features as ocelli. Moreover, the median eye may have been marginal or ventral, and therefore would not be seen in the fossil, which preserves only the dorsal crust.

(4) Paired eyes are not present, or at least not visible in the protaspis stages of primitive trilobites. They may through acceleration appear in the protaspis stages of later genera, as they do in the nauplius embryos of certain modern decapods.

I do not believe that the nauplius has any great phylogenetic significance, and have considered it "as a derived larva modified by adaptation" (*l. c.*, p. 190), and as a "modified crustacean larva" (*ibid.*, p. 191).

It does not seem necessary to correlate the post-oral second pair of trilobite appendages with the mandibles of higher Crustacea. The second pair in the nauplius is also post-oral and manducatory, though they later develop into the antennæ and are pre-oral.

As to the cephalon of a primitive crustacean, I have merely accepted the conclusion approved by Claus, as stated by Lang, in his reconstruction of the original crustacean, which is as follows: "The head segment was fused with the four subsequent trunk segments to form a cephalic region" (Comparative Anatomy, p. 406).

Similarly in regard to the interpretation of the biramous appendages, I have adopted the statements and conclusions of a large number of zoölogists who consider the most primitive appendages as branched or consisting of a dorsal and a ventral member, and I have followed them in thus interpreting the trilobite appendages, which are clearly of this nature.

3. THE LARVAL STAGES OF TRILOBITES [*]

(Plates III–V)

Introduction.

It is now generally known that the youngest stages of trilobites found as fossils are minute ovate or discoid bodies, not more than one millimetre in length, in which the head portion greatly predominates. Altogether they present very little likeness to the adult form, to which, however, they are traceable through a longer or shorter series of modifications.

Since Barrande[2] first demonstrated the metamorphoses of trilobites, in 1849, similar observations have been made upon a number of different genera by Ford,[22] Walcott,[34, 35, 36] Matthew,[26, 27, 28] Salter,[32] Callaway,[13] and the writer[4, 5, 7]. The general facts in the ontogeny have thus become well established, and the main features of the larval form are fairly well understood.

Before the recognition of the progressive transformation undergone by trilobites in their development, it was the custom to apply a name to each variation in the number of thoracic segments and in other features of the test. The most notable example of this is seen in the trilobite now commonly known as *Sao hirsuta* Barrande. It was shown by Barrande[3] that Corda[17] had given no less than ten generic and eighteen specific names to different stages in the growth of this species alone.

The changes taking place in the growth of an individual

---

[*] *American Geologist*, XVI, 166–197, pls. viii–x, 1895.

are chiefly : the elongation of the body through the gradual addition of the free thoracic segments ; the translation of the eyes, when present; the modifications in the axis of the glabella ; the growth of the free-cheeks ; and the final assumption of the mature specific characters of pygidium and ornamentation.

In the present paper the larval stages of several species are described and illustrated for the first time, and a review is undertaken of all the known early larval stages thus far described. This work would have no special interest in itself were it not for the fact that, with our present understanding of trilobite morphology, it is possible to reach some conclusions of general importance which have a direct bearing on the significance and interpretation of several of the leading features of the trilobite carapace, and incidentally upon the structure and relations of the nauplius of the higher Crustacea.

The Protaspis.

Barrande[3] recognized four orders of development in the trilobites, as follows : —

TYPES.

I. { Head predominating, incomplete. / Thorax nothing or rudimentary. / Pygidium nothing. } *Sao hirsuta.*

II. { Head distinct, incomplete. / Thorax nothing. / Pygidium distinct, incomplete. } *Trinculeus ornatus* and all *Agnostus.*

III. { Head complete. / Thorax distinct, incomplete. / Pygidium distinct, incomplete. } *Arethusina Konincki.*

IV. { Head complete. / Thorax complete. / Pygidium distinct, incomplete. } *Dalmanites Hausmanni.*

A study of these groups shows at once that they form a progressive series in which the first alone is primitive.

The others are more advanced stages of development, as shown by the larger size of the individuals, and their having characters which appear successively in the ontogeny of a species belonging to the first order of development. To attain the stage which is represented by actual specimens, they must have passed through earlier stages, which as yet have not been found. Furthermore, it is evident that Barrande did not consider the orders after the first as primitive, and characteristic of the genera cited, for, in some remarks under the third order, he says:[3] "Il est très-vraisemblable, que la plupart des Trilobites de cette section, si ce n'est tous, devront être un jour transférés dans la première, par suite de la découverte probable d'embryons sans segmens thoracique."

The geological conditions necessary for the fossilization of the minute larval forms of trilobites are such that only in comparatively rare instances are any of the immature stages preserved. Larval specimens are doubtless often overlooked or neglected by collectors, but generally the sediments are too coarse for the preservation of these small and delicate organisms. In certain horizons and rocks, however, such remains are quite abundant, and complete ontological series may be obtained. Yet it is not strange that series of equal completeness have not been found in all Paleozoic horizons.

The abbreviated or accelerated development of many of the higher Crustacea has resulted in pushing the typical free-swimming, larval nauplius so far forward in the ontogeny that this stage is either eliminated or passed through while the animal is still within the egg, so that when hatched it is much advanced. Although the trilobites show distinct evidence of accelerated development through the earlier inheritance of certain characters which will be taken up later, yet it is not believed that the normal series or periods of transformation were to any degree disturbed, since both the simplest and most primitive genera whose ontogeny is known and the most highly specialized forms agree in having a common early larval type. This would be expected from their great antiquity, their comparatively generalized and uniform structure, and from the fact that no sessile, attached, parasitic, land, or freshwater species are known. These conditions, by introducing new elements into the ontogeny, would tend to modify or abbreviate it in various ways, especially among the higher genera.

Before discussing any of the various philosophical and theoretical problems involved in an attempt to correlate the larval forms of Crustacea, a brief consideration of the known facts relating to the larvæ of trilobites will be presented.

Minute spherical or ovoid fossils associated with trilobites have been described as possible trilobite eggs, by Barrande[3] and Walcott,[32] but nothing is known, of course, of the embryonic stages of the animals themselves. The smallest and most primitive organisms that have been detected, and traced by means of series of specimens through successive changes into adult trilobites, are, as stated above, little discoid or ovate bodies not more than one millimetre in length, as shown on Plates III and IV. It is fair to assume that we have here a general exhibition of trilobite larval stages, since the ten species represented are from various geological horizons belonging to the Cambrian, Ordovician, and Silurian sediments, with Devonian types, and showing the simple as well as the highly specialized forms.

All the facts in the ontogeny of trilobites point to one type of larval structure. This is even more noticeable than among recent Crustacea, in which the nauplius is considered as the characteristic larval form. It is desirable to give a name to this early larval type apparently so characteristic of all trilobites, and among different genera varying only in features of secondary importance. This stage may therefore be called the *protaspis* (πρῶτος *primus*, ἀσπìς *scutum*).

The principal characters of the protaspis are the following: Dorsal shield minute, varying in observed species from .4 to 1 mm. in length; circular or ovoid in form; axis distinct, more or less strongly annulated; head portion predominating;

glabella with five annulations; abdominal portion usually less than one-third the whole length of the shield, axis with from one to several annulations; pleural portion smooth or grooved; eyes when present anterior, marginal, or sub-marginal; free-cheeks when present very narrow, marginal.

Several moults took place during this stage before the complete separation of the pygidium or the introduction of thoracic segments. When such moults are recognized they may be considered as early, middle, and late protaspis stages, and designated respectively as anaprotaspis, metaprotaspis, and paraprotaspis. They introduced various changes, such as the stronger annulation of the axis, the beginning of the free-cheeks, and the growth of the pygidial portion from the introduction of new appendages and segments, as indicated by additional grooves on the axis and pleura. Similar ecdyses occur during the nauplius stage of many living Crustacea before a decided transformation is brought about. Certain of these later stages have received a distinctive name, and are called the metanauplius.

It is believed that the protaspis is homologous with the nauplius or metanauplius of the higher Crustacea. Most of the reasons for this belief will appear later in the present paper; some which may be stated now are as follows:—

(1) The size of the protaspis does not differ greatly from that of many nauplii, and represents as large an animal as could be hatched from the bodies considered as the eggs of trilobites.

(2) Some of the sediments carefully examined by the writer could preserve smaller larval trilobites were such originally present and provided with a chitinous test, as shown by the abundance of minute ostracodes, and the perfection of detail in these and other fossils.

(3) The protaspis can be shown to be structurally closely related to the nauplius, and in a more marked degree possesses some characters required in the theoretical crustacean ancestor.

Review of Larval Stages of Trilobites.

Matthew [27, 28] has carefully described several early larval (protaspis) stages of trilobites from the Cambrian rocks of New Brunswick, which are very simple and primitive, and will be noticed first.

Solenopleura Robbi Hartt; Plate III, figure 1; from the Cambrian of New Brunswick; after Matthew.[27] This larva is very minute and circular in outline; the glabella is obscurely annulated and extends to the anterior margin, where it is expanded; the neck ring is the only one well defined; the abdominal portion is less than one-third the whole length, and is limited by a slight transverse furrow; no traces of eyes or free-cheeks discernible.

Liostracus onangondianus Hartt; Plate III, figure 2; from the Cambrian of New Brunswick; after Matthew.[27] This form is similar to the preceding, though larger, and with the glabella more rapidly expanding in front. The neck segment is the only one which is distinct.

It should be mentioned that most of the larval specimens here described and figured are preserved in fine shales and slates, as casts of the interior of the dorsal shield, so that some features are not as emphatic as on the exterior of the test. When well preserved, the axis always shows the typical five annulations on the cephalon.

Ptychoparia Linnarssoni Walcott; Plate III, figures 3 and 4; from the Cambrian of New Brunswick; after Matthew.[28] The earliest stage is slightly more elongate than the preceding forms. The axis is narrow, expanding in front and obscurely annulated, five annulations belonging to the cephalon, and one to the pygidium, which is very short and separated from the cephalon by a distinct groove.

The second stage (figure 4) is decidedly more elongate; the axis is more distinctly annulated; the occipital pleura defined; and the pygidium is larger and has an additional segment.

Ptychoparia Kingi Meek; Plate III, figures 5, 6, and 7;

from the Cambrian of Nevada and Utah. Figure 5 represents a cast of the protaspis, and shows a defined occipital ring, with the axis slightly expanded and undefined in front; pygidium truncate behind. Figure 6, which is referred to a later stage (metaprotaspis) of the same species, shows the inception of several characters that have not as yet appeared in the previous larvæ. The axis is very strongly annulated; the anterior lobe is nearly as long as the four posterior annulations of the cephalon, and on each side there is a furrow representing the eye-line of the adult; the free-cheeks are present as narrow marginal plates, including the genal spines; the pygidium shows two segments separated by a furrow.

An adult *Ptychoparia Kingi* is shown in figure 7, and may be taken as representing the sum of the changes passed through in the development of larvæ like the preceding, belonging to the genera *Solenopleura*, *Liostracus*, and *Ptychoparia*. The introduction and growth of the segments of the thorax are perhaps the most marked changes, but other points of importance to be noted are: the comparatively smaller size of the cephalon and its transverse form; the limitation and recession of the glabella, which is now rounded in front, and only extends about two-thirds the length of the cephalon; the growth of the eyes and free-cheeks at the expense of the fixed-cheeks; the increased segmentation of the abdomen, shown in the axial and pleural grooves on the pygidium.

Sao hirsuta Barrande; Plate III, figures 8, 9, 10, and 11; from the Cambrian of Bohemia; after Barrande.[3] The specimens of this species are preserved as casts, and several of the features are therefore somewhat subdued. The earliest or anaprotaspis stage, represented in figure 8, is quite as primitive in most respects as any of the preceding. It is circular in outline, the annulations of the axis are distinctly shown only in the neck segment and pygidial portion, and the eye-line is present. In figure 9 of the metaprotaspis quite an advance is seen in the development of the free-cheeks and the more pronounced annulation of the glabella, together with pleural grooves from the neck segment and those of the pygidium. The next stage (figure 10) probably represents the close of the protaspis stage (paraprotaspis) and the inception of the nepionic condition, when the cephalon and pygidium are distinct and before the development of the free thoracic segments.

In considering the changes necessarily passed through by these larvæ previous to attaining their adult characters (figure 11) the most notable, aside from increase in size and addition of the sixteen thoracic segments, are: the appearance and translation of the eyes *pari passu* with the growth of the free-cheeks; the growth of the border in front of the glabella, which now narrows anteriorly, and terminates about one-third the length of the cephalon within the margin; the less distinct annulation of the glabella; and the development of the spines and tubercles ornamenting the test.

Triarthrus Becki Green; Plate III, figures 12, 13, and 14; from the Ordovician, Utica slate, near Rome, New York. A larval form of this species was figured by the writer[5] in 1893. At this time the eye-line was confused with the anterior annulation of the axis, making the cephalon appear to have six instead of five annulations. A recent examination of a large number of specimens shows that five is the invariable number, as here represented. Two protaspidian stages of this species have been noticed, differing chiefly in the size of the pygidium. Both agree in showing a strongly annulated axis, not expanded in front and terminating some distance within the margin. From the first annulation a slightly elevated ridge on each side indicates the eye-line, and extends to the marginal eye-lobe. The adult form (figure 14) shows, in addition to several characters noted in the previous species, the nearly complete loss of the two anterior annulations of the glabella; the disappearance of the eye-line; and the development of a row of nodes along the axis, from the neck segment to the proximal segment of the pygidium.

Acidaspis tuberculata Conrad; Plate IV, figures 1, 2, and

3; from the Lower Helderberg group, Albany county, New York.[4] Several of these remarkable larvæ have been found perfectly silicified in a limestone from which they have been freed by etching. In general form they resemble the second larval stage of *Sao* (Plate III, figure 9), but the pygidium is shorter and the glabella does not expand and terminate in the anterior margin. No eye-line is present, but the eye-lobes may be seen a little within the margin. The glabella has the characteristic number of annulations; margin provided with a row of denticles; genal angles extended into spines; pygidium with four spines.

The adult condition (figure 3) shows that the eyes have moved inward and backward to near the neck segment. The glabella has lost its annulations and is broken up into a median lobe with two smaller ones on each side, while the neck ring is projected into a spine. The changes noted here are much more profound than in any of the preceding genera, since *Acidaspis* is one of the most highly specialized of trilobites in its glabellar structure and elaborate ornamentation. The protaspis, too, partakes of this specialization, and although the general form of the shield and the annulation of the axis are as primitive as in *Triarthrus*, yet the characteristic spinosity of the genus appears even at this early stage and is a marked instance of acceleration of development.

Arges consanguineus Clarke; Plate IV, figure 4 ; from the Lower Helderberg group, Albany county, New York. A single larval form of this type has been found, and at first was provisionally referred to *Phaëthonides*.[4] The recent publication by Clarke[14] of *Arges consanguineus* from the same horizon and a comparison of the larva with the description and with considerable additional material, renders it now possible to determine definitely the relations of this interesting form. As the main details of structure in *Acidaspis* and *Arges* are so similar, the transformations undergone by the larva are much alike in each case. The young *Arges* likewise shows the same acceleration in the development of the spines and surface ornamentation, and the retention of the primitive features of the glabella. The specimen seen in figure 4 represents a late larval stage (paraprotaspis), as shown by the transverse form of the cephalon and the large size of the pygidium.

Proëtus parviusculus Hall; Plate IV, figures 5, 6, and 7; Utica slate, near Rome, New York. Two larval stages of this species have been found. The younger (figure 5) is smooth, broadly ovate, .72 mm. long, and widest in front; axis distinctly annulated, cylindrical on the cephalon, tapering on the pygidium; eyes nearly transverse to the axis, very large and prominent, situated on the anterior margin, separated only by the axis. The specimen represented in figure 6 is in the paraprotaspis stage, and measures .96 mm. in length. It shows an advance over the other in its size, its larger pygidium with grooved pleura, and the beginning of the recession of the eyes.

The adult of this small species is shown in outline enlarged two diameters, in figure 7. The principal changes from the larva which should be noticed are: the loss of the four anterior annulations of the glabella, the neck segment being the only one wholly defined, although the basal lobes represent remnants of the next anterior; the translation of the eyes backward as far as the pleura of the neck segment, and the change from a transverse to a parallel position with respect to the axis.

In the original description of this species,[23] no mention was made of fine undulating striæ ornamenting the entire dorsal surface of the test, nor of the basal lobes of the glabella. Both these features are present in the type specimen, which is from Cincinnati, Ohio, as well as in all the specimens from the Utica slate, near Rome, New York. With these additional characters the species is very closely related to *Proëtus decorus* Barrande.

Dalmanites socialis Barrande; Plate IV, figures 8–11; from the Ordovician of Bohemia; after Barrande.[3] A nearly complete series of the growth stages of this species is given by Barrande. The earliest, or anaprotaspis, stage found (figure 8) exhibits an outline and axis similar to *Acidaspis*. The eyes are quite large and situated, as in the same stage of *Proëtus*, transverse to the axis, on the anterior border. Genal angles present, but in this case not produced by the free-cheeks as in *Sao* and *Ptychoparia;* glabella strongly annulated, increasing in diameter anteriorly, although not expanding at the frontal margin as in *Sao*, etc. In the two following stages (figures 9, 10), the pygidium increases in size, and the pleura are defined. To reach maturity (figure 11), eleven segments are developed in the thorax, the glabella becomes more prominently developed in front, but the five annulations are maintained. The eyes have travelled in and back as far as the third cephalic segment, and their longer axes have swung around into a position parallel with the axial line, as in *Proëtus*. The pygidium has added many new segments, and the extremity is prolonged into a spine.

Before proceeding further in the discussion of the protaspis, it is necessary to notice a number of forms of young trilobites which have heretofore been referred to the embryonic and larval stages, but which are now believed to belong to stages later than the protaspis.

Besides the truly elementary forms described by Barrande and already noticed (*Sao hirsuta* and *Dalmanites socialis*), there are others which he referred to his second, third, and fourth orders of development.[3] Among these *Agnostus* may be taken first. The youngest forms of *Agnostus nudus* and *A. rex* (figures 76, 77) measure respectively 2 and 1.3 mm. in length, and the adults 13 and 15 mm. The earliest stages of the genera shown on Plates III and IV measure less than 1 mm., while the adults are more than 25 mm., with the exception of *Proëtus parviusculus*, which is seldom more than 10 mm. long, though this species has a protaspis .72 mm. in length. The cephalon and pygidium of the youngest known *Agnostus* are quite separate and distinct, which is not the case with the typical protaspis stage. It therefore seems probable that on account of the comparatively large size and advanced structure of the youngest stages observed, the elementary forms of this genus are as yet unknown, and possibly the extreme tenuity of the test in the protaspis has prevented their preservation. In the same way the young of *Trinucleus* (figure 78) show a separate cephalon and pygidium, and the specimens are in a much more advanced stage of development than the protaspis of *Proëtus,* shown on Plate IV, figure 5. An evidence of age is furnished, also, in the transverse shape of the head, which, in typical elementary forms, is longer than wide instead of wider than long.

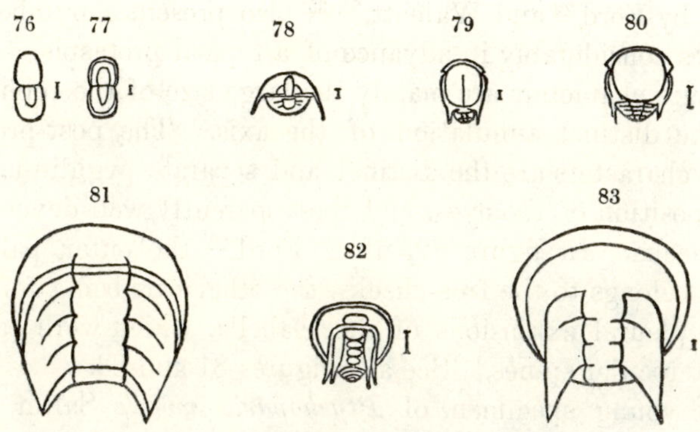

FIGURE 76. — *Agnostus nudus* Beyrich.   (After Barrande.)
FIGURE 77. — *Agnostus rex* Barrande.   (After Barrande.)
FIGURE 78. — *Trinucleus ornatus* Sternberg.   (After Barrande.)
FIGURE 79. — *Hydrocephalus saturnoides* Barrande.   (After Barrande.)
FIGURE 80. — *Hydrocephalus carens* Barrande.   (After Barrande.)
FIGURE 81. — *Olenellus* (*Mesonacis*) *asaphoides* Emmons ; Ford collection. (Original × 30.)
FIGURE 82. — *Olenellus* (*Mesonacis*) *asaphoides* Emmons.   (After Ford.)
FIGURE 83. — *Olenellus* (*Mesonacis*) *asaphoides* Emmons.   (After Walcott.)

The youngest specimens of *Arethusina Konincki,* figured by Barrande,[3] are 2 mm. or upward in length, and have seven or more free thoracic segments, with the cephalon wider than long. The facts of ontogeny show that younger stages must be admitted in which the number of segments diminishes to nothing, continuing down to a form agreeing with the protaspis of other genera.

It has already been suggested [4] that the species described by Barrande [3] under the generic name of *Hydrocephalus* are probably the young of *Paradoxides*. This conclusion receives further support from the undoubted young of *Olenellus*, a related genus, which in its immature stages bears a strong resemblance to *Hydrocephalus*. The youngest examples of the latter have a distinct pygidium, a well-developed cephalon, and large eye-lobes at the sides of the glabella, as in adult forms. Free-cheeks were evidently present, though not generally preserved. See figures 79 and 80.

The young of *Olenellus asaphoides*, described and illustrated by Ford [22] and Walcott,[35, 36] also present a number of features considerably in advance of a typical protaspis. The immature characters are mainly the large size of the cephalon and the distinct annulation of the axis. The post-protaspidian characters are the distinct and separate pygidium, the adult position of the eyes, and the apparently well-developed free-cheeks. In figure 82, after Ford,[22] the outer pair of spines belongs to the free-cheeks, the other pair being formed by the pleural extensions of the glabella, which were called the interocular spines. See also figures 81 and 83.

The young specimen of *Ptychoparia monile* Salter sp., figured and noticed by Callaway,[13] is 1.5 mm. in length, and agrees, as far as can be determined without seeing the original, with what is known of other species of the same genus. It probably belongs to a stage later than the protaspis.

Matthew [26] has carefully described some small cephala of *Ctenocephalus* (*Hartella*) *Matthewi* and *Conocoryphe* (*Bailiella*) *Baileyi*, from the Cambrian of New Brunswick. The fact of their being separate cephala, transverse in form, and from 2 to 3 mm. in length, is sufficient to show that they do not represent the youngest stages of these species.

The immature examples of *Agnostus, Trinucleus, Arethusina, Paradoxides, Olenellus, Ctenocephalus,* and *Conocoryphe,* here briefly noticed, are of great interest in a study of the ontogeny of the various species to which they pertain. In the present paper, however, it is intended chiefly to establish the primary larval characters of the trilobites, and therefore only the earliest stages are considered. Under the genera just mentioned the writer has endeavored to show that as yet their ontogeny cannot be traced as far back as the stage which has been defined as the *protaspis*. Therefore any general notions of first larval forms must at present be based on the genera *Solenopleura*, *Liostracus*, *Ptychoparia*, *Sao*, *Triarthrus*, *Acidaspis*, *Proëtus*, and *Dalmanites*.

Analysis of Variations in Trilobite Larvæ.

After taking a general survey of the earliest known larval stages of trilobites figured on Plates III and IV, it is evident that an accurate and detailed description of any one would not apply to any other except in certain broad characters. To formulate a definition of the protaspis applicable to all, as has been done previously (pp. 169 and 170), it is necessary to neglect or eliminate some rather striking characters which should now be mentioned. A few features thus omitted are considered as very primitive larval characters, while others are modifications introduced in higher or later genera through the operation of the law of earlier inheritance.

From the best evidence now obtainable, the eyes have migrated from the ventral side, first forward to the margin and then backward over the cephalon to their adult position, thus agreeing with Bernard's conclusions.[12] Therefore the most primitive larvæ should present no evidence of eyes on the dorsal shield, and naturally there would be no free-cheeks visible. Just such conditions are satisfied in the youngest larva of *Ptychoparia*, *Solenopleura*, and *Liostracus*, which are the most primitive genera whose protaspis is known. The eye-line is present in the later larval and adolescent stages of these genera, and persists to the adult condition. In *Sao* it has been pushed forward to the earliest protaspis, and is also found in the two known larval stages of *Triarthrus*. *Sao* retains the eye-line throughout life, but in *Triarthrus* the adult has no traces of it, and none of the higher and later genera studied has an eye-line at any stage of development. Matthew has considered this feature as especially characteristic of most of the Cambrian genera, and now it is further shown to be a character first appearing in the later larval stages of certain genera (*Ptychoparia*, etc.), next in the larval stages (*Sao*), then disappearing from adult stages (*Triarthrus*), and finally pushed out of the ontogeny altogether (*Acidaspis, Dalmanites*, etc.). The eyes are visible on the margin of the dorsal shield after the paraprotaspis stage, later than the eye-line in *Ptychoparia, Solenopleura, Liostracus, Sao*, and *Triarthrus;* but in the other genera through acceleration they are present in all the protaspis stages, and persist to the mature, or ephebic, condition, moving in from the margin to near the sides of the glabella.

The changes in the glabella are equally important and interesting. Throughout the larval stages the axis of the cephalon is five-segmented or annulated, indicating the presence of as many paired appendages on the ventral side. In its simplest and most primitive state it expands in front, joining and forming the anterior margin of the cephalon (larval *Ptychoparia, Sao*). During later growth it becomes rounded in front and terminates within the margin. In higher genera through acceleration it is rounded and well-defined in front even in the earliest larval stages, and often ends within the margin (larval *Triarthrus, Acidaspis*). From these common types of simple, pentamerous glabellæ, all the diverse forms among adult individuals of various genera have been derived, through changes affecting any or all of the lobes. The modifications usually take place in the anterior lobes first, and gradually involve the others, though rarely disturbing the neck segment which is the most persistent of all. Six lobes are occasionally found in the glabellæ of some species. They do not indicate an additional pair of limbs, for the extra lobe is produced (*a*) by division of the anterior lobe through the greater or less extent of the eye-line across the axis, as in *Olenellus, Paradoxides* and *Ogygia;* or (*b*) by the

marked development of muscular fulcra, which are supposed
to be connected with the hypostoma.

The next structures not especially noticeable in all stages
of the protaspis are the free-cheeks, which usually manifest
themselves in the meta- or paraprotaspis stages, though sometimes even later. Since they bear the visual areas of the
eyes, their appearance on the dorsal shield is practically
simultaneous with these organs; and before the eyes have
travelled over the margin the free-cheeks must be wholly
ventral in position. They are very narrow when first discernible (Plate III, figures 6, 9, and 10), and in *Ptychoparia*,
Sao, etc., include the genal angles, but in *Dalmanites* they
extend only a short distance below the eyes.

The remaining features of the protaspis which here require
notice are the pleural furrows and the pygidium. The pleura
from the anterior segments of the glabella are occasionally
shown, as in the young of *Olenellus* (figure 81), but usually
the pleura of the neck segment are the first and only ones to
be distinguished on the cephalon, the others being so intimately coalesced as to lose all traces of their individuality.
This makes the cranidium, or head shield, exclusive of the
free-cheeks, consist of the fused lateral extensions or pleura
of the head segments, as already noticed by Bernard.[12] The
possible pleural or segmental nature of the free-cheeks will
be noticed later.

The distinct pleura of the pygidium appear soon after the
anaprotaspis stage, and in some genera (*Sao*, *Dalmanites*) are
even more marked than in the adult state, much resembling
separate segments. The growth of the pygidium is very
considerable through the protaspis stages. At first it is less
than one-third the length of the dorsal shield, but by the
successive addition of segments, it soon becomes nearly one-
half as long. In some genera it is completed before the
appearance of the free thoracic segments, though usually new
segments are added during the adolescence of the animal.

A number of genera present adult characters, which agree
closely with some of the larval features noticed in this

section, and are important in a phylogenetic study of the trilobites.   The main features of the cephalon in the simple protaspis forms of *Solenopleura*, *Liostracus*, and *Ptychoparia* are retained to maturity in such genera as *Carausia* and *Acontheus*, which have the glabella expanded in front, joining and forming the anterior margin.   They are also without eyes or eye-line.   *Ctenocephalus* retains the archaic glabella nearly to maturity, and likewise shows eye-lines and the beginnings of the free-cheeks (larval *Sao*).   *Conocoryphe* and *Ptychoparia* are still further advanced in having the glabella rounded in front, and terminated within the margin (larva of *Triarthrus*).   These facts and others of a similar nature show that there are characters appearing in the adults of later and higher genera, which successively make their appearance in the protaspis stage, sometimes to the exclusion or modification of structures present in the most primitive larvæ.   Thus the larvæ of *Dalmanites* or *Proëtus*, with their prominent eyes, and glabella distinctly terminated and rounded in front, have characters which do not appear in the larval stages of ancient genera, but which may appear in their adult stages. Evidently such modifications have been acquired by the action of the law of earlier inheritance or tachygenesis. Altogether it seems that we have represented on Plates III and IV a progressive series of first larval stages in exact correlation with adult forms, the latter also constituting a progressive series, structurally and geologically.

A summary of the features added to the dorsal shield of the anaprotaspis stage of acceleration during the evolution of the class, from the simpler forms of Cambrian times to the later and more highly differentiated *Dalmanites*, *Proëtus*, and *Acidaspis*, would include: the free-cheeks; the eyes; the more strongly lobed glabella, rounded in front; the transient eye-line; the genal angles; and the ornaments of the test.

These additions, as may be seen by reference to Plates III and IV, considerably complicate and modify the primitive protaspis, but, as previously mentioned, it does not lose any of its essential structures.   Besides, it is possible to trace the origin and significance of the acquired characters, and thus to assign to each its true value.

Antiquity of the Trilobites.

The superlative age of the trilobites has been generally recognized, and is too well known to require more than a passing notice. Even in the earliest Cambrian they bear evidence of great antiquity in their diversified form, their larval modifications, and their polymerous cephalon and caudal shield, all of which features show that trilobite phylogeny must reach far back into pre-Cambrian times.

Not only are the smallest species (*Agnostus*) found in the Cambrian, but also many of the largest (*Paradoxides*). There is a great range of variation in the number of free thoracic segments, varying from two in *Agnostus* to twenty in *Paradoxides*. The pygidium likewise shows extreme variation of from two to upward of ten ankylosed segments. The eyes may be absent as in *Agnostus* and *Microdiscus*, or very large as in *Paradoxides*, though both in this respect and in the number of somites, free or fused, the Cambrian genera are exceeded in later deposits. In ornamentation and spiniform processes the Cambrian species show considerable development, though not as great as others since that time. However, the wide variation they do present in this particular indicates differentiation and specialization considerably removed from the beginning of the trilobite phylum.

The acquisition of distinct larval stages could only have been reached through a long series of changes in ancestral forms. The composition of the cephalon and caudal shield indicates a derivation from some primitive form, probably annelidan, in which, through adaptation to special requirements, certain polar segments became fused, forming very distinct terminal body regions. Furthermore, the trilobites are the only large division of the Arthropoda which has become extinct. The Merostomata and Phyllocarida culminated a little later, though still represented by living species;

but all the other divisions apparently have continued to increase since their inception during Paleozoic time. The only known arthropod contemporaries of the trilobites in the Cambrian are the Merostomata, Ostracoda, Phyllopoda, and Phyllocarida, all of the higher forms apparently having developed since that time. A more graphic view of the geological range and distribution of the arthropods is represented in the following table: —

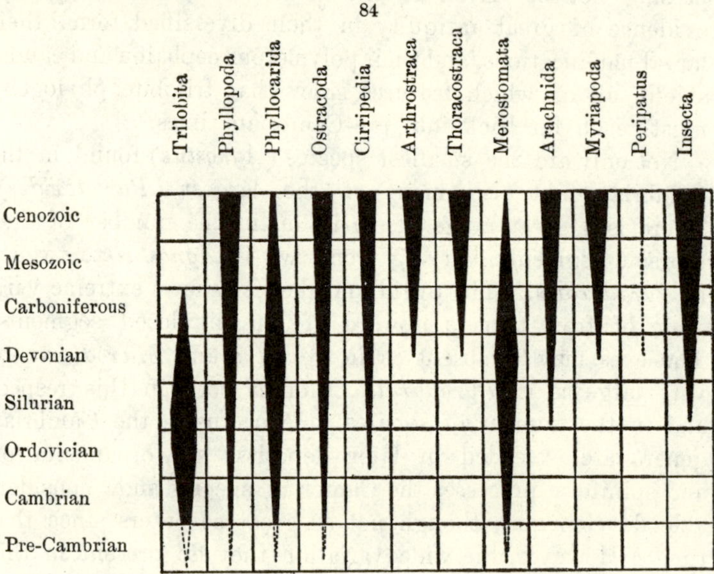

FIGURE 84. — Geological Range and Distribution of Arthropoda.

Having thus far reviewed the features of the primitive protaspis and some of the characters it acquired through earlier inheritance, together with the comparative age of the different groups of arthropods, it must be conceded that, in interpreting crustacean phylogeny from the facts of ontogeny, the trilobites, so far as they show structure, are entitled to first place. Moreover, since the appendages are quite fully known and from them the trilobite proves to be a most generalized and primitive crustacean, still greater reliance can be placed on deductions based upon a study of this type. The recent discoveries of the antennæ and the exact details of trilobite structure, together with the larval homologies here made and the concordance of trilobites with the theoretical original crustacean, leave almost no doubt as to their true crustacean affinities. Woodward,[37] from another point of view, reaches the same opinion by saying: " The trilobita, being certainly amongst the earliest forms of crustacea with which we are acquainted, cannot be removed from that class without destroying its ancestral record."

Restoration of the Protaspis.

At first thought the attempt to reconstruct the ventral side of the trilobite protaspis may seem a little hazardous or premature, but a careful consideration of all the data leads the writer to undertake this with some confidence.

The genus *Triarthrus* is taken for the basis of this restoration, as it is to-day the best known of all the trilobites, and its ventral structure has been ascertained to a degree of perfection of detail which compares favorably with many of the recent crustaceans.[6, 7, 8, 9] The writer has studied the structure of many adult and immature specimens, some of them not more than 5 mm. in length, so that fortunately the appendages are known at many stages of growth. Especially are the young and rudimentary limbs near the extremity of the pygidium in adolescent individuals of considerable morphological interest, for they agree closely with the phyllopodiform trunk appendages in the metanauplius of *Apus*, and protozoëa of *Euphausia*, or, in a general way, with the still more rudimentary trunk limbs in the nauplius stages of these and other forms.

It has been definitely ascertained that the cephalon in trilobites bears five pairs of jointed appendages or limbs.[9] In larval or immature specimens, and in adults in which the glabella retains its primitive structure, this number is indicated on the dorsal shield by the five lobes or annulations of the glabella,

including the neck ring. These may therefore be taken as
representing, in so far, the original segmentation of the cephalon, and agree with what is generally accepted as the primitive
structure in modern true Crustacea. The head portion of the
protaspis clearly shows this pentasomitic structure, and evidently carried a corresponding number of paired limbs on the
ventral side. It has also been demonstrated that the annulations on the axis of the pygidium correspond to the number
of paired limbs beneath, exclusive, of course, of the anal segment. Here, too, it is possible to tell from the pygidial portion of the protaspis the number of limbs present during life.
The protaspis of *Triarthrus*, represented in Plate III, figure
13, on this basis had five pairs of limbs attached to the head
portion, and two pairs to the pygidium.

Next, as to the composition and form of these elementary
protaspis limbs, it is safe to assume that the anterior pair,
corresponding to the antennules, must be uniramous, since
they are so during all the young and adult stages observed,
and since this form is common to all nauplius stages of modern
Crustacea, and is recognized as primitive and elementary for
the class. There is apparently a greater similarity in the
larval antennules than between any other appendages, and as
Apus and *Euphausia* have these in a very generalized form,
they are taken as types of the first pair of limbs of the trilobite protaspis, as shown in Plate V, figure 1 (*I*). It should
be noted, too, that the antennules of the trilobites arise from
the sides of the upper lip or hypostoma, as in the nauplius.

The other head appendages are typically branched, though
in many of the recent Crustacea they lose this character after
the larval stages. Especially is this true of the third pair of
limbs, which become modified into the mandibles. In trilobites the primitive biramous structure of the head limbs persists to adult stages, occurring also in limbs of all the posterior
segments where they become more and more phyllopodiform.[8]
In the restoration of the protaspis it seems only necessary to
append this archaic type of limb to each segment, agreeing as
it does in form and structure with the rudimentary limbs of

older stages and with the nauplius and metanauplius stages of *Apus*.

It cannot be doubted that the protaspis had five pairs of limbs on the head portion and one or more on the pygidium, and although these are the main points necessary to prove the argument in the next section, on the nauplius, yet it seems perfectly warrantable and better for graphic purposes to attach the required number of elementary limbs to the ventral side of the protaspis, as represented in Plate V, figure 1.

There are other organs and structural details occurring in the nauplius and in adult trilobites, which deserve recognition in a restoration of the protaspis stage. First among these is the labrum, or upper lip. Nowhere is this plate so well developed and so striking a ventral feature as among the trilobites. There can be no hesitation, therefore, in accepting this as characteristic of the protaspis.

The trilobites and most recent crustaceans have a metastoma, or lower lip. This is already developed in the nauplius stage of some Crustacea, as *Euphausia* and *Peneus*, and probably represents an early larval character. It usually appears as a median plate divided into two small plates, or lappets, on each side of the median line, posterior to the mouth, and is thus represented in the restored protaspis. As it occurs on a segment bearing also a pair of legs and has no separate neuromere, it cannot well be considered as representing a somite.

An anal opening is found in most nauplii, especially in those of the non-parasitic Crustacea, and in those in which this stage is normal and free-swimming. The protaspis, as representing a free-swimming larval stage of trilobites, therefore probably possessed an anal opening.

The only character represented in the restoration which is accepted purely from analogy is the median unpaired eye. This organ is almost universally present in the nauplius, and is regarded as a very primitive character wherever found.

The next and last structures to be noticed are the freecheeks and the beginnings of the paired eyes, as shown in Plate V, figure 1 (*g, oc*). Their existence has already been indicated in the descriptions and observations of the protaspis and its derived characters, and need not be repeated here. Apparently the nauplius presents nothing homologous, unless possibly the frontal sensory organs of *Apus, Balanus, Peneus,* etc., may be taken as such. The paired eyes and frontal sensory organs are close together and seem to have some intimate connection, for, as the paired eyes develop, the latter dwindle and disappeear. Likewise in the trilobites the free-cheeks bear the visual areas, and may be almost wholly converted into eyes as in *Æglina* (*Cyclopyge*).

The greater or less separation of the cerebral ganglia in the chætopods and in some of the lower Crustacea leads to the idea that the free-cheeks in trilobites are the pleura of an occuliferous head segment, which otherwise is lost. If the hypostoma is homologous with the annelid prostomium, as urged by Bernard,[11] then the free-cheeks may be considered as representing the second procephalic segment, which is the number required on the supposition that each neuromere corresponds to a somite. There is a separate neuromere to each mesodermic metamere posterior to the head, and from analogy we should expect that each neuromere in the cephalon would represent an original segment, especially as it can be demonstrated that the head is composed of consolidated or fused segments (Kingsley[24]).

Having thus shown the probable ventral structure of the protaspis, we are prepared to make some general observations on the larval type of modern Crustacea known as the *nauplius*. Before doing this it is well to emphasize again that there is very positive evidence, amounting virtually to certainty, that the protaspis had five pairs of limbs attached to the cephalic portion, behind which was an abdominal portion containing the formative elements out of which all the posterior somites and appendages were developed.

The Crustacean Nauplius.

The name *Nauplius* was first used by O. F. Müller[29] to designate a minute crustacean believed to represent an adult animal. Afterwards it was found to be a larval stage of *Cyclops*, but because it agreed in structure with the larvæ of many other Crustacea the name was retained for that type of larval form and is now in general use. Primarily it is supposed to represent the first free-swimming stage after the escape of the animal from the egg. However, many species are quite fully developed when leaving the egg, and undergo comparatively slight subsequent metamorphoses, and in these and other species there may be developed in the egg an embryo having some of the characters of the nauplius. Therefore the term is also applied to all cases where a certain assemblage of nauplian characters occurs in the development of any crustacean. Thus it may be considered as a stage of development not restricted to a definite period of ontogeny.

The adult *Apus* possesses so many nauplian features, and in its development passes through such simple metamorphoses, that it has been aptly considered by Bernard [11] as a nauplius grown to maturity. Balfour [1] also states that the chief point of interest in the development of *Apus* "is the fact of the primitive Nauplius form becoming gradually converted without any special metamorphoses into the adult condition." * This form, together with the nauplii of other crustaceans and the study of the larval and adult characters of the trilobites, ought to afford definite knowledge of the characters possessed by the ancestral forms of the Crustacea.

Before further examining the nauplius it may be well to state the characters which, on the grounds of comparative anatomy and phylogeny, are believed to represent the primitive adult crustacean. It will be seen that in many respects the trilobite recalls this type, but, as already suggested, is removed some distance from the prototype, although in itself a most primitive crustacean. Lang [25] gives a very comprehensive description of the racial form, as follows: "The original Crustacean was an elongated animal, consisting of numerous

---

* The adult *Apus* properly has five pairs of cephalic limbs. A sixth pair of appendages has been correlated as maxillipedes, though from their innervation they seem to be metastomic and homologous with the chilaria of *Limulus*.

and tolerably homonomous segments. The head segment was fused with the 4 subsequent trunk segments to form a cephalic region, and carried a median frontal eye, a pair of simple anterior antennæ, a second pair of biramose antennæ and 3 pairs of biramose oral limbs, which already served to some extent for taking food. From the posterior cephalic region proceeded an integumental fold which, as dorsal shield, covered a larger or smaller portion of the trunk. The trunk segments were each provided with one pair of biramose limbs. Besides the median eye there were 2 frontal sensory organs. The nervous system consisted of brain, œsophageal commissures and segmental ventral chord, with a double ganglion for each segment and pair of limbs. The heart was a long contractile dorsal vessel with numerous pairs of ostia segmentally arranged. In the racial form the sexes were separate, the male with a pair of testes, the female with a pair of ovaries, both with paired ducts emerging externally at the bases of a pair of trunk limbs. The excretory function was carried on by at least 2 pairs of glands, the anterior pair (antennal glands) emerging at the base of the second pair of antennæ, the posterior (shell glands) at the base of the second pair of maxillæ. The mid-gut possibly had segmentally arranged diverticula (hepatic invaginations)."

The characters ascribed to the typical nauplius have been selected mainly on the principle of general average. They do not satisfy the theoretical demands resulting from a comparative morphological study, nor are they consistent with the accepted requirements of an ancestral type of the Crustacea. Claus [16] urges that the nauplius is a modified or secondary larval form, and the writer now hopes to further substantiate this view, and partly to reconstruct the nauplius from internal evidence and from its more primitive representative, the protaspis of the trilobites.

The usual features attributed to the nauplius are: three pairs of appendages, afterwards forming two pairs of antennæ and the mandibles; the first pair is uniramous and sensory in function; the second and third pairs are biramous, swimming appendages; body usually unsegmented; anteriorly there is a single median eye, and a large labrum, or upper lip; an alimentary canal bent anteriorly, and ending in an anus near the posterior end of the body; a dorsal shield; the second pair of antennæ are innervated from a sub-œsophageal ganglion. Frontal sense organs and a rudimentary metastoma are sometimes present. The trunk and abdominal regions are not generally differentiated.

Balfour[1] remarks of the nauplius that: "In most instances it does not *exactly* conform to the above type, and the divergences are more considerable in the Phyllopods than in most other groups." This variation is indeed quite marked among nearly all the groups besides the phyllopods, and furnishes the facts for the conclusion that the hexapodous condition is not primitive.

On Plate V are represented some of the leading types of nauplius structure, taken chiefly from the excellent compilation by Faxon.[20] Bearing in mind the typical and average characters of this larva, some of the variations will be briefly reviewed.

The nauplius of *Apus*, represented in Plate V, figure 2, shows the rudiments of five trunk segments, which in a later stage (figure 3) develop phyllopodiform appendages belonging to the sixth, seventh, and eighth pairs of limbs. They are the anterior trunk appendages, and appear at a time when the fourth cephalic pair is a mere rudiment while the fifth is entirely undeveloped. The fourth and fifth pairs of head appendages evidently must have some existence, though undeveloped in the nauplius. The physical conditions of nauplius life probably do not require them, and they therefore remain for a time quiescent or undeveloped.

In figures 4, 5, 8, and 6, respectively, of *Branchipus*, *Artemia*, *Leptodora*, and *Limnaida*, the first pair of appendages becomes progressively shortened, until, in the last, they almost disappear. *Leptodora* (figure 8) and *Lepidurus* (figure 7) also have rudimentary trunk segments and appendages (*y*). Figures 9 and 10, of *Daphnia* and *Moina* (from summer eggs), show how rudimentary the nauplius appendages may become when this stage is passed within the egg. Even a more marked reduction is exhibited in the embryos of *Palæmon* and *Astacus* (figures 25 and 26). *Cyclops* is a very normal form, though even here in a second nauplius stage (figure 12) a fourth pair of limbs is developed.

Examples have been cited showing the reduction and obsolescence of the anterior antennæ, or first pair of nauplius limbs, and some cases will now be cited in which the third pair also becomes reduced and rudimentary. *Achtheres* (figure 14) and *Mysis* (figure 22) afford instances of this variation. The former is of additional interest, as showing that the appendages from the fourth to the eighth may be developed, while the third remains quiescent, and that the second pair, typically biramous, is here unbranched. Similarly, in *Mysis*, *Nebalia* (figure 19), and especially in *Cypris* (figure 18), the nauplius limbs are simple. The embryo of *Lucifer* (figure 24) and a late nauplius stage of *Euphausia* (figure 21) are also of moment in showing the beginnings of the metastoma (*mt*) with the two maxillæ and first maxillipedes.

It appears from the foregoing facts that enough has been shown to prove the marked variations in the number and state of development of the nauplius appendages, and to reach the conclusion that potentially five pairs of cephalic appendages are present. The two posterior pairs are the ones usually not developed until after some of the trunk limbs appear. Very satisfactory explanations have been offered as to why the first three pairs have been selected by the larva, although it does not seem to have been recognized that the fourth and fifth have been more or less suppressed during the evolution of the class. Lang[25] accounts for the three pairs of nauplian limbs by saying that: "In a young larva which, like the *Nauplius*, is hatched early from the egg, only a few of the organs most necessary for independent life and independent acquisition of food can be developed. The 3 most anterior pairs of limbs which serve for swimming may be described as such most necessary organs. The third pair perhaps belongs to this category, because as mouth parts, generally provided with masticatory processes, they serve not only with the others for locomotion, but also for conducting food to the oral aperture."

Another point in favor of the original pentamerous composition of the cephalic portion of the nauplius or protonauplius is the dorsal shield which is present in many forms, and is considered (*vide* Bernard[11]) as a dorsal fold of the fifth segment. So that, in reviewing the nauplius structures, we find here and there evidences of the entire series of head segments.

Now, since the protaspis fulfils the requirements by having five well-developed cephalic segments, and is besides the oldest crustacean larva known, it is believed that, in so far, at least, it represents the primitive ancestral larval form for the class.

The nauplius, therefore, is to be considered as a derived larva modified by adaptation.

Other variations in the characters of the nauplius occur, but as they have clearly originated (*a*) from the parasitic habits of the adult, (*b*) from embryonic conditions, or (*c*) from earlier inheritance, they need not enter into consideration here. Such, for example, are (*a*) the absence of an intestine in *Sacculina*, (*b*) the absence of the median eye in *Daphnia* and *Moina*, and (*c*) the bivalve shell in *Cypris*. The larval stages of other, and especially later and higher groups of arthropods, offer more considerable differences and need not enter into this discussion, which is aimed chiefly to establish the genetic relationship between the protaspis of trilobites and the nauplius of recent Crustacea.

Summary.

Barrande first demonstrated the metamorphoses of trilobites in 1849, and recognized four orders of development, which are now shown to be stages of growth of a single larval form.

A common early larval form is recognized and called the *protaspis*.

The protaspis has a dorsal shield, a cephalic portion composed of five fused segments and a pygidial portion consisting of the anal segment with one or more fused segments.

The simplest protaspis stage is found in the Cambrian genera of trilobites. During later geological time it acquired additional characters by earlier inheritance and became modified, though retaining its pentamerous glabella and small abdominal portion.

Some of these acquired characters of the dorsal shield are the free-cheeks, the eyes, the eye-line, the genal angles, and the ornaments of the test. The free-cheeks and eyes moved to the dorsum from the ventrum.

The history of the acquired characters is traced by means of comparisons between larval and adult trilobites, through Paleozoic time, and a progressive series of larval forms established in exact correlation with adult forms, which themselves constitute a progressive series, chronologically and structurally.

The antiquity of trilobites is indicated by their remains in the oldest Paleozoic rocks, and especially by the fact that in the early Cambrian they are already much specialized and differentiated in number of genera. The age of the trilobite or crustacean phylum is further shown from the distinct larval stages of trilobites and their having a cephalon and pygidium of consolidated segments.

Since the trilobites are among the oldest and most generalized of Crustacea, their ontogeny is of considerable importance in interpreting crustacean phylogeny.

The protaspis in its segmentation shows that the cephalon had five pairs of appendages as in the adult.

The crustacean nauplius is shown to be homologous with the protaspis and to have potentially five cephalic segments bearing appendages, which should therefore be taken as characteristic of a protonauplius.

The nauplius is a modified crustacean larva. The protaspis more nearly represents the primitive ancestral larval form for the class, and approximates the protonauplius.

References.

1. Balfour, F. M., 1885. — A Treatise on Comparative Embryology, memorial edition.

2. Barrande, J., 1849. — *Sao hirsuta* Barrande, ein Bruchstück aus dem " Système silurien du centre de la Bohême." *Neues Jahrbuch für Mineralogie*, etc.

3. —— 1852. — Système silurien du centre de la Bohême, 1er partie.

4. Beecher, C. E., 1893. — Larval forms of Trilobites from the Lower Helderberg Group. *Amer. Jour. Sci.* (3), vol. xlvi.

5. —— 1893. — A Larval Form of Triarthrus. *Amer. Jour. Sci.* (3), vol. xlvi.

6. —— 1893. — On the Thoracic Legs of Triarthrus. *Amer. Jour. Sci.* (3), vol. xlvi.

7. —— 1894. — On the Mode of Occurrence, and the Structure and Development of Triarthrus Becki. *American Geologist*, vol. xiii.

8. —— 1894. — The Appendages of the Pygidium of Triarthrus. *Amer. Jour. Sci.* (3), vol. xlvii.

9. —— 1895. — Further Observations on the Ventral Structure of Triarthrus. *American Geologist*, vol. xv.

10. —— 1895. — Structure and Appendages of Trinucleus. *Amer. Jour. Sci.* (3), vol. xlix.

11. Bernard, H. M., 1892. — The Apodidæ. A Morphological Study. *Nature Series.*

12. —— 1894. — The Systematic Position of the Trilobites. *Quar. Jour. Geol. Soc. London*, vol. l.

13. Callaway, C., 1877. — On a new area of Upper Cambrian rocks in South Shropshire, with a description of a new fauna. *Quar. Jour. Geol. Soc. London*, vol. xxxiii.

14. Clarke, J. M., 1894. — The Lower Silurian Trilobites of Minnesota. In advance of vol. iii, Pt. ii, *Geol. and Nat. Hist. Surv. of Minn.*

15. Claus, C., 1876. — Untersuchungen zur Erforschung der genealogischen Grundlage des Crustaceen-Systems.

16. —— 1885. — Neue Beiträge zur Morphologie der Crustaceen. *Arb. zu. Inst. Wien*, vi.

17. Corda, A. J. C. [and I. Hawle], 1847. — Prodrom einer Monographie der bömischen Trilobiten. *Abhandl. böhm. Gesell. Wiss.*, Prag, vol. v.

18. Dohrn, Anton, 1870. — Untersuchungen über Bau und Entwickelung der Arthropoden, *Zeitsch. für Wiss. Zool.*, Bd. XXI.

19. Dohrn, Anton, 1870. — Geschichte des Krebs-Stammes nach embryologischen, anatomischen und palæontologischen Quellen. *Jenaische Zeitsch.*, vol. vi.

20. Faxon, W., 1882. — Crustacea. Selections from Embryological Monographs. *Mem. Mus. Comp. Zoöl.*, vol. ix, No. 1.

21. Fernald, H. T., 1890. — The Relationships of Arthropods. *Studies from the Biological Laboratory, Johns Hopkins Univ.*, vol. iv, No. 7.

22. Ford, S. W., 1877. — On Some Embryonic Forms of Trilobites. *Amer. Jour. Sci.* (3), vol. xiii.

23. Hall, James, 1860. — New Species of Fossils from the Hudson-River Group of Ohio and other Western States. Appendix, *Thirteenth Ann. Rept. N. Y. State Cabinet.*

24. Kingsley, J. S., 1894. — The Classification of the Arthropoda. *American Naturalist*, vol. xxviii.

25. Lang, Arnold, 1891. — Text-Book of Comparative Anatomy. English translation by H. M. and M. Bernard.

26. Matthew, G. F., 1884. — Illustrations of the Fauna of the St. John Group continued : On the Conocoryphea, with further remarks on Paradoxides. *Trans. Roy. Soc. Canada*, vol. ii, section iv.

27. —— 1887. Illustrations of the Fauna of the St. John Group. No. IV. — Part II. The Smaller Trilobites with Eyes (Ptychoparidæ and Ellipsocephalidæ). *Trans. Roy. Soc. Canada*, vol. v, section iv.

28. —— 1889. — Sur le Développement des premiers Trilobites. *Ann. Soc. Roy. Malac. de Belgique.*

29. Müller, O. F., 1785. — Entomostraca, seu Insecta testacea, quae in aquis Daniae et Norvegia reperit, etc.

30. Müller, Fritz, 1864. — Für Darwin.

31. Packard, A. S., Jr., 1883. — A Monograph of North American Phyllopod Crustacea. *Twelfth Ann. Rept. U. S. Geol. and Geol. Surv.*

32. Salter, J. W., 1866. — A Monograph of British Trilobites. Part III. *Pal. Soc.*, London, vol. xviii.

33. Walcott, C. D., 1877. — Notes upon the Eggs of Trilobites. Published in advance of *Thirty-first Rept. N. Y. State Mus. Nat. Hist.*

34. —— 1879. — Fossils of the Utica Slate and Metamorphoses of Triarthrus Becki. Printed in advance of *Trans. Albany Inst.*, vol. x.

35. —— 1886. — Second Contribution to the Studies on the Cambrian Faunas of North America. *Bull. U. S. Geol. Surv.*, No. 30.

36. —— 1890. — The Fauna of the Lower Cambrian or Olenellus Zone. *Tenth Ann. Rept. Director U. S. Geol. Surv.*, 1888–89.

37. Woodward, Henry, 1895. — Some Points in the Life-history of the Crustacea in Early Palæozoic Times. Anniversary Address of the President. *Quar. Jour. Geol. Soc. London*, vol. li.

4. ON THE MODE OF OCCURRENCE AND THE STRUCTURE AND DEVELOPMENT OF TRIARTHRUS BECKI[*]

(PLATE VI)

THE presence of antennæ and other appendages on a trilobite from the Utica Slate was announced in May, 1893, by W. D. Matthew.[†] The specimens were discovered by W. S. Valiant,[‡] near Rome, New York, where they occur in a fine-grained carbonaceous shale. It was apparent that specimens preserving organs so delicate as antennæ ought to show, in addition, other anatomical features which would be of great assistance in determining the zoölogical position of the Trilobita. With this object in view, and with the assistance of Professor Marsh, a collection was made for the Yale University Museum. From this material it is hoped that the remaining details in the structure of this important fossil may be made out. The preliminary examination of the specimens shows a number of new and remarkable structural features, some of which will be briefly noticed here. It was also possible for the writer to make observations in the field, which furnish interesting facts as to the mode of occurrence and to the habits of the trilobite.

---

[*] Abstract of a paper "On the Structure and Development of Trilobites," read before the National Academy of Sciences, November 8, 1893. *American Geologist*, XIII, 38–43, pl. iii, 1894.

[†] On Antennæ and other Appendages of Triarthrus Beckii. Read before the N. Y. Academy of Sciences, May, 1893. Published in *Amer. Jour. Sci.* (3), XLVI, 121–125, August, 1893.

[‡] Mr. Valiant informs me that he found the first specimen showing antennæ in 1884, but it was not until 1892 that other specimens were obtained by him and M. Sid. Mitchell fully establishing the discovery. The specimens sent to Columbia College were collected by W. S. Valiant, of Rutgers College.

In their present condition the specimens contain very little calcite matter, and nearly the entire calcareous and chitinous portions of the animal are represented by a thin film of iron pyrite. To this kind of fossilization is doubtless due the preservation of delicate organs and structures which otherwise would have been destroyed. For, as is well known, pyrite may replace such organic tissues as chitine or even soft dermal structures, the change occurring by the slow decomposition of these tissues in the presence of iron sulphate in solution, or from the action of hydrogen sulphide as a result of decomposition in a chalybeate water.

From the mode of occurrence of the specimens it is evident that some physical change of a rather sudden nature must be inferred to explain the facts. This is shown from the following considerations: (1) Their restricted vertical distribution; (2) nearly all specimens are complete and preserve their appendages; (3) they are of all ages, from larval forms up to full-grown individuals; (4) the rock has a characteristic structure and composition; and (5) the adjacent strata contain a rather sparse fauna in which the trilobites are generally fragmentary, or usually without appendages.

It does not require a violent catastrophe to account for these peculiarities, and, as in the case of the recent destruction of the tile-fish off the eastern coast of the United States, it is possible that a temporary change in the direction of an ocean current, with the consequent variation of temperature, would be amply sufficient. Just what occurred in the present instance has not been determined. Throughout the trilobite-bearing rocks generally, young and larval forms are extremely rare, while, of full-grown examples, fragments are the rule and entire specimens the exception. Therefore it is believed that the remains commonly found represent sheddings or moults, and not in each case the death of a separate individual. In the present material, however, the almost invariable perfection of the specimens precludes this view. Moreover, the appendages are apparently in the position held in life, and not such as obtain in the cast-off shells of recent Crustacea.

Another feature noticed in the field is that the specimens nearly all lie with the back down. The same thing has been observed by other investigators, and has been accounted for by the assumption that in being drifted about along the bottom such a position would be assumed from the centre of gravity being on the convex side. This idea does not seem tenable, because, while on their backs, the trilobites would be most easily rocked by the currents of water, and eventually be turned over or dismembered. A further explanation has been offered by Hicks and accepted by Walcott,* to the effect that trilobites probably lived with the ventral side down, and the accumulation of gases in the viscera during decomposition was sufficient to overturn the animal and allow it to be buried by the deposition of sediments in the position now found. This theory, also, does not meet the facts as here observed, for in turning over a dead and limp animal provided with long and slender antennæ, delicate jointed legs, and fringed appendages, the legs would be either folded under the carapace on one side, or displaced from their natural position. But, as has been already noticed, the present material generally shows the legs extended on both sides of the body and the antennæ in a very lifelike position. (Plate VI, figures 3–7.)

It seems most probable that trilobites could both swim freely and crawl along the bottom, and that, on dying, they coiled themselves up in the same manner as the recent isopods. Then upon unrolling they would necessarily lie on their backs. Even if they did not coil up, any swimming animal having a boat-shaped form would settle downward through the water with the concave side up.

The definite structure of the legs of *Triarthrus* is now for the first time clearly shown, and is of much interest. Furthermore, a difference can be seen in the appendages of the pygidium, thorax, and cephalon. Those of the caudal

---

* The Trilobite: new and old evidence relating to its organization. *Bull. Mus. Comp. Zoöl.*, VIII, No. 10, 1881.

region overlap each other, and are furnished with very long hairs, or setæ. The appendages of the head include the antennæ and the mouth parts, the latter consisting of the mandibles and maxillæ bearing palps and setæ.

The legs of the thorax have been worked out in detail, and are shown on Plate VI, figures 8, 9. No essential differences have been observed in the series attached to the free segments. Each segment bears a pair of biramous appendages, originating at the sides of the axis, as in other trilobites (Walcott, *l. c.*). The anterior legs are the longest, and the others gradually become shorter towards the pygidium. Those which are here taken for description are the legs of the second and third free thoracic segments. The entire length of the legs has been exposed from the dorsal side, by removing the overlying pleuræ of the thorax, which concealed nearly half their length. Each limb consists of two nearly equal members, one of which was evidently used for crawling and the other for swimming. These two members and their joints may be correlated with certain typical forms of crustacean legs among the Schizopoda, Cumacea, and Decapoda, and may be described in the same terms. Therefore each limb is composed of a stem, or shaft, with an outer branch (exopodite) and an inner branch (endopodite). Plate VI, figure 9, shows the joints of the stem (*6, 7*), the exopodite (*ex, 1* and *2*), and the endopodite (*en, 1–5*). The precise form of the coxal joint of the stem (coxopodite) has not yet been clearly made out. It is followed by a broad joint about twice as long as wide, which may be referred to the protopodite.

The endopodite (figure 9, *en*) was the member used for crawling, as in the Schizopoda. The three proximal joints (*5, 4, 3*) are similar in form to *6*, and taper gradually outward. The distal portion is completed by two slender cylindrical joints (*2, 1*), the latter bearing at its extremity short setæ, or bristles, of which three are commonly to be seen.

The other member, the exopodite (*ex*), lies over the endopodite. It apparently articulates with the protopodite, but may spring from what is here referred to the coxopodite, as its basal portion is very broad and originates close to the articulation of the protopodite with the coxal joint. The proximal joint of the exopodite (*2*) is somewhat arched and tapers rapidly. It extends to the ends of the pleuræ, and is the longest joint of either branch. The posterior edge is finely denticulate, and carries a row of long setæ. The distal portion (*1*) is multiarticulate, being composed of ten or more joints. In general form it is slightly crescentic, with the margins thickened, the anterior one being strongly crenulated. Long setæ extend posteriorly from the crenulations on the dorsal side of the leg, making a conspicuous fringe along the distal half of the exopodite.

Plate VI, figure 7, represents a dorsal view of *Triarthrus Becki*, showing the antennæ and the exposed portions of the appendages. The antennæ and legs on the right side are drawn from one specimen, and the legs on the left side are as shown in another individual. The biramous character of the entire series of thoracic legs is very evident, as is also the distinction between the crawling and swimming members. Figure 8 shows the right second and third legs of the free thoracic segments. In figure 9 the upper exopodite is represented without setæ, so as to bring out the structure in greater detail. On the lower leg the setæ are shown.

The antennæ are about as long as the head, and are composed of short conical joints. They usually occur in the position shown in figures 5 and 7, but occasionally lie close to the margin, figures 3 and 4, and sometimes curve backward over the head, as in figure 6.

It is not necessary in this place to describe in detail the development of *Triarthrus Becki*, but attention may be called to two early larval forms. The youngest is shown on Plate VI, figure 1, and may be compared with the first segmented stage, figure 2, and with the adult, figure 7. At this early stage the animal is less than one millimetre in length (.63 mm.), and has no distinct separation into parts. The division into a cephalic and a caudal region is indicated by a transverse groove, but as yet the body segments are undeveloped. After the separation of the head and pygidium the thoracic segments are introduced successively between the head and abdomen until the full number is reached, and the animal measures from 10 to 55 millimetres in length. The segmented stages have been described fully by Walcott,[*] and an outline figure of the stage with one thoracic segment is given in figure 2.

The final conclusions to be reached from a complete study of the development and structure of these animals can as yet be only surmised. It is quite evident, however, that they are related to the true Crustacea. The Trilobita are shown to be a primitive type in (1) their multiple segmentation, (2) the irregular number of thoracic legs, and (3) the biramous structure of the legs. They therefore present characters common to the Entomostraca and Malacostraca.

[*] *Trans. Albany Inst.,* X.

5. FURTHER OBSERVATIONS ON THE VEN-TRAL STRUCTURE OF TRIARTHRUS *

(Plates VII and VIII)

In previous papers on the ventral structure of *Triarthrus,* the anterior antennæ, thoracic legs, and appendages of the pygidium have been described.† There yet remain for investigation the appendages of the head and additional details of other parts of the animal. These characters have not been easily obtained on account of the labor of removing the rock from such delicate structures. Moreover, but few specimens are in the proper position or condition to yield satisfactory results. The appendages of the head either suffered greater decomposition than those of the thorax, before mineralization, or were so tenuous as to be easily obliterated, and are now seldom sufficiently well preserved for study. Further, the number and compact arrangement of such complicated organs, even when present, make it difficult to trace their precise form. A similar difficulty would be experienced were one to endeavor to describe the appendages of *Apus* by examining the ventral side without cutting away some of the limbs or at least unfolding or bending them around.

---

* *American Geologist,* XV, 91–100, pls. iv and v, 1895.

† W. D. Matthew.— On Antennæ and other appendages of Triarthrus Beckii. *N. Y. Academy of Sciences,* May, 1893 ; *Amer. Jour. Sci.,* August, 1893.

C. D. Walcott.— Note on some Appendages of the Trilobites. *Proc. Biol. Soc. Washington,* March, 1894.

C. E. Beecher.— On the Thoracic Legs of Triarthrus. *Amer. Jour. Sci.,* December, 1893.

—— On the mode of Occurrence, and the Structure and Development of Triarthrus Becki. *American Geologist,* January, 1894.

—— The Appendages of the Pygidium of Triarthrus. *Amer. Jour. Sci.,* April, 1894.

The features described in the present paper have been obtained by further work on the material secured for the Yale Museum by Professor Marsh. No detailed review of the ventral anatomy of *Triarthrus* will be given at this time, only such additional characters as have been observed since the publication of the last paper on this trilobite. The precise structure and relations of the organs here described must also be left subject to slight modifications required by researches which are still in progress. The writer has carefully prepared the specimens and made the drawings from camera-lucida outlines. The appendages, however, are often so faintly preserved or so obscure that in order to represent them in a pen-drawing it is necessary to emphasize their limits and their prominence, and this may sometimes lead to errors of interpretation. It seems almost unnecessary to state that errors are not due to any preconceived notions of trilobite anatomy, since from the beginning of these investigations it has not been possible to predict with safety the exact form and details of any of the appendages. Even their presence has been more or less doubtful until revealed by a fortunate discovery.

The paired appendages of the cephalon will be taken up in their order, beginning with the most anterior; next the newly observed characters of thoracic legs; then the organs in the median line, — the hypostoma, mouth, metastoma, and anal opening.

Close observation of the specimens thus far prepared for the purpose of showing the under side of the head fails to detect more than five pairs of appendages attached to the cephalon, apparently corresponding to the five typical limbs of the crustacean head. Considerable difficulty is experienced in determining from the ventral side of the specimens the posterior limit of the cephalon. The ventral membrane, which alone is usually visible, does not show marked evidence of segmentation, and the observer is guided chiefly by the margin of the cephalon, the extremities of the pleura, and obscure transverse lines on the axial membrane. In a few cases the evident sliding or displacement of the dorsal and ventral surfaces further complicates the attempt to refer the appendages to definite divisions of the animal.

Paired Uniramose Appendages.

Anterior Antennæ, or Antennules. — These have been described by Matthew, Walcott, and the writer (*l. c.*). Walcott showed their proximal extremities and their mode of attachment at the side of the hypostoma. Little more can now be added except that they are evidently the first pair of antennal organs, and correspond to the antennules of other Crustacea. The strong basal joint or shaft is shown in Plate VIII, figures 9, 10, 11, attached to the ventral side of the head at each side of the hypostoma, near the middle of its length. The shaft carries a single flagellum, and thus agrees with the typical uniramose antennule of the nauplius of Crustacea. This simple antennule is still present in the Isopoda, as in *Mannuopsis typica*. The flagella curve forward and approach, nearly touching as they cross the doublure. Beyond the limits of the head they are variously disposed, though usually extending forward, at first diverging for half their length and then slightly converging (Plate VIII, figures 5, 6, 7).

Paired Biramous Appendages.

The remaining paired appendages of the trilobite all seem to be biramous, and agree closely in their general features. Adjacent members of the series present very slight differences. It is only when the primitive and simple phyllopodous legs of the pygidium are compared with the anterior thoracic or cephalic appendages that variations of note can be observed, although these are of form and not of structure. On this account there is no well-defined separation into posterior antennæ, mandibles, maxillæ, maxillipeds, thoracic, and pleopodal appendages. It is most convenient, therefore, to number them from before backward, and to indicate homologous positions with other Crustacea only when there is some evident reason for so doing.

First Pair of Biramous Appendages, or Posterior Antennæ. — The second pair of appendages, corresponding to the posterior antennæ, are attached to the head at each side of the glabella, on a line with the extremity of the hypostoma. They are apparently biramous, and thus agree with the second pair of nauplian limbs and with the typical posterior antennæ of many Entomostraca and Malacostraca. They may be compared with the posterior antennæ in *Euphausia pellucida*, one of the schizopods, especially with the *Furcilia* and *Cyrtopia* stages. The details of the endopodite and exopodite are not clearly shown. The former is more commonly preserved, and its distal joint extends just beyond the edge of the carapace. The coxopodite is developed into a triangular plate, the inner angle carrying a masticatory ridge, the whole extending about three-fourths the distance from the side of the glabella to the median line, just below the hypostoma, and directed obliquely backward (Plate VIII, figures 8–11).

Second Pair of Biramous Appendages, or Mandibles. — The appendages here correlated with the mandibles are immediately behind the first pair of biramous limbs. The proximal portion, or coxopodite, is similar in form to the preceding, though somewhat smaller, and overlapping its basal part. The palps, or endopodial and exopodial branches, have not been distinctly traced, though their presence is indicated on Plate VII, figure 1, where, on the left side, there are endopodites and exopodites in sufficient number for each appendage of the head. That these should be referred to the cephalic limbs is further indicated by their being in advance of the endopodite, which manifestly pertains to the first thoracic segment. The inner edge of the mandibles as well as that of the other gnathobases of the head is apparently finely denticulate, as shown on Plate VII, figure 1, and Plate VIII, figure 2.

Third and Fourth Biramous Appendages, or Maxillæ. — Following the appendages referred to the mandibles are two pairs of strong limbs, with broad plate-like basal portions, or

coxopodites, serving as gnathites (Plate VIII, figures 8–11).
They resemble each other, and are similar in form to the two
preceding limbs, though somewhat larger.   They are usually
fairly well preserved, and their form and structure can be
approximately made out.   The endopodites are composed of
stout joints, and could be extended but a short distance
beyond the margin of the head.   The exopodites are more
slender, and carry an abundance of stiff setæ, which often
diverge in a fan-like manner from their line of attachment.
These brushes of setæ occupying the cavities of the cheeks
are often preserved in specimens where the other details of
the limbs are obscure or obliterated.   In *Triarthrus* they are
evidently homologous with similar brushes observed by Walcott in *Calymmene.**

This completes the number of paired appendages which can
be definitely referred to the head.   It is evident they do not
differ conspicuously from each other, and, as will be presently
shown, they closely resemble the thoracic legs in all essential
structural characters.

Thoracic Legs. — In the paper by the writer (*l. c.*) describing
the structure of the thoracic legs, the endopodites and exopodites of the second and third pairs were illustrated, together
with their points of attachment.   The form of the coxopodite,
or basal portion, was at that time unknown.   With the
present material it is possible to add several details.   The
most important are the inward prolongation of the coxopodite
of each limb toward the axial line, forming a gnathobase, and
the progressive development of this member.   First it has
a slender cylindrical form in the posterior half of the series,
then becomes flattened and denticulate, and finally widens,
until on the head it forms the triangular plate-like coxopodite,
with masticatory ridge and functioning as a gnathite (Plate
VII, figure 1 ; Plate VIII, figures 1–4, 8–11).

The large basal portions of the limbs of *Asaphus*, in the
specimen illustrated by Walcott (*Science*, March, 1884), are

* The Trilobite : New and old Evidence relating to its Organization.   *Bull.
Mus. Comp. Zoöl.*, VIII, No. 10, 1881.

evidently the gnathobases, as will be seen at once from a comparison with *Triarthrus* (Plate VIII, figure 1).

Organs in the Median Line.

The Hypostoma. — There is nothing peculiar in the hypostoma of *Triarthrus*, since it presents features commonly found in many other genera. It is longer than wide, and extends more than half the length of the head. The posterior end is narrowly rounded, margined by a slight doublure, and often presents a transverse elevation near the apex, as shown on Plate VIII, figure 9. This may represent a corresponding hollow on the inner side to allow for movements of the manducatory organs.

In considering the exact location of the appendages of the head, it must be understood that in their present positions they are probably somewhat displaced. During the process of fossilization the whole inner tissues of the animal were removed without replacement, allowing the ventral membrane to come more or less in contact with the under side of the dorsal crust, and thus causing some stretching of the membrane and consequent displacement of the organs. The hypostoma, being more rigid and attached in front to the margin of the head, doubtless was not shifted, but dropped down into the cavity of the glabella. When raised to its natural position, it probably extended a little over the mouth parts. The fact that the mouth and lower lip do not come opposite the organs correlated as mandibles may be due in part to the unequal stretching of the membrane over the uneven inner surface of the dorsal crust. The extended gnathobases directed obliquely backward and lying in the axial hollow cause the appendages to appear as though originating further back than is really the case. Nevertheless the posterior position of the second and third pairs of appendages, or the antennæ and mandibles, with respect to the mouth, does not offer any serious difficulty. As shown by Lankester,[*]

* Observations and Reflections on the Appendages and on the Nervous System of Apus cancriformis. *Quar. Jour. Mic. Sci.*, XXI, 1881.

they were doubtless originally post-oral in the Crustacea, as is indicated from their innervation from the ventral nerve ganglion chain and not from the archicerebrum of the prostomium, or cephalic lobe. Besides, in the embryo of *Limulus* all the appendages are post-oral, as shown by Packard.[*]

The Mouth. — Although the opening of the mouth itself has not been observed in *Triarthrus*, there can be little doubt as to its exact location, since it must have been situated in the median line between the hypostoma and metastoma. As both these organs are quite close together, the place of the mouth would be as indicated on Plate VIII, figure 11, *m*. Further corroborative evidence may be had from the genus *Calymmene*, in which the mouth was determined by Walcott (*l. c.*) to lie at the end of the hypostoma, opening obliquely backward.

The Metastoma. — The lower lip, or metastoma, here represented for the first time, is generally clearly shown as a convex arcuate plate just posterior to the extremity of the hypostoma. On each side, at the angles, are two small elevations, or lappets, which suggest similar structures in many higher Crustacea, and apparently represent the entire metastoma in *Apus* and some other forms (Plate VIII, figures 9 and 11, *met*).

The Anal Opening. — In tracing the intestinal canal as preserved in *Trinucleus*, Barrande determined its posterior termination to be at the extremity of the pygidium, and Bernard[†] has recently succeeded in reaching a similar conclusion, from his studies of *Calymmene*, in which the anal opening was found just at the inner margin of the doublure of the pygidium, in the median line. Plate VII, figure 1, of *Triarthrus* shows the anus in the same position, outlined by a slightly elevated wrinkled ring.

---

[*] The Development of Limulus polyphemus. *Mem. Boston Soc. Nat. Hist.,* II, 1872.

[†] The Systematic Position of the Trilobites. *Quar. Jour. Geol. Soc. London,* L, 1894.

Observations.

With these additional discoveries relating to *Triarthrus*, several observations upon its general organization and comparisons with other Crustacea may be made. This cannot be done exhaustively or comprehensively at this time, and only a few points will be touched upon. The simplicity and primitiveness of the trilobite structure will first impress the student. The variable number of segments in the thorax and pygidium in the different genera shows the unstable metameric condition of the class. The head alone seems to have a permanent number of segments and appendages. Even this is not often apparent, but the constant number of five head segments in larval trilobites shows this to be the true number, although subsequent growth may obscure or obliterate this pentasomitic character, as has been shown by the writer in *Acidaspis* (*Amer. Jour. Sci.*, August, 1893) and observed in other genera.

With the exception of the antennules, all other paired appendages of the animal seem to agree in every point of structure, and vary only in the relative development of certain parts. The appendages of the pygidium are ontogenetically the youngest, and express the typical phyllopodiform structure. Passing anteriorly, the joints become less leaf-like, until in the anterior thoracic legs they are quite slender, and the limbs resemble those of schizopods. Corresponding to this, there is, through the whole series, a gradual development of a process from the coxopodite, forming a gnathobase to the limb. On the head these serve as true manducatory organs. Posteriorly they were like the basal endites of *Apus*, and enabled the trilobite to convey food along the entire length of the axis to the mouth.

Bernard (*l. c.*) has made a strong exposition of the evidence in favor of the phyllopod affinities of the Trilobita, and especially of their relations to *Apus*. A portion of the under side of the head of *Apus* is introduced for comparison on Plate VIII, figure 12. Both pairs of antennal organs

(*1, 2*) are rather rudimentary in this genus, and are situated further forward than in *Triarthrus*. The powerful mandibles (*3*) are partly covered by the labrum, or hypostoma (*hy*). Then follow two well-developed gnathobases, representing the maxillæ (*4, 5*), the more slender maxilliped (*6*), and the large first thoracic limbs (*7*), behind which are the basal endites, or gnathobases, of two of the phyllopodous append-ages (*8, 9*). The general similarity of the cephalic organs of *Apus* and *Triarthrus* is quite apparent. The most con-spicuous differences, as the absence of normal endopodial and exopodial elements, disappear in a study of the ontogeny of the limbs of *Apus*, thus bringing these organs in the two groups into nearly exact correlation.

There are, however, important structural features of other parts of the body, which are quite dissimilar from *Apus* and the higher Crustacea, and the exact relations of the trilobite with any one group cannot be considered as fixed. Points of likeness may be established with almost every order, showing chiefly the relationship between the trilobite and the ancestors of modern Crustacea.

Summary of Ventral Organs of Triarthrus.

A pair of appendages to each potential segment of the animal, all of which are biramous except the anterior pair.

Five pairs of appendages on the cephalon.

Anterior antennæ, or antennules, attached at the sides of the hypostoma; simple, with a single many-jointed flagellum.

First pair of biramous limbs, or posterior antennæ, with endopodite and exopodite; basal portion manducatory in function.

Second pair of biramous limbs, or mandibles, similar to preceding.

Third and fourth pairs of biramous limbs, or maxillæ, same as preceding, with large gnathobases, well-developed endopo-dites, and fringed exopodites.

Thoracic limbs biramous; endopodite a jointed crawling leg; posteriorly the joints become flattened and leaf-like;

exopodite fringed with setæ, and developed into a swimming organ; coxopodite with gnathobase.

Appendages of the pygidium, true phyllopodous limbs.

Hypostoma.

Mouth between hypostoma and metastoma.

Metastoma, a convex crescentic plate, with side lappets.

Anus in median line, near ventral extremity of pygidium.

6. THE MORPHOLOGY OF TRIARTHRUS *

(PLATE IX)

MOST of the recent advances in the knowledge of trilobite structure have come from the study of *Triarthrus*. Since Valiant's discovery of the antennæ, and its announcement by Matthew in 1893, the writer has published a series of papers on the detailed structure of this trilobite. Much time has also been spent in carefully working out the numerous specimens from the abundant material in the Yale Museum. Altogether upward of five hundred individuals with appendages more or less complete have been investigated, and at the present time it may be safely said that the important exoskeletal features have been seen and described. †

Notwithstanding the amount of information regarding the details of the various organs, very little has been shown illustrating the general appearance of the animal with the appendages in a natural and lifelike position, and it is one object of the present article to supply this deficiency.

Several specimens have been lately developed which preserve not only the appendages in great perfection, but also show them extended and disposed in a very lifelike manner. No new structural points are here brought out, yet the representation of the complete animal serves as a summary of present knowledge, and also gives a definite picture of

---

* *Amer. Jour. Sci.* (4), L, 251–256, pl. viii, 1896. Reprinted in *Geological Magazine* (London), dec. iv, III, 193–197, pl. ix, 1896.

† The more important literature relating to the structure of the genus *Triarthrus* is given at the end of the present article; numbers in the text refer to this.

great assistance in forming a conception of general trilobite morphology.

The dorsal view represented on Plate IX is from a camera drawing based upon three specimens of about the same size. One gives the entire series of legs down to the ninth free segment, with the exception of the exopodites of the head, which are supplied from a second individual. In the third specimen the anterior appendages are bent and irregularly arranged, while from the ninth backward to the end of the pygidium they are complete and uniformly extended. The figure is, therefore, a restoration only in so far as representing the best portions of three individuals.

The ventral view (Plate IX) is based mainly upon two very excellent specimens. One was figured on Plate IV, vol. xv, of the *American Geologist*, and another, since found, nearly completes the ventral aspect. The under side of the head and pygidium was carefully compared with all the available material, and no attempt was made to supply any characters except as to the exact number of joints in the endopodial cephalic elements and the precise form of the cephalic exopodites, which, from every character observed, and from analogy with similar structures elsewhere, were as represented.

So many specimens preserve the appendages in the position shown in the figures, that this must be recognized as natural, and one likely to have been assumed by the living animal when extended. Few, however, show the details of the limbs with sufficient clearness to enable one to make out all their joints and more minute characters.

In comparison with what is now known of the appendages of several other genera of trilobites, especially *Trinucleus*,[*] those of *Triarthrus* seem to have been exceptionally long. On this point Bernard, in a letter to the writer, suggests that " *Triarthrus* must have been a sort of ' Daddy longlegs ' among the Trilobites, as *Scutigera* is among the Myriapoda."

[*] Structure and Appendages of Trinucleus. *Amer. Jour. Sci.* (3), XLIX, April, 1895.

The entire length of a thoracic leg, including the coxal joint, is nearly equal to the width of the body at that point, and about half the length projects beyond the pleura.

The limbs of the head diminish in length forward until the anterior pair scarcely extends beyond the border of the cephalon. The anterior thoracic legs are the longest, and there is a gradual shortening backward in the series, especially noticeable after passing the fifth, those at the extremity of the pygidium being about one-ninth the length of the first thoracic leg. Their position is also of interest. At the posterior extremity they point almost directly backward, while those on the head are directed more or less forward. Between these two extremes all the intermediate positions occur in regular order.

The gnathobases, or coxopodites, become more and more specialized anteriorly, growing broader and having their inner edge denticulate, until on the head they function as true manducatory organs. The second pair, however, corresponding to the mandibles of higher Crustacea, has not become clearly differentiated from the rest of the series, and apparently has not lost the exo- and endopodial branches.

Few changes of importance can be traced in the exopodites, though the latter are considerably reduced in size on the cephalon. Over the anterior half of the thorax they functioned as vigorous paddles, and on the pygidium their length and compact arrangement made them overlap each other, thus producing two broad flaps, or fin-like organs. The conclusion cannot be avoided that *Triarthrus* must have been an active creature, and with its rows of endopodites and exopodites it was as fully equipped as the bireme in classic navigation. The form of the animal and the multiplicity of locomotor organs were well adapted for rapid motion either along the sea-bottom or through the water.

The youngest and most immature limbs are on the pygidium, and in a young trilobite they are very much like those in the larval *Apus*[4] and are typically phyllopodiform. According to the law of morphogenesis, these limbs may be taken as of phylogenetic value and indicative of the primitive type of limb structure.

The whole series of endopodites anterior to the last two or three show modifications from the phyllopodous type, the change involving progressively from one to all of the endites. The endopodites of the pygidium have a true phyllopodiform structure, and are composed of broad leaf-like joints, wider than long. This character is gradually lost in passing anteriorly, the distal endites being the ones first affected. By the time the anterior pygidial limb is reached, the three distal joints are longitudinally cylindrical. The ninth thoracic endopodite shows a fourth endite becoming cylindrical, and on the first and second thoracic legs even the proximal ones are thus modified, making all the endites of these limbs slender in form.

This gradual modification of a phyllopodiform swimming member into a long, jointed, cylindrical, crawling leg deserves more than passing notice, for here, probably, better than in any known recent form, can the process and its significance be studied. No living type of crustacean more nearly conforms to the theoretical archetype of the class than do the trilobites, and as *Triarthrus* belongs to an ancient Cambrian family, it may be expected to retain very primitive characters.

In this genus several causes evidently influenced the modification of the appendages. First may be mentioned the specialization into oral organs of the gnathobases of the head, which would tend toward a reduction of the other portions of the limbs. Next, the assumption of a walking habit would gradually lead to a corresponding adaptation of the anterior thoracic endopodites, this region of the body being naturally the place where they would be most operative. Lastly, any tendency to change the form of the anterior limbs would be accelerated through the greater number of moults they undergo as compared with the abdominal appendages.

Since the anal segment of Crustacea contains the formative elements out of which all the trunk segments are successively developed, it may be considered as the same segment in all

Crustacea, no matter how many nor what kinds of segments may intervene between it and the head. The youngest segment, therefore, is always in the budding zone, just in front of the telson, or terminal somite, and those further anterior and more differentiated are older. This sequential order in the age of the segments and appendages may be greatly obscured in higher forms, so that, as in the Thoracostraca, the last pair of pleopods, forming with the telson the caudal fin, appears at an early stage of the ontogeny. In such cases, as Lang says, "the grade of development and physiological importance of a section of the body or of a pair of limbs in the adult animal may be recognized by the earlier or later appearance of their rudiments."*

In *Triarthrus* these disturbing factors are hardly to be recognized, for no pair of limbs had an excessive physiological importance over any other pair or series of pairs, and increase progressed regularly by the addition of new members in front of the anal segment. The pygidium being formed of fused segments accommodated itself to this kind of growth by pushing forward the series of limbs and by the formation of a new free segment at the posterior end of the thorax. This process of metameric growth continued from the protaspis stage with no free thoracic segments, and successively added segment after segment with corresponding moults, until the full complement was reached, after which the moulting resulted mainly in increase in size. The repetition of moults afforded the chief means by which modifications in the appendages could be brought about.

The earliest protaspis stage shows, from the segmentation of the axis, that there were present five pairs of appendages on the head and two on the pygidium.[6] The adult animal has thirteen or fourteen free thoracic segments and six pygidial. † Now, so far as is known of trilobite ontogeny, there was never more than one segment added at a single

---

* Text-Book of Comparative Anatomy, English edition (Bernard), p. 410.

† A few individuals of this species (*T. Becki*) have been observed with one or two additional thoracic segments. Walcott.[11]

moult, though there is no evidence that there may not have been more moults than segments between the protaspis stage and the finished segmentation. In *Triarthrus* the average full number of segments was attained by the time the animal reached a length of about 7 mm. So that the limbs of the anterior thoracic segment in an individual 7 mm. in length, and containing the full complement of fourteen free and six pygidial segments, must have undergone at least seventeen moults. The second thoracic segment, therefore, at this stage of growth would have been moulted sixteen times, the fifth thirteen times, the tenth eight times, and the fourteenth four times. The length of full-grown individuals is from 25 to 40 mm., and to have reached this size a considerable number of additional moults must have occurred in which all the segments participated alike.

Some mention should be made of the probable method of respiration of *Triarthrus*. No traces of any special organs for this purpose have been found in this genus, and their former existence is very doubtful, especially in view of the perfection of details preserved in various parts of the animal.

The delicacy of the appendages and ventral membrane of trilobites and their rarity of preservation are sufficient demonstration that these portions of the outer integument were of extreme thinness, and therefore perfectly capable of performing the function of respiration. Similar conditions occur in most of the Ostracoda and Copepoda, and also in many of the Cladocera and Cirrepedia, where no special respiratory organs are developed.

The fringes on the exopodites in *Triarthrus* and *Trinucleus* are made up of narrow, oblique, lamellar elements becoming filiform at the ends. Thus they presented a large surface to the external medium, and partook of the nature of gills. But, as Gegenbaur says, "the functions of respiration and of locomotion are often so closely united that it is difficult to say whether certain forms of these appendages should be regarded as gills, or feet, or both combined." * For purposes

---

* Elements of Comparative Anatomy, English edition (Bell and Lankester), p. 241.

of locomotion, the limbs of the cephalon and pygidium were of feeble assistance compared with those on the thorax, and in the higher Crustacea these two regions are the ones where the greatest branchial specialization takes place.

References.

1. Beecher, C. E., 1893. — A Larval Form of Triarthrus. *Amer. Jour. Sci.* (3), vol. xlvi, pp. 361, 362, November.

2. —— 1893. — On the Thoracic Legs of Triarthrus. *Amer. Jour. Sci.* (3), vol. xlvi, pp. 467–470, December. Abstract of a paper " On the Structure and Development of Trilobites," read before the National Academy of Sciences, November 8.

3. —— 1894. — On the mode of occurrence, and the structure and development of Triarthrus Becki. *American Geologist*, vol. xiii, pp. 38–43, pl. iii, January. Abstract of a paper " On the Structure and Development of Trilobites," read before the National Academy of Sciences, November 8, 1893.

4. —— 1894. — The Appendages of the Pygidium of Triarthrus. *Amer. Jour. Sci.* (3), vol. xlvii, pp. 298–300, pl. vii, April. Read before the Connecticut Academy of Arts and Sciences, March 21.

5. —— 1895. — Further observations on the ventral structure of Triarthrus. *American Geologist*, vol. xv, pp. 91–100, pls. iv, v, February.

6. —— 1895. — The Larval Stages of Trilobites. *American Geologist*, vol. xvi, pp. 166–197, pls. viii–x, September.

7. Bernard, H. M., 1894. — The Systematic Position of the Trilobites. *Quar. Jour. Geol. Soc. London*, vol. 1, pp. 411–432, August. Read March 7.

8. —— 1895. — Supplementary notes on the Systematic Position of the Trilobites. *Quar. Jour. Geol. Soc. London*, vol. li, pp. 352–359, August. Read April 24.

9. —— 1895. — The Zoölogical Position of the Trilobites. *Science Progress*, vol. iv, pp. 33–49, September.

10. Matthew, W. D., 1893. — On Antennæ and other Appendages of Triarthrus Beckii. *Amer. Jour. Sci.* (3), vol. xlvi, pp. 121–125, pl. i, August. Read before the New York Academy of Sciences, May, and published in *Trans. N. Y. Acad. Sci.*, vol. xii, pp. 237–241, pl. viii, July 22.

11. Walcott, C. D., 1879. — Fossils of the Utica Slate. *Trans. Albany Inst.*, vol. x, pp. 18–38, pls. i, ii, 1883. Author's extras printed in advance, June, 1879.

12. —— 1894. — Note on some Appendages of the Trilobites. *Proc. Biol. Soc. Washington*, vol. ix, pp. 89–97, pl. i, March 30. Read March 24. *Geological Magazine*, N. S. dec. iv, vol. i, pp. 246–251, pl. viii, June.

7. STRUCTURE AND APPENDAGES OF TRINUCLEUS *

(PLATE X)

TRINUCLEUS departs so widely from the common type of trilobite form that any contribution of new facts regarding its structure and appendages is a matter of interest. Moreover, this added information will be of assistance in interpreting some peculiar and striking features in the natural group of genera of which *Trinucleus* is evidently a member.

For the present it is convenient to consider in this group such forms as *Trinucleus, Harpes, Harpides, Dionide,* and *Ampyx.* Most of these have the genal angles extending to or beyond the pygidium, with a broad, finely perforated or punctate margin around the head. They are further characterized by the absence or obsolescence of visual organs, while the facial sutures are either peripheral, as in *Harpes,* or in addition include the genal spines, as in *Trinucleus, Dionide,* and *Ampyx.* Several other genera have been recognized as having affinities with those mentioned, but they are imperfectly known, and will be merely noticed here. *Harpina* Novák, based upon the features of the hypostoma, is probably of only sub-generic value under *Harpes. Arraphus* Angelin is apparently based upon a specimen of *Harpes* denuded of the punctate border. *Salteria* of W. Thompson and *Endymionia* of Billings, both generally considered as closely related to *Dionide,* were founded upon too imperfect material to afford decisive data as to their affinities. Angelin's sub-genera of *Ampyx* (*Lonchodomus, Raphiophorus,* and

* Amer. Jour. Sci. (3), XLIX, 307–311, pl. iii, 1895.

Ampyx) are based upon the length of the glabellar spine, and the possession of five or six free thoracic segments. Similar characters in *Trinucleus* are not considered as worthy of such marked distinction.

In 1847 Salter * illustrated and described an eye-tubercle on each cheek of *Trinucleus*, from which there was a raised line extending obliquely upward to a punctum or spot on each side of the glabella. He considered this line as a discontinuous facial suture, but the true suture was afterward correctly determined by Barrande, † and in well-preserved specimens may be easily observed extending around the entire frontal and lateral border of the head, and including the genal spines. The "eye-line" was further recognized by McCoy, ‡ and made one of the bases for a division of the genus into two sections or genera, — *Trinucleus* proper and *Tetraspis*. These divisions were accepted by Salter, but later were thoroughly discussed, and rejected by Barrande (*l. c.*, p. 617), upon valid grounds. Nicholson and Etheridge, § in 1879, reviewed these facts at some length, and gave original figures illustrating the ocular tubercle and eye-line. They also agree with Barrande in recognizing them as clearly adolescent characters.

The justice of these conclusions is substantiated, and additional results are reached, from the study of a series of *Trinucleus concentricus* Eaton, found associated with *Triarthrus Becki* Green, in the Utica slate, near Rome, New York. The remarkable preservation of the fossils at this locality has already afforded a means of determining all the principal details of the ventral structure of the trilobite genus *Triarthrus*, and there is now distinct evidence as to the nature of the appendages in another type, — *Trinucleus*, as well as to the probable significance of the so-called "eye-tubercle."

---

* On the structure of Trinucleus, with Remarks on the Species. *Quar. Jour. Geol. Soc.*, III, 251–254.

† *Syst. Sil. Bohême*, I, 1852.

‡ *Ann. Mag. Nat. Hist.*, 2d Series, IV, 1849.

§ Monograph of the Silurian Fossils of the Girvan District in Ayrshire, fasc. ii, 1879.

As compared with *Triarthrus*, specimens of *Trinucleus* are not very common at this locality, and although more than fifty individuals of the latter have been obtained from the collections presented to the Yale Museum by Professor Marsh, not more than half a dozen of these are adult specimens, and but three show any appendages. Young specimens of all ages occur, from about 1 mm. across the cephalon upward, and in all the eye-line and eye-tubercle are present until a width of nearly 5 mm. is attained, when in the present species these features dwindle and disappear, leaving no discoverable traces in the adult.

Two cephala of young individuals, without the free-cheeks, are shown enlarged in figures 1 and 2 of Plate X. Figure 2 represents a specimen before the appearance of the perforate border, and figure 1 gives a later stage, having two rows of perforations around the head. On both specimens the eye-line is clearly shown, extending somewhat obliquely backward from the anterior lobe of the glabella to the central area of the fixed-cheeks, enlarging slightly, and terminating in a rounded node or tubercle (*a, a,* figure 2).

In seeking for homologous features in other trilobites, the genera *Harpes* and *Harpides* are immediately suggested, since they have similar ocular ridges extending from the sides of the glabella, and ending in a tubercle, which, in *Harpes*, contains from one to three eye-spots, as determined by Barrande. They further agree in having these visual organs on the fixed-cheeks, while in all other trilobites with distinct eyes the free-cheeks carry the visual areas. This type of eye is thus quite different in its relations to the parts of the cephalon, from that of *Phacops* or *Asaphus*, and more nearly resembles the eyes of some of the Merostomata (*Bellinurus*), as do also the triangular areas in the young *Trinucleus*, so distinctly marked off from the fixed-cheeks on each side of the glabella behind the eye-line. Adult *Trinucleus* and *Harpes* have these areas much reduced and often obsolescent. A spot or node in the median line on the glabella has been noticed by many observers, and although its nature has not been demonstrated, it has generally been called an ocellus. It is more clearly preserved in adult specimens, though it can be detected in young examples, as indicated in figures 1, 2, Plate X.

An eye-line occurs in many early trilobite genera, and is well marked in *Conocoryphe*, *Olenus*, *Ptychoparia*, and *Arethusina*. At least four-fifths of the Cambrian forms preserve this feature, which is almost entirely eliminated before Devonian time. It differs in extent, but not necessarily in nature, from the eye-line of *Trinucleus* and *Harpes* in running entirely across the fixed-cheeks to the free-cheeks, ending in the palpebral lobe in eyed forms. It is evidently a larval character in the trilobites, as shown from its geological history and the ontogeny of *Trinucleus*. From the direction of the optic nerve in *Limulus*, and its relations to the surface features of the cephalothorax, the eye-line probably represents the course of that nerve, and is of much less morphological importance than the different types and arrangement of visual organs.

The pygidium of young *T. concentricus* (Plate X, figure 3) is remarkable for the lack of definition between the axis and pleura. In later and adult stages the number of ridges on the pleura and axis do not correspond, and from figures 4, 5, and 6 it is evident that in this genus the number of pleura is no indication of the number of pygidial segments or pairs of appendages, which, however, may be shown, as in this case, by the annulations of the axis. In this respect the pygidia in *Encrinurus*, *Cybele*, and *Dindymene* are of the same nature. Figure 6 also shows a narrow, striated doublure, a character generally overlooked in descriptions of *Trinucleus*.

Appendages.

Three specimens have been thus far observed which show the nature of the appendages in *Trinucleus*. Two of these are illustrated in figures 4, 5, and 6 of Plate X. Figure 4 represents the thorax and pygidium viewed from the dorsal side. In this specimen the pyrite which replaced the chitinous remains of the animal has decomposed, and the dorsal crust weathered away, exposing below the stems of the exopodites, with their fringes extending over the entire pleural areas on both sides. A pygidium, with three attached thoracic segments, from another entire specimen (figures 5 and 6), preserves the details of the appendages in the most perfect and satisfactory manner. As both halves showed essentially the same extent and disposition of the fringes on the dorsal side, the specimen was cut in two along the centre of the axis, and the left side was then embedded in paraffine. By careful preparation the appendages were exposed from the ventral side.

The cephala of the three specimens described are considerably compressed, and from them a very imperfect knowledge of the mouth parts could be obtained, so that this information must be left to future discovery.

Endopodites. — The three posterior thoracic endopodites are very similar, and in a general way closely resemble those of *Triarthrus* from the same region of the thorax. They are, however, comparatively shorter and stouter, and could not be extended beyond the ends of the pleura. The two distal joints are cylindrical, with well-marked articular surfaces and ridges. The joints preceding these proximally become much wider, flattened, and produced into transverse extensions which carry large tufts of setæ at the end, as also does the end of the last joint of the limb (dactylopodite).

The endopodites on the pygidium offer no conspicuous differences from those just described, except that a gradual change in form is manifest as the terminal limbs are reached. The separate endites become more and more transversely cylindrical, until the whole limb appears to be made up of cylindrical segments transverse to its length. A similar condition was observed in the young of *Triarthrus*.*

Exopodites. — These seem to be composed of slender joints, the distal exites being long and slightly curved outward.

---

* *Amer. Jour. Sci.* (3), XLVII, pl. vii, fig. 3, April, 1894.

They carry very long, close-set, overlapping, lamellose fringes, which evidently had a *branchial* function. Some of the lamellæ are spiniferous. The exopodites become shorter on the pygidium, and apparently are represented near the end of the series of limbs by the oval plates indicated at *c*, figure 6. If this interpretation is correct, the posterior exopodites are simple flabella attached to the limbs, as in *Apus*.

Both Professors A. E. Verrill and S. I. Smith agree that the characters of the appendages in *Trinucleus* indicate an animal of burrowing habit, which probably lived in the soft mud of the sea-bottom, much after the fashion of the modern *Limulus*. In addition to its limuloid form, the absence of eyes seems to favor this assumption. So does the fact that many specimens have been found preserving the cast of the alimentary canal, showing that the animal gorged itself with mud like many other sea-bottom animals.

III

STUDIES IN THE DEVELOPMENT OF THE BRACHIOPODA

III

STUDIES IN THE DEVELOPMENT OF THE BRACHIOPODA *

1. DEVELOPMENT OF THE BRACHIOPODA

Part I. Introduction †

(Plate XI)

THE Brachiopoda have been so carefully studied that any new general conclusions regarding them must be naturally based upon features not heretofore considered. In other classes of animals such important results have been recently reached by the application of the law of morphogenesis as defined by Hyatt, that the writer was led to study the Brachiopoda from this standpoint. The facts observed by this method are mainly new to the class, and considerably affect the taxonomic positions and affinities of the various families and genera.

The value of the stages of growth and decline in work relating to phylogeny and classification is now generally admitted. The memoirs of Hyatt, Jackson, and others amply show that the clearest and simplest understanding of a group may be thus reached. The application of the principles of growth, acceleration of development, and mechan-

---

* No revision of these papers has been undertaken further than to bring the nomenclature of genera and species up to date; also the current terms of auxology, as agreed upon by Hyatt, Buckman, and Bather, have been substituted for those first proposed.

† *Amer. Jour. Sci.* (3), XLI, 343–357, pl. xvii, 1891.

ical genesis form the main factors in the studies here made. The geologic sequence of genera and species in this connection is also of the greatest importance, for in this way the development of ancient species may be studied, which in their adult condition represent neanic or nepionic stages of later forms.

The prolific development of the Brachiopoda, both in point of numbers and variety of genera and species, together with their geological history, mark this group as one which should furnish important data for the study of its genesis and of the limits of a specialized variation in a single class. Moreover, as its culmination was reached in Paleozoic time, the group should afford illustration of many principles of evolution.

The main characters common to the class of Brachiopoda are as follows: The bivalve shell; the pedicled or fixed condition; the animal composed of two pallial membranes intimately related to the shell; a visceral sac; and two arms or appendages near the mouth. The extreme range of variation does not eliminate any of these features, and, consequently, no univalve or multivalve forms are found, nor any strictly free-swimming species, nor growths or modifications adapting the organism to a pelagic life. Thus the limits of modification are narrowly restricted as compared with those of several other classes; namely, the Echinodermata and the Pelecypoda, but the thousands of known species of Brachiopoda show what differentiation has taken place within these limits.

The Protegulum.

The first important observation to be noted is that all brachiopods, so far as studied by the writer, have a common form of embryonic shell, which may be termed the *protegulum.*[*] The protegulum is semi-circular or semi-elliptical in outline, with a straight or arcuate hinge-line, and no hinge-area. A slight posterior gaping is produced by the pedicle valve being usually more convex than the brachial. The

* From πρώ early, and τέγος a covering.

modifications noted are apparently due to accelerated growth, by which characters primarily neanic become so advanced in the development of the individual as to be impressed finally upon the embryonic shell.   This feature is well shown in the development of *Orbiculoidea* and *Discinisca*, and is reserved for discussion under these genera.

As the protegulum has been observed in about forty genera[*] representing nearly all the leading families of the class, its general presence may be safely assumed.   In size it varies in different genera and species.   The range is from .05 to .60 mm.   A similar range in the prodissoconch of pelecypods has been noticed by Dr. Robert T. Jackson.   The protoconch of cephalopods and gastropods also varies greatly.   In all these classes the size of the initial shell has no special relation to the mature form, and it seems to have little significance in related genera or species.

The structure of the protegulum has been described as corneous and imperforate.   In all probability it is the same for the entire class, whether among the corneous and phosphatic linguloids and discinoids, or the terebratuloids and other forms having carbonate of calcium shells.   Professor E. S. Morse, in describing the early stages of *Terebratulina*,[†] says: "A heart-shaped corneous shell is formed even at this early stage, for in several cases I met with it where the softer portions had been removed by Paramæcia."   Similarly, in the genus *Cistella* according to Kovalevski:[‡] "En même temps la coquille se forme, par suite du dépôt sur la cuticule chitineuse des minces couches de calcaire, dans lesquelles on

---

[*] *Atretia* (*Cryptopora*), *Chonetes, Cistella, Conotreta, Crania, Craniella, Discina, Discinisca, Glottidia, Gwynia, Kraussina* (*Megerlina*), *Laqueus, Leptæna, Lingula, Lingulops, Linnarssonia, Liothyrina, Magellania* (*Macandrevia*), *Martinia, Mühlfeldtia, Obolus? Œhlertella, Orbiculoidea, Orthis* group, *Pholidops, Productella, Rhynchonella* (*Hemithyris*), *Schizambon, Schizobolus, Schizocrania, Schizotreta, Spirifer, Streptorhynchus* (*Orthothetes*), *Stropheodonta, Strophomena, Terebratella, Terebratulina, Thecidium,* (*Lacazella*), *Trematis, Tropidoleptus, Zygospira.*

[†] Embryology of Terebratulina.   *Mem. Boston Soc. Nat. Hist.*, II, 257, *vide* figs. 68, 76, pl. viii, 1873.

[‡] Developpement des Brachiopodes, Kovalevski.   Analyse par MM. Œhlert et Deniker, 65, 67, 1883.

ne voit point encore les perforations tubulaire." Previous to this stage, "Les lobes du manteau commencent alors à se recouvrir d'une cuticule épaisse et rigide que ne leur permet plus de se mouvoir que dans le sens vertical."

From the minuteness and the tenuous nature of the protegulum, its fossil preservation in an unaltered condition would not be expected. Neither would it be found on the beaks of mature shells, whether recent or fossil. In rare cases of unusually perfect conservation of the beaks the protegulum is retained, but frequently its form and characters are exhibited, after its removal, by the impression left in the surrounding calcareous test. To study the features of the protegulum, and the early stages in the growth of the shell, it is very desirable and often necessary to have young and well-preserved specimens. The rapid encroachment of the pedicle on the ventral beak commonly obliterates, at an early period, all traces of the protegulum and early nepionic stages; while, in the dorsal valve, abrasion from foreign objects, or against the deltidial covering, or the pedicle itself usually removes all early lines of growth or nepionic characters. In general, fully matured shells, recent or fossil, do not furnish material for a study of the incipient growth-stages.

Affinities. — In looking for a prototype preserving throughout its development the main features of the protegulum, and showing no separate or distinct stages of growth, the early primordial form hitherto known as *Kutorgina* Billings is at once suggested. This genus, as shown below, includes two distinct types, for one of which the name *Paterina* [= *Iphidea* Walcott] is proposed.*

---

* The strict definition of *Kutorgina* limits it to calcareous shells such as are found near Swanton, Vermont, often occurring as casts in the limestone. The original description of *Obolella cingulata* by Billings (*Geology of Vermont*, II. 948, figs. 347–349, 1861) seems to include two species. One, represented by figures 347 and 349 (*loc. cit.*), agrees with phosphatic species having a straight hinge-line as long as the width of the shell. The other, shown in figure 348, has a calcareous test, shorter hinge, flattened brachial valve, and convex pedicle valve with arching beak. Upon the latter species the genus was founded, and it

The valves of *Paterina* [= *Iphidea*] are sub-equal, the pedicle valve being a little more elevated than the brachial. They are semi-elliptical in outline. In mature specimens all lines of growth, from the nucleal shell to the margin, are unvaryingly parallel and concentric, terminating abruptly at the cardinal line. In other words, no changes occur in the outlines or proportions of the shell during growth, through the nepionic and neanic stages up to and including the completed ephebic condition. The resemblance of this form to the protegulum of other brachiopods is very marked and significant, as it represents a mature type having only the common embryonal features of other genera. It is of further importance as representing, in many species, an early condition of nepionic growth subsequent to the pro-tegulum, during which the proportions and features of the shell undergo no modification except increase in size. This is termed the *Paterina stage*. It is well shown in the brachial valve *Orbiculoidea minuta* Hall (Plate XI, figure 5).

Modifications from Acceleration. — The modifications in the form of the protegulum are due to the influence of accelerated

has been recognized as the type by C. D. Walcott (*Bull. U. S. Geol. Surv.*, No. 30, 102, pl. ix, figs. 1, 1 *a*, *b*, 1886). The species represented by Billings in figures 347 and 349 resembles *Obolus labradoricus* (fig. 345, *loc. cit.*), and is repre-sented by Walcott (*loc. cit.*, pl. ix, figs. 2, 2 *a*, *b*) and referred by him also to *Kutorgina*. Mr. Walcott recognizes two groups of species, which are classified (p. 102) as : " shell structure calcareous (*K. cingulata, K. Whitfieldi*) or horny (*K. Labradorica, K. sculptilis*)."

An examination of specimens representing both groups leads the writer to consider *Kutorgina cingulata* and *Obolus labradoricus* of Billings as generically distinct. Therefore the name *Paterina* is here proposed to include species of the type of *Obolus labradoricus*, var. *swantonensis*. This name is intended to express the primitive ancestral characters which it possesses (Plate XI, figs. 1, 2). Exfoliated specimens of *Paterina labradorica* show a roughened area on the cast, each side of the median line near the beak. These probably represent muscular attach-ments. Sections of the shell show no hinge-area as described in *K. cingulata*. A study of the latter would doubtless present distinct stages of growth. The dissimilar valves, arcuate ventral beak, and mesial depression could be developed only by passing through several well-marked phases. This in itself seems sufficient for a separation were no other characters present. [Walcott has reinvestigated the genus *Iphidea*, and has decided that it includes the form above designated as *Paterina*. *Proc. U. S. Nat. Mus.*, XIX, 707, 1897.]

growth, by which nepionic and sometimes neanic features are pushed forward, or appear earlier in the history of the individual, so as to become impressed upon the early embryonic shell. Only a brief review of these changes will be noted here, as a fuller description properly belongs under the discussions of the various genera and families. Naturally, the greatest departure from the normal protegulum is exhibited in the most variable and specialized valve, the ventral valve. The nearly equivalve genera, as *Lingula* and *Glottidia*, present almost no modification. In the ventral valve of *Linnarssonia* and *Orbiculoidea* (Plate XI, figure 7), the protegulum has a hinge more or less arcuate. *Discinisca* shows a subcircular ventral protegulum with a pedicle-notch, and the evidence of any hinge in the dorsal valve is very slight (Plate XI, figures 8, 9). The discinoid character appearing in the second and third nepionic stage of the Paleozoic *Orbiculoidea* (Plate XI, figure 6) has become so accelerated in neozoic and recent *Discinisca* as to produce a discinoid protegulum.

The strophomenoid shells usually retain a normal protegulum in the dorsal valve, but in the ventral valve the protegulum has an abbreviated hinge and arcuate hinge-line (Plate XI, figures 13, 14, 15).

No marked variation has been yet observed among the spire-bearing genera, nor has any been seen in the terebratuloids or rhynchonelloids further than the radii on the protegulum of *Atretia* (*Cryptopora*). Possibly this feature in *Atretia* is an inheritance from the radiate character of the shell in the Rhynchonellidæ. It may be, however, one of the features consequent upon its fragile nature and deep-sea habitat, as observed among other abyssal shells.

Differences in the Valves.

The dissimilarity in the form and relations of the two valves progressively increases in the following genera: *Lingula, Terebratulina, Cistella, Discinisca, Thecidium* (*Lacazella*), and *Crania*. *Lingula* is nearly equivalve, both valves bearing a close resemblance to each other. In *Terebratulina* and

Cistella the two valves are more strongly specialized, while in *Discinisca, Thecidium,* and *Crania* they are quite unlike.

Two important organic characters accompany and partake of a similar amount of variation; (*a*) the length and direction of the pedicle, and (*b*) the position and structure of the pedicle-opening. *Lingula* with a long, fleshy, mobile pedicle receives uniformly disposed axil impacts on the valves, and therefore, with equal physiological reactions, equality in size and form is produced. *Terebratulina* and most of the other terebratuloids and rhynchonelloids have a shorter and less flexible pedicle. As a whole the motions of the animal are more restricted; the pedicle-opening is confined mainly to one valve; the valves, consequently, are differently related to the environment, and express this difference in their dissimilarity. In these examples, also, the inclination of the pedicle to the longitudinal axis, or of the shell to the surface of support, agrees, *pari passu*, with the amount of unlikeness in the valves, except when the pedicle is so shortened as to interfere with their free movement. To this inclination is probably due the difference in the action of the forces from without.

Normally, in *Lingula*, the pedicle is in direct linear continuation with the axis of the shell. *Terebratulina* and *Magellania* are inclined at an angle of 40° to the surface of support, but in *Cistella* and *Mühlfeldtia* this is increased to about 70°. In the latter genera, although the position of the axis is nearly vertical, the shortening of the pedicle precludes more than a slight elevation and rotation of the organism. The more the pedicle-opening is confined to one valve the greater is the difference between both.

Passing to *Discinisca*, the pedicle is found to be at right angles to the longitudinal axis, and the valves become strictly an upper and a lower. The lower rests upon the object of support, and the animal is capable of raising and rotating it only to a slight degree. Under such circumstances the lower valve is wholly different in its relations to the environment, and, naturally, it expresses the greatest dissimilarity in the two valves of any genus yet discussed. In some allied genera, as *Discina* (type *D. striata*) and *Schizotreta*, where the pedicle is small and the lower valve rises above the object of support, a similar form in both valves is again produced by the conical growth of the lower valve.

More primitive types, as *Acrotreta* and *Acrothele*, having the plane of the dorsal valve at right angles to the direction of the pedicle, retain a marginal upper beak, while the lower is elevated, sub-central, and perforate. These features in *Acrotreta* and *Discina* resemble, in a measure, those in the *Rudistes*. In *Acrotreta*, as in *Caprotina*, the upper valve shows its normal affinities, while the other has become highly modified and dissimilar. But in *Discina* and *Hippurites* the hinge-line is lost, and the apex of the upper valve is sub-central. This conical habit of growth in erect attached organisms has been explained as the physiological reaction from equal radial exposure to the environment. It constitutes the law of radial symmetry, ably discussed by Haeckel, Jackson, Korshelt, and Heider. Its application to the Brachiopoda can be made mainly in forms having the pedicle perforation sub-centrally located in the lower valve.

In *Thecidium* and *Crania* the calcareous union of the lower valve to the object of support represents the extreme of unlike conditioning, and such forms exhibit the greatest difference in the features of the opposite valves. *Crania*, being probably derived from discinoid stock, is without proper hinge. In the history of its development, so far as known, it does not show beyond the protegulum an early hinged condition. Hence there is no indication of direct derivation from hinged forms. A false hinge is sometimes present, but it clearly shows a secondary mechanical adaptation, and not a phylogenetic character. On the other hand, true hinged attached genera, such as *Thecidium* (*Lacazella*), *Davidsonia*, and *Strophalosia*, possess this feature as a later ancestral character, and in their chronological history tend to shorten and gradually eliminate it. An illustration of this is seen in the succession of the species in *Strophalosia*, or in the ontogeny of one of the Permian species. *Strophalosia Gold-fussi*, in early neanic stages, has a hinge-line about equal to the width of the shell, but in mature individuals it is usually less than one-half the width. This reduction of the hinge and ostrean form of growth are in accordance with the deductions and observations made upon the Oyster and its allies by Jackson, and the mechanical principles are evidently the same in both cases.

One of the most conspicuous examples of a difference in the form of the valves is shown in the abnormal genus *Pro-boscidella*. In early neanic stages it resembles an ordinary *Productus*. Afterward, probably from burrowing in the mud, the ventral valve becomes extravagantly developed anteriorly into a calcareous tube. This is accomplished by the excessive growth of the anterior and lateral margins. Then an infolding takes place until the lateral edges unite, after which the tube is built up by concentric increment around the free end. The resemblance of *Proboscidella* to *Aspergillum* is quite marked, except that, in the latter genus, the tube is formed from the growth and union of two valves instead of one.

From the morphological differences of the ventral and dorsal valves it will be seen that the highest modifications occur in the former; while the variations in the latter are expressed mainly as adaptive reactions or accommodations to these changes. The explanation of the fact that greater alteration takes place in the ventral valve evidently lies not in the greater plasticity of this member, but in its more highly specialized and differentiated external form, and mainly in its being the lower and attached valve.

No account is taken here of the crura, loops, and spires of the brachial valve, so characteristic and important in many families and genera. These are evidently processes devel-oped by the internal requirements of the animal, and are not affected by the environment. Therefore they are internal calcified organs independent of the form or manner of growth of the external covering. This is shown by the fact that in each group there is a frequent recurrence of similar general external features, whether in crurate, looped, or spire-bearing genera.

Genesis of Form.

The principal characters shared by the two valves are the general outline and the hinge. In typical and generalized forms, as *Lingula*, *Terebratulina*, *Cistella*, and *Discinisca*, considered as before in regard to length of pedicle, freedom of movement, and direction of longitudinal axis to the object of support, we find a key to these types of structure. In the individual development of *Terebratulina*, as shown by Morse, we first have the early embryonic shell (protegulum), with a short pedicle and straight hinge. The next stage retains both these characters, but the valves have become more unequal and the pedicle-opening confined to the delthyrium of one valve. The result is a shell very much like *Argiope* or *Megerlia* (*Megathyris* and *Mühlfeldtia*), to which Professor Morse also called attention. The same author next showed that the succeeding stage had a comparatively long pedicle, and a shell linguloid in form. Afterward the defining of the pedicle-opening, shortening of the pedicle, and truncation of the ventral beak produced the final characteristic external features of *Terebratulina*. The deduction from this example and from *Lingula* is that genera having pedicles sufficiently long to admit of freedom of axial movement have elongate and rostrate shells. The shortening of the pedicle brings the posterior part of the shell in more or less close proximity to the object of support, and, as growth cannot take place in that direction, it increases laterally, resulting in broader forms with extended hinge-areas, as in many species of *Cistella*, *Scenidium*, *Mühlfeldtia*, *Terebratella*, *Kraussina*, etc.

The variety known as *Mühlfeldtia truncata*, var. *monstruosa* Davidson, further shows how discinoid characters may be produced in an entirely different type of shell. A specimen was found by the writer in a position which readily gave the solution to its variation from the normal species. It was attached to a foreign object under the hinge-line of a large mature specimen of *M. truncata*, thus forcing the axis and plane of the valves into parallelism with the object of support. In this way the pedicle emerged at right angles to the axis. The growth of the shell and the increase in the size of the pedicle caused the latter to encroach on the substance of the lower beak, forming a dorsal perforation or pedicle-notch, which in this example amounted to an arc of 180°. As the ventral valve was the upper and the dorsal the lower, with the pedicle-opening through the latter, only the abnormal position of the shell can account for this anomalous discinoid condition. In the development of *Orbiculoidea*, a true discinoid genus, it will be seen that during the early stages it had a straight hinge and marginal beaks (Plate XI, figures 5, 6, 7). Then, from its procumbent position and peripheral growth, the pedicle became more and more enclosed by the lower valve, until in *Schizotreta* (figure 11) and *Acrothele* (figure 12), the opening finally became sub-central.

The resemblance between this form of growth and habit and *Anomia* is very suggestive. Morse and Jackson have shown that from an early normal, bivalve, hinged shell the right valve in its subsequent growth surrounds the byssus, which occupies much the same position and performs a function similar to the pedicle of *Discinisca* and *Orbiculoidea*. Peripheral growth also causes the initial shell to recede from the margin. Another instance is thus furnished of a discinoid habit in an organism otherwise entirely different. It is therefore evident that the discinoid form is purely due to the mechanical conditions of growth. Hence the writer believes that any bivalve shell with the plane parallel to the object of support, and attached by a more or less flexible, very short organ, as a byssus or a pedicle, without calcareous cementation, assumes a discinoid mode of growth.

The conditions of radial symmetry and ostrean growth were briefly mentioned in a preceding section, and need only be cited here as resulting from the cemented state of fixation, as shown in species of *Thecidium*, *Strophalosia*, and *Crania*.

A long pedicle accompanies elongate shells with short hinges. A short pedicle causes extended hinge growth when the plane of the valves is ascending or vertical, but a discinoid form results when the plane of the valves is horizontal.

Types of Pedicle-openings.

M. Deslongchamps is one of the few writers who have given much consideration to the characters of the pedicle-opening. His studies, although mostly confined to the terebratuloids and later spire-bearing genera, conclusively show the importance of this feature.* In a recent paper by the writer † attention was called to the persistence and embryonic features of this portion of the shell. "It has been shown by J. M. Clarke and the writer that all species, so far as examined, possessing a true deltidium [= deltidial plate] in the adult state, show that it was gradually developed in early stages of growth, by concrescence along the lateral margins of an open triangular area. Also, that all species furnished with a pedicle-sheath [= deltidium] have it fully developed in the earliest growth-stages which have been observed for these species, and the subsequent growth of the individual does not materially alter its general characters, except that it is sometimes retrogressive, the parts becoming atrophied or functionally obsolete. A feature of such importance, and so intimately connected with the embryonal growth of the shell, must be given considerable significance in discussing the various genera in which it is present or absent." At that time the development and true interpretation of these different features of the pedicle-opening and the early stages of the shell had not been studied sufficiently, and a more general application of the principles involved could not then be made. The results of later studies give prominence to these characters, and show that they furnish a method for an ordinal grouping of the genera of brachiopods. This is

---

* Note sur le développement du deltidium chez les brachiopodes articulés. *Bull. Soc. Géol. France* (2), XIX, 409–413, pl. ix, 1862.

† *Amer. Jour. Sci.* (3), XL, 217, September, 1890.

found to agree with the chronological history of the class, as well as with the anatomical and shell characters, and therefore it is believed to be a natural and reliable sub-division.

The first and simplest type of pedicle-opening is in shells with a posterior gaping of the valves, through which the pedicle protrudes in line with the axis. It is shared more or less by both valves, although, generally, the greater portion of the periphery is included by the pedicle valve. The genus *Lingula* and related genera afford types of this form of pedicle-opening.

The second type is characterized by a pedicle wholly confined to the lower valve, and emerging at right angles to the plane of the valves. In primary forms it is not entirely surrounded by shell growth, but occupies a sinus, slit, or fissure. A further specialization carries it quite within the periphery, and it finally becomes sub-central. A serial illustration of this type is presented in the genera *Schizocrania*, *Orbiculoidea*, *Discinisca*, *Schizotreta*, and *Acrothele*. The group probably terminates with forms like *Crania* and *Pholidops*, as shown by the development of the dorsal valve and from internal characters. The development of the lower valve, however, has not been observed as yet in either of these genera.

The third form somewhat resembles the second. During the first nepionic stage of shell growth the pedicle is enclosed by the ventral valve and the pro-deltidium. The perforation remains sub-marginal, and does not tend to become centralized as in the preceding group. The initial pedicle-opening may be maintained by further growth, forming a deltidium; or it may be merged into the hinge opening by resorption of the shell or by pedicle abrasion. *Orthisina*, *Leptæna*, *Strophomena*, *Chonetes*, and *Stropheodonta* furnish illustrations of the first condition, and the second is represented in the groups of *Orthis*.

The fourth type in its incipient stage marks a return to the simple conditions of the first, but in early neanic stages the pedicle is confined to the ventral beak, and deltidial plates are developed in the majority of species. These plates at maturity may entirely limit the pedicle-opening below, so that the pedicle emerges immediately under the beak, or encroaches upon the substance of the beak itself. This type of opening is shown by *Zygospira, Spirifer, Rhynchonella, Terebratulina, Magellania*, etc.

The only divisions of the class which have had continued recognition are the Arthropomata and Lyopomata, proposed by Owen in 1858.* Subsequently various authors gave names to express other characters, but all included the same elements in the two divisions. Professor Huxley's terms, the Articulata and Inarticulata, have also come into current use, and are convenient to express the nature of the union of the valves. All the names proposed for these divisions by Owen, Bronn, Huxley, Gill, and King, are based upon (1) the intestinal canal whether ending in an anus or in a blind sac, (2) the relative proportions of the viscera and brachia to the shell cavity, and (3) the character of the union of the valves.

If, as Agassiz has said,† orders should be founded upon facts of development or embryology, the ordinal division into groups expressing the genesis of an important common character should furnish a satisfactory classification. The Articulata and Inarticulata do not appear to have a primary developmental basis in nature. These names may be conveniently retained as two divisions or sub-classes, but they fail to express the true relationships of the various groups included in them.

In 1883 Dr. Waagen (Palæontologia Indica) proposed a classification comprising six sub-orders, founded partly on the pedicle-opening and on the form of the brachial supports. Two of his groups, the Mesokaulia and Aphaneropegmata, are nearly equivalent in extent to the Atremata and Protremata now proposed. Daikaulia and Gasteropegmata of Waagen are here included in the Neotremata, and the Telo-

---

* Encycl. Brit., 8th ed., XV, 301, 1858.
† Methods of Study in Natural History, L. Agassiz, 8th ed., 76, 1873.

tremata comprises the Kampylopegmata and Helicopegmata of the same author. With the transfer of some genera in his sub-orders, they may be properly recognized and serve further to differentiate the class into comprehensive groups.

After this preliminary discussion the four groups proposed can be defined and understood. The special details with full illustration and demonstration of the development and affinities in each group are left for future consideration. At present it is aimed to give only the general results which have been reached through the study of individual development (ontogeny) among various species representing the families of nearly the entire class. Of the sixteen families of Brachiopoda recognized by Œhlert in Fischer's "Manuel de Conchyliologie," fifteen have thus been studied and determined. The genera marked by an asterisk have been examined somewhat in detail. The others have been investigated partly from adult specimens, and from the published descriptions of the genera.

ATREMATA.

(*a* priv., and τρῆμα perforation.)

(Plate XI, figures 1–4.)

Protegulum semi-circular or semi-elliptical; hinge-line straight or slightly arcuate. Growth taking place mainly around the anterior and lateral margins, never enclosing or surrounding the pedicle, which in all stages emerges freely between the two valves, the opening being more or less shared by both. Valves inarticulate.

Including the genera:

Dignomia.	**Leptobolus.*	*Obolus.*
Dinobulus.	**Lingula.*	**Iphidea.*
Elkania.	*Lingulasma.*	*Paterula.*
Glossina.	**Lingulops.*	*Rhynobolus.*
**Glottidia.*	*Monomerella.*	*Trimerella.*
Lakhmina.	*Obolella.*	

NEOTREMATA.

(νεός young, and τρῆμα perforation.)

(Plate XI, figures 5–12.)

Protegulum as in the preceding order in primitive forms, becoming more circular and with shorter and more arcuate hinge in the pedicle valve of derived types. Growth of the dorsal valve tending to become peripheral. In the opposite valve the pedicle more or less surrounded by progressive neanic growth posterior to the initial hinge. Pedicle fissure remaining open in primitive mature forms, becoming enclosed in secondary forms during neanic stages, and in derived types enclosed in early neanic or nepionic stages. Valves inarticulate.

Including the genera:

Ancistocrania.	*Discinopsis.*	*Pholidops.*
Acrothele.	*Helmersenia.*	*Pseudocrania.*
Acrotreta.	*Kayserlingia.*	*Rœmerella.*
Conotreta.	*Lindstrœmella.*	*Schizambon.*
Crania.	*Linnarssonia.*	*Schizobolus.*
Craniella.	*Mesotreta.*	*Schizocrania.*
Craniscus.	*Œhlertella.*	*Siphonotreta.*
Discina.	*Orbiculoidea.*	*Trematis.*
Discinisca.		

PROTREMATA.

(πρώ early, and τρῆμα perforation.)

(Plate XI, figures 13–21.)

Protegulum of the dorsal valve as in the Atremata. In the ventral valve it has become modified to an elliptical or circular form with arcuate hinge. Pedicle enclosed in early nepionic stages by a pro-deltidium; posterior covering (deltidium) retained at maturity, or resorbed or abraded in neanic stages, so that the pedicle protrudes between the two valves. Valves articulate.

Including the genera:

Amphigenia.	*Hipparionyx.*	**Productella.*
Aulosteges.	**Lacazella.*	*Productus.*
Bactrynium.	**Leptœna.*	**Rhipidomella.*
Bilobites.	*Leptœnisca.*	*Schizophoria.*
Camarella (group).	*Lyttonia.*	*Sieberella.*
Camarophoria.	*Meekella.*	*Streptis.*
**Chonetes.*	*Mimulus.*	**Streptorhynchus.*
Clitambonites.	*Oldhamina.*	*Stricklandinia.*
Conchidium.	**Orthis* (group).	*Strophalosia.*
Davidsonella.	*Orthisina.*	**Stropheodonta.*
Davidsonia.	**Orthothetes.*	**Strophomena.*
Daviesiella.	*Pentamerella.*	**Strophonella.*
Derbya.	*Platystrophia.*	*Thecidella.*
Enteletes.	**Plectambonites.*	**Thecidium.*
Eudesella.	*Porambonites?*	*Thecidopsis.*
Hemipronites.	*Proboscidella.*	*Triplecia.*

TELOTREMATA.

(τέλος last, and τρῆμα perforation.)

(Plate XI, figures 22–28.)

Protegulum as in Atremata. Pedicle-opening shared by both valves in nepionic stages, usually confined to one valve in later stages, and becoming more or less limited by two deltidial plates in ephebic stages. Arms supported by calcareous crura, spirals, or loops. Valves articulate.

Including the genera:

Acanthothyris.	*Cœnothyris.*	*Glassia.*
Ambocœlia.	*Cryptonella.*	*Grünewaldtia.*
Amphiclina.	*Cyrtia.*	**Hemithyris.*
**Athyris.*	*Cyrtina.*	*Hindella.*
**Atretia* (*Cryptopora*).	*Dayia.*	*Ismenia.*
**Atrypa.*	*Dictyothyris.*	*Karpinskya.*
Bifida.	*Dielasma.*	*Kayseria.*
Bouchardia.	*Dimerella.*	*Kingena.*
Centronella.	*Disculina.*	*Koninckella.*
**Cistella.*	*Eatonia.*	**Koninckina.*
Clorinda.	*Eudesia.*	**Kraussina.*
**Cœlospira.*	*Eumetria.*	**Laqueus.*

Leptocœlia.	*Nucleospira.*	*Stringocephalus.*
Leiorhynchus.	*Pentagonia.*	*Suessia.*
Liothyrina.	*Peregrinella.*	*Syringothyris.*
Macandrevia.	*Platydia.*	*Terebratella.*
Magas.	*Rensselœria.*	*Terebratula.*
Magellania.	*Reticularia.*	*Terebratulina.*
Martinia.	*Retzia.*	*Terebratuloidea.*
Martinopsis.	*Rhynchonella.*	*Thecospira.*
Megathyris.	*Rhynchonellina*	*Trematospira.*
Megalanteris.	*Rhynchoporina.*	*Trigonosemus.*
Megerlina.	*Rhynchotrema.*	*Tropidoleptus.*
Merista.	*Rhynchotreta.*	*Uncinulus.*
Meristella.	*Spirifer.*	*Uncites.*
Meristina.	*Spiriferina.*	*Zellania.*
Mühlfeldtia.	*Spirigerella.*	*Zygospira.*

PART II. CLASSIFICATION OF THE STAGES OF GROWTH AND DECLINE *

(PLATE XII)

A BRIEF review of the known embryology of the Brachiopoda is desirable, in order to account for some of the differences presented by adult forms in the several divisions of the class. This knowledge is far from complete, and is confined to a few species, but much of interest bearing on the later development of the organism may be obtained.

The important memoirs † of Morse,[18, 19] Kovalevski,[15] Lacaze-Duthiers,[16] and Shipley[22] contain nearly all that is known regarding the early embryology of brachiopods. The genera included in the works of these authors comprise *Cistella, Terebratulina, Liothyrina,* and *Lacazella.* Later larval stages of the genus *Glottidia* have been fully described by Brooks.[4] Müller,[20] also, has given a description and figures of a larval form doubtfully referred to *Discinisca.* The results of these observers must at present be taken without reservation, and are thus made use of in the present paper.

---

* *Amer. Jour. Sci.* (3), XLIV, 133–155, pl. i, 1892.

† The works referred to by numbers are cited in full in the list appended to this article.

Something is known, therefore, of the early stages in each of the four groups or orders proposed by the writer.[2] The Atremata, Neotremata, and Protremata are represented by a single genus only in each; *Glottidia*, *Discinisca*, and *Lacazella*, respectively; and the Telotremata, by *Cistella*, *Terebratulina*, and *Liothyrina*. Were *Glottidia* and *Discinisca* as well known as *Cistella*, *Terebratulina*, and *Lacazella*, some comparisons could undoubtedly be made which would enlighten many obscure points of anatomy and morphology, as well as give clearer insight into the history and origin of each group.

Cistella and *Terebratulina* are taken as standards of the embryological development on account of the completeness with which they have been studied, and because their points of difference are not great. *Lacazella* shows such peculiar features that its history must be discussed separately. The nepionic *Glottidia* and *Discinisca*, too, present characters which evidently had an early history somewhat different from *Cistella* or *Terebratulina*.

In taking up the review of the observed stages of growth, an attempt will be made to fix their limitations. To this end the admirable nomenclature proposed by Hyatt[9, 10] is here adopted, as it is more convenient and of wider application and significance than the terms heretofore used. Thus far this system has been employed principally in studies relating to the Mollusca, and its application to the Brachiopoda will necessarily require some illustration and explanation. In the preface to "Genesis of the Arietidæ" Hyatt has presented a summary of the theoretical opinions resulting mainly from his studies in the Cephalopoda. It is believed that nearly the same ground may be covered in the Brachiopoda, and thus the truth of these deductions will receive further evidence from another class of organisms.

Embryonic Stages.

The true embryonic stages are classified by Hyatt as *Protembryo*, *Mesembryo*, *Metembryo*, *Neoembryo*, and *Typembryo*.

To these Jackson [11] has added the *Phylembryo*, taking it from the later stages of the *Typembryo* to represent the period when the animal can be referred definitely to the class to which it belongs.

The succeeding stages in the growth of the animal to maturity are termed by Hyatt [and emended by Buckman and Bather] *nepionic* (young), *neanic* (adolescent), and *ephebic* (mature), while old-age characters are called *gerontic*. The stages are further divided by using the prefixes *ana*, *meta*, and *para;* as *anagerontic*, *metagerontic*, and *paragerontic*.

The application of this nomenclature of the stages of growth and decline to the Brachiopoda is shown on the following pages.

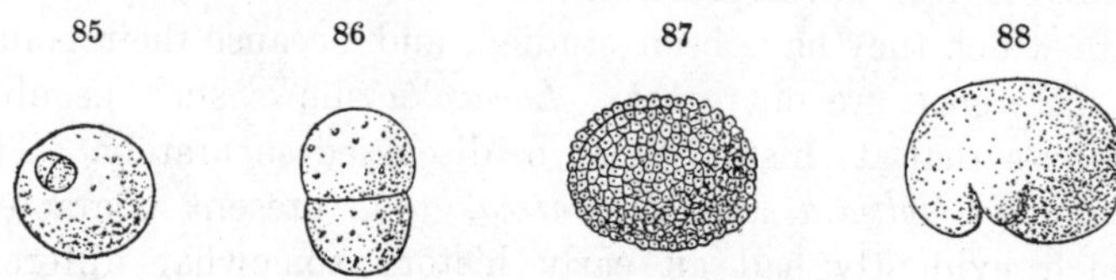

Cistella neapolitana Scacchi.

FIGURE 85. — Protembryo; unsegmented ovum.
FIGURE 86. — Protembryo; ovum composed of two spheres.
FIGURE 87. — Mesembryo; blastosphere.
FIGURE 88. — Metembryo; gastrula. (Figures 85–88, after Shipley.)

The *Protembryo*, as in other groups of organisms, includes the ovum and its segmented stages preceding the formation of a blastula cavity. Figures 85 and 86 show protembryonic stages of *Cistella*. The eggs are spherical, pyriform, or ovoid, and the segmentation proceeds in a regular manner, resulting in a blastosphere composed of equal parts.

The *Mesembryo*, or blastosphere (figure 87), has been observed in *Cistella*, *Terebratulina*, and *Lacazella*. The blastula cavity is small.

The *Metembryo*, or gastrula stage (figure 88), is developed from the blastosphere in two ways: (*a*) by embolic invagination in *Cistella* and *Terebratulina* (Kovalevski and Shipley), and (*b*) by delamination in *Lacazella* (Kovalevski). At the close of this stage the archenteron in *Cistella* is trilobed, consisting of a central cavity, or mesenteron, connecting on each side with the body cavity.

The *Neoembryo*, represented by the trochosphere and segmented, ciliated, cephalula stages, has been more fully observed than any of the preceding. The first advance from the completed gastrula is in the separation of the mesenteron from the body cavity, and the division of the organism into two segments or lobes, the cephalic and caudal (figure 89). Later a third or thoracic segment is developed and carries four bundles of stiff barbed setæ (figure 90). The cephalic and caudal lobes are densely ciliated. During the subsequent cephalula period two eyes, then two others, appear in *Cistella*, and at the same time the dorsal and ventral sides of the thoracic segment become extended over the caudal, and are progressively defined as two lobes (figures 89–93, 108, 109).

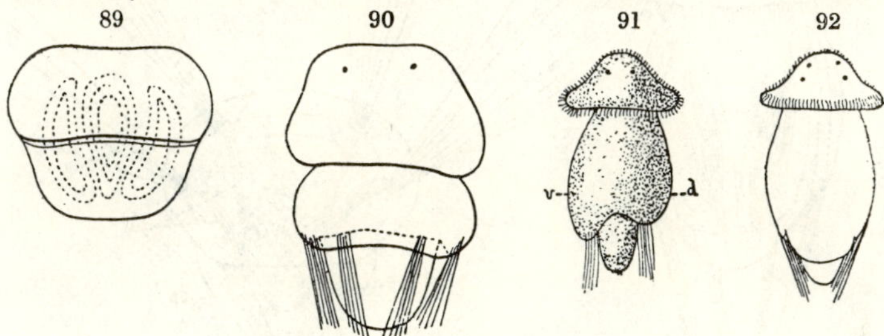

Cistella neapolitana Scacchi.

FIGURE 89. — Neoembryo; embryo of two segments.

FIGURE 90. — Neoembryo; cephalula; ventral side; showing cephalic, thoracic, and caudal segments, eye-spots, and bundles of setæ. (Figures 89 and 90, after Kovalevski.)

FIGURE 91. — Neoembryo; lateral view of completed cephalula stage; showing extent of dorsal (*d*) and ventral (*v*) mantle lobes, and umbrella-like cephalic segment.

FIGURE 92. — Neoembryo; same stage; ventral view. (Figures 91 and 92, after Shipley.)

Terebratulina has a tuft of bristles on the top of the cephalic segment. In *Lacazella* the bundles of setæ are absent, and the head is more distinctly differentiated from the anterior segment than in *Cistella*. The closing cephalula stage in *Cistella* has an umbrella-like expansion of the cephalic border, and the organism becomes a free-swimming larva (figures 91–93).

Larval Stages.

The *Typembryo* is the larval stage at which some distinctive features make their appearance, but before the special characters of the class are to be found (figure 94). It is analogous to the molluscan embryo in which a shell gland and plate-like initial shell are developed. There is, however, no homology of parts or organs between the typembryonic mollusk and brachiopod.

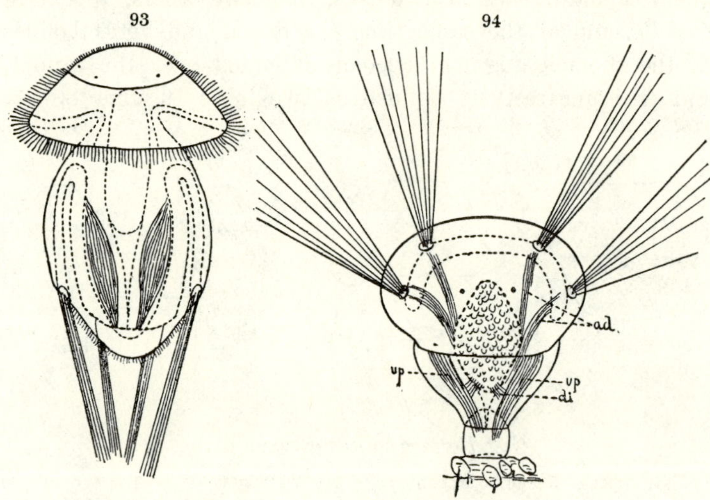

Cistella neapolitana Scacchi.

FIGURE 93. — Neoembryo; completed cephalula stage.

FIGURE 94. — Typembryo; transformed larva resulting from folding upward of mantle lobes over cephalic segment. *ad*, muscles from bundles of setæ to sides of body cavity; *di*, muscles from dorsal to ventral sides of body; *vp*, muscles from ventral side of body to caudal segment or pedicle. (Figures 93 and 94, after Kovalevski.)

In *Cistella* and *Terebratulina* the development of the typembryo has been observed, and consists of the folding upward of the lobes which have been developed from the thoracic segment to form the mantle, so that they gradually

enclose the anterior end (figures 108–111).   The surfaces of
the mantle which were exterior in the cephalula have now
become inner and the bundles of setæ have revolved 180°,
changing their direction from posterior to anterior.   This
leaves the lower part of the thoracic, and the whole of the
caudal, segment exposed.   The outer surface of the mantle is
invested with a hard integument, which, upon completion
and before the growth of the true shell, forms the protegulum.   The pedicle at this stage is also defined, being a modification of the caudal segment.   It may serve to attach the
larva to foreign objects, as in *Cistella* (figure 94) and *Terebratulina*, or it may remain undeveloped for a time, as in
Glottidia and *Discinisca*.   A rudimentary digestive tract is
present.

The body muscles which have been developed thus far
consist of four distinct pairs.   Two pairs lie close to the
sides of the body cavity, and extend to the points of inser-
tion of the bundles of bristles (figure 94, *ad*).   They become
after transformation the four adductor muscles of the valves.
The third pair extends from the ventral side of the body to
the caudal segment, and is converted into the ventral pedicle
muscles (figures 94, 99, 100, *vp*).   The fourth pair is situated
posterior to the digestive tract, and extends from the dorsal
to the ventral wall of the body (figure 94, *di*).   They form
the divaricator muscles in the mature brachiopod (figure 100,
di), and are divided into or duplicated by a pair of dorsal and
a pair of ventral divaricators.   There is also a pair of dorsal
pedicle muscles in the larva of *Liothyrina* and *Terebratulina*.

The folding upward of the mantle lobes forms the first
hinge-line of the future valves (*hl*, figures 110, 111).   Thus
its origin is not, as in pelecypods, a line produced by the
bending of a single plate (Jackson), but is the line along
which the two mantle lobes are bent against the body.
Between them projects posteriorly nearly half the body of
the animal, and the whole opening corresponds to the pedicle-
opening of later stages of growth.   The hinge of brachiopods,

therefore, is not primarily a line of articulation of the valves, but the limiting borders between the body and the attached edges of the mantle. Secondarily, and during later growth, the extension of the valves along a line of apposition forms a true hinge-line.

The first points of contact of the valves to form the true hinge lie adjacent to the right and left sides of the body of the animal, at the cardinal extremities (figure 99, *t*). Here naturally the first hinge-teeth are formed, and their position corresponds to that in adult individuals; namely, on each side of the cardinal opening. The enlarging of the cardinal opening by shell growth results in the gradual divergence or separation of the teeth, as in *Terebratulina*. In species with extended hinge-lines, as in many forms of *Spirifer*, *Orthis*, and *Strophomena*, the teeth still lie in their original position on each side of the cardinal opening, and the elongation of the hinge has come not only from the enlargement of the opening by growth, but by additions at the hinge extremities, so that the teeth are situated on each side of the central area, below the beak, and not at the cardinal angles. The young of these genera, however, all have the hinge-teeth at the extremities of the hinge, as the cardinal opening then occupies the whole posterior area of the shell.

Adult specimens of *Kutorgina* (*K. cingulata* Billings) have a deltidium as in *Strophomena*. The cardinal opening including the deltidium occupies the whole posterior end of the shell, and according to a statement made to the writer by Mr. Charles Schuchert, there are rudimentary teeth at the cardinal extremities. Therefore this genus represents a nepionic condition of later forms, and, on account of these and other characters, it is believed to be related to *Orthisina* and *Strophomena*, of which it is the ancestral type. It consequently belongs to the articulate brachiopods.

The embryonic stages up to this point have frequently been compared to similar stages in other organisms, especially in the Annelida and Polyzoa. Without repeating these comparisons, which may be consulted elsewhere,[4, 12, 15, 16-19, 21] attention is called to the similarity of development of the brachiopod typembryo to the larval stages of *Spirorbis*. There are, however, important structural differences. An article by J. W. Fewkes, "On the Larval Forms of *Spirorbis borealis* Daudin,"[7] contains a nearly complete and very interesting account of the development of this chætopod. There is a striking resemblance in the characters of the cephalula

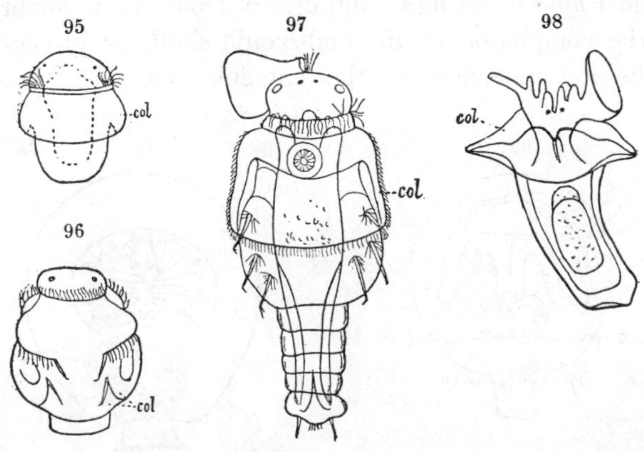

Spirorbis borealis Daudin.

FIGURE 95. — Cephalula, developing lobe from the body (*col*).
FIGURE 96. — More advanced stage.
FIGURE 97. — Larval form before transformation; showing posteriorly directed expansion (*col*) from thoracic segment.
FIGURE 98. — Transformed *Spirorbis*; showing folding upward of collar partially enclosing head. (Figures 95–98, after Fewkes.)

stages in both organisms, as may be seen on comparison (figures 95 and 96). *Spirorbis* develops a posteriorly directed extension from the middle segment, called a collar, which in later stages is reflexed anteriorly so as to cover more or less the cephalic portion, thus agreeing with the growth and change in position of the mantle in *Cistella*. The ventral lobe is also the larger in both. Many other comparisons and homologies have been made by Morse,[19] and the one here described is even more marked than his reference to the lobation of the cephalic collar in *Sabella*. Four figures are introduced illustrating the principal changes in *Spirorbis*. They may be compared with the development of *Cistella* shown in figures 90–94.

It is not intended by this to indicate a close relationship with the chætopods, for the writer is inclined to accept the opinion of Joubin,[12] that the brachiopods constitute a distinct and independent class.

The *Phylembryo* (figure 99) differs from the typembryo in (*a*) the completion of the embryonic shell, or protegulum; (*b*) the first appearance of the tentacular lobes of the lopho-

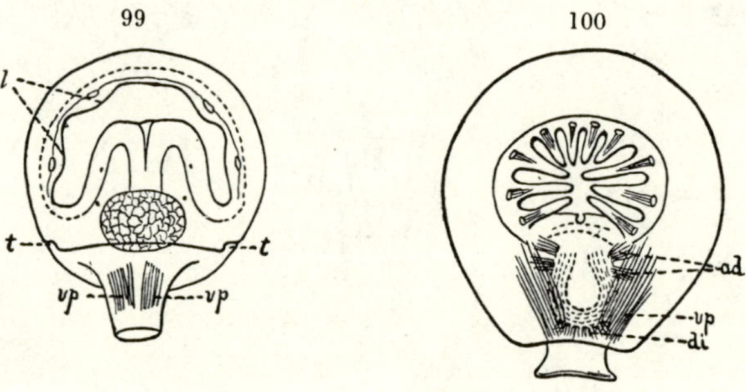

Cistella neapolitana Scacchi.

Figure 99. — Phylembryo ; brachiopod ; showing shell (protegulum), beginning of tentacles of lophophore (*l*), obsolescence of eye-spots, and formation of œsophagus ; *t*, hinge-teeth ; *vp*, ventral pedicle muscles.

Figure 100. — Nepionic brachiopod ; showing distinct tentacles of lophophore, mouth, and stomach, and transformation of muscles from typembryo (figure 94). *ad*, adductors ; *di*, divaricators; *vp*, ventral pedicle muscles. (Figures 99 and 100, after Kovalevski.)

phore, or arms; (*c*) the usual dehiscence of the four bundles of setæ; (*d*) the obsolescence of the eyes; (*e*) the definition of the œsophagus and stomach, and (*f*) the agreement of the muscular system with that in adult forms. These features, with the pedicle which appeared in a preceding stage, represent the brachiopod phylum, and are properly referred to the phylembryonic period of Jackson. Although the molluscan stage called the prodissoconch in pelecypods, the protoconch in cephalopods and gastropods, and the periconch in scaphopods, represents the completed phylembryo of these groups, as the protegulum represents a like period in the developing brachiopod, yet there is no homology of distinctive organs.

The mantle of mollusks is first formed on the posterior dorsal side, and is in the shape of a disk, which gradually envelops the animal to a greater or less extent, and may become distinctly lobed. As has been shown, this organ in the brachiopods develops simultaneously from the dorsal and ventral sides of the thoracic segment of the cephalula, and is primarily bilobed.

The initial shell of brachiopods is not produced from a distinct shell gland, as in the Mollusca, but is an integument of the surface of the mantle lobes, and intimately connected with them. The position of the valves is dorsal and ventral. The pedicle has no organic similarity with either a foot or a byssus.

The mouth of mollusks (and annelids) is formed below the base of the cephalic lobe of the cephalula, and may be the blastopore, while in the brachiopods it is near the anterior pole within the cephalic segment. Notwithstanding these differences, so many parts are functional equivalents that their growth and development may be discussed and interpreted in the same terms.

Before passing to later stages of growth which become more and more divergent from a common simple type, some points previously omitted, relating to *Thecidium* (*Lacazella*), *Lingula* (*Glottidia*), and *Discinisca*, should be here noted. As *Lacazella* is a form in which the ventral valve in the neanic and ephebic stages is cemented to foreign objects by calcareous fixation, it bears about the same relation to other brachiopods that *Ostrea* bears to *Avicula*, among the pelecypods, and a corresponding early absence or modification of many features present in adult individuals should be looked for. From what is known of the geological history of *Thecidium*, and if the interpretations of its phylogeny by the writer are correct, it is derived from an ancestry which had a similar condition of fixation as early as the Upper Silurian. *Thecidium* is apparently not a terebratuloid genus. Its structural affinities are evidently with the strophomenoids, especially such forms as *Plectambonites, Leptœnisca,* etc. Briefly the reasons for this statement are (*a*) the presence of a deltidium of one plate; (*b*) the absence of a true loop supporting the arms (the internal calcification or spiculization is confined wholly to the mantle, and does not extend to the arms [16]); (*c*) a concave plate in the cavity of the ventral beak, bearing the divaricator muscles; (*d*) the attached ventral valve, and (*e*) the cardinal processes in the dorsal valve.* The first character is of prime importance, because all the strophomenoids and none of the terebratuloids have a deltidium of one plate.

It would appear, therefore, that the early, free-swimming, larval state, and the later pediculate stage have become lost by acceleration, thus accounting for the very unequal development of the mantle lobes in the cephalula stage, and the non-active and early sedentary larvæ as described by Kovalevski and Lacaze-Duthiers.

The young *Lingula* (*Glottidia*) described by Brooks, and the *Discinisca* by Müller,[20] both representing the phylembryonic stage, were active and free-swimming animals, with rudimentary pedicles. *Terebratulina* becomes attached or rests on the caudal segment during the cephalula stage (Morse), while at the end of this period in *Cistella* (Kovalevski and Shipley) there is an active, swimming, ciliated organism, which later attaches itself by the pedicle in the typembryonic period.

From the facts that young individuals of Paleozoic species belonging to such genera as *Zygospira, Spirifer, Orthis, Rhynchonella,* and *Scenidium,* have been observed by the

---

* Dall in 1870 (*Amer. Jour. Conchology*) made a clear statement of the characters of *Thecidium* and of many of its radical points of difference with the Terebratulidæ, showing that it was entitled to rank as the type of a distinct family.

writer to retain their original relations to the objects of support, and that casts of the pedicles of fossil Lingulæ and *Eichwaldia* have been described (Davidson,[5] Walcott[23]), it cannot be assumed that the free-swimming condition was ever present in neanic or ephebic individuals. Evidently it has always been a larval character.

Origin of the Deltidium and Deltidial Plates.

The origin and significance of the deltidium * (= " pseudo-deltidium ") are made apparent in the development of *Thecidium*, and it may be well in this place to make a few observations on the genesis of this important character, and its relations to the deltidial plates of other genera, as *Rhynchonella* and *Terebratula*. It has been already noted (Part 1), that the deltidium in all species possessing it (the Protremata) is an embryological, or nepionic feature, which may or may not continue to the ephebic period; while the deltidial plates in other brachiopods (the Telotremata) appear later during the neanic and ephebic periods, or may never be developed. The detailed researches of Kovalevski on *Cistella* and *Thecidium*, together with other observations now first made, furnish data for a clear understanding of these differences.‡

Figure 102 represents a dorso-ventral section of a ripe cephalula just before the transformation, and shows the un-

---

* The single plate or covering to the triangular opening beneath the ventral beak should be termed the *deltidium*, as it was thus extensively used by Davidson. When it consists of two plates, they may be called *deltidial plates*. These names have been loosely used. In Part I of this paper the deltidium proper is referred to as pedicle covering, pedicle-sheath, and pseudo-deltidium. Hall and Clarke have proposed to call the triangular opening in the beaks of brachiopods the *delthyrium*, and the concave plate in the ventral beak of *Pentamerus*, *Orthisina*, etc., they have termed the *spondylium*. There yet remains a term for the convex plate covering the opening below the beak of the dorsal valve, and resembling the deltidium of the opposite valve. For this feature the name *chilidium* (χεῖλος) is here proposed.

‡ Kovalevski[15] For *Thecidium* consult the explanation of pl. iv, figs. 15–26 ; for *Cistella*, pl. i, figs. 13–15 ; pl. ii, figs. 17, 19–21.

equal lobes of the mantle, *v* being the ventral lobe, and *d* the dorsal; *h* is the head, and *p* the caudal segment developing into a pedicle.   A deposit of integument representing

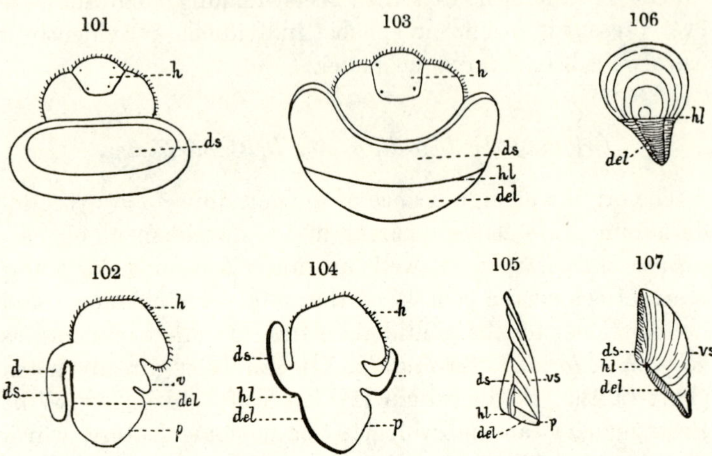

Thecidium (Lacazella) mediterraneum Risso.

Figure 101.—Cephalula; dorsal side. *ds*, dorsal shell plate; *h*, head. (After Kovalevski.)

Figure 102.—Dorso-ventral longitudinal section of cephalula of about same age as preceding. *h*, head; *d*, dorsal mantle lobe; *v*, ventral mantle lobe; *ds*, beginning of dorsal valve; *del*, shell plate forming on dorsal side of body; *p*, pedicle. (After Kovalevski.)

Figure 103.—Typembryo; larva transformed from folding upward of mantle lobes. *h*, head; *ds*, dorsal valve; *hl*, hinge-line of dorsal valve; *del*, shell plate on body and pedicle posterior to hinge-line of dorsal valve. (After Kovalevski.)

Figure 104.—Dorso-ventral longitudinal section of preceding. References as in figure 103. *vs*, ventral valve; *p*, pedicle.

Figure 105.—Profile view of neanic *Leptæna rhomboidalis*. The features of the shell are placed and lettered as in figure 104. *ds*, dorsal valve; *hl*, hinge-line; *del*, deltidium; *p*, pedicle-opening; *vs*, ventral valve.

Figure 106.—Adult *Thecidium (Lacazella) mediterraneum*; dorsal side; showing ventral area and deltidium.

Figure 107.—Profile of same. References as in figures 104 and 105.

the shell has formed on the inner side of the dorsal mantle lobe (*ds*), and also on the adjacent dorsal side of the body lobe (*del*).   A larva somewhat more advanced is represented in figure 101, as viewed from the dorsal side.   The mantle lobe is still directed posteriorly, as in the preceding figure, and the underlying shell plate is shown at *ds*. In the process of transformation (figures 103, 104), the mantle lobe is turned forward in the usual manner, bringing the shell on the outside of the animal, so that both dorsal plates are now exposed, *ds* being the dorsal valve, and *del* the shell developed on the dorsal side of the walls of the body and caudal segments. As this plate (*del*) is below or posterior to the hinge-line (*hl*), and extends down over the pedicle, it is evidently the beginning of the deltidium. At the same time there is an extension of the edges of the mantle and pedicle on the ventral, or lower, side and shelly matter is deposited, forming the ventral valve (*vs*, figure 104). At this stage the hinge-line (figures 103, 104, *hl*) is the line between the dorsal mantle shell (*ds*) and the dorsal body shell plate (*del*). The beak of the ventral valve is separated from the dorsal beak by the pedicle and the shell covering to the pedicle and body lobe, or the deltidium. The valves afterward meet at their peripheries; the hinge is extended beyond the deltidium, forming the true hinge of articulate brachiopods. As there is no motion between the ventral valve and the deltidium, the two become ankylosed. Figures 106 and 107, showing an adult *Thecidium*, are lettered in the same manner as the preceding, and express the same relation of parts.

The deltidium is not, therefore, primarily, on account of its manner of origin, an integral part of the ventral valve, but is a shell growth from the dorsal side of the body, which afterward becomes attached to the ventral valve, and is then considered as belonging to it.

The further growth of the deltidium around the body and pedicle, and its consequent extension into the cavity of the ventral umbo, may explain the origin of the spondylium.

Kovalevski[15] believed the ventral valve in *Thecidium* was secreted by the expanded edges of the pedicle and the body walls; whether or not this is so does not affect the interpretation of the origin of the deltidium. From the observations of Lacaze-Duthiers,[16] it seems, however, as though the ventral mantle lobe must have formed the shell in the usual way. This appears all the more probable from the fact that the lower or ventral valve is punctate, and, so far as known, the mantle contains all the cæcal prolongations, which alone could produce the punctate structure. Careful microscopic examination has failed to detect punctæ in the deltidia of *Thecidium*, *Strophomena*, *Leptæna*, and other punctate genera belonging to the Protremata.

It is true that *Aulosteges* has spines on the deltidium, but spines even when tubular are not equivalent to punctæ, as shown in *Productus*, *Strophalosia*, and some species of *Spirifer*. *Aulosteges* is a gerontic genus, which has become excessively spinose, and has also reverted to ancestral characters in its high hinge-area and conspicuous deltidium. It is well known that even the spires of *Spiriferina* and the loop of *Macandrevia* are spinose.

Turning now to *Cistella* as a representative of the Telotremata, a different process obtains.

Figure 108 represents the fully developed, free-swimming cephalula of *Cistella*, and shows the extent of the folds of the mantle and their posterior direction. Figure 109 represents the same in section. The inner sides of the mantle lobes are to form the future valves, the dorsal (*ds*), and the ventral (*vs*). The transformed larva or typembryo is represented in figure 110 and in section in figure 111. It is seen that the transformation consists in the folding forward of the mantle lobes over the head segment (*h*). Now the shell-secreting layers of the mantle are exterior, and the two valves begin to form, the dorsal shell (*ds*), and the ventral (*vs*). The pedicle and posterior portion of the body come out freely between the valves and mantle lobes and limit the hinge-areas of both (*hl* and *hl'*).

The further process of growth increases the distance between the initial dorsal and ventral hinges, for while the original dorsal beak is usually maintained at the hinge-line, the ventral beak is progressively removed and the ventral hinge travels from its first position at the beak, along the

edges of the umbo, leaving an open triangular area or delthyrium in the ventral valve occupied by the pedicle.   This

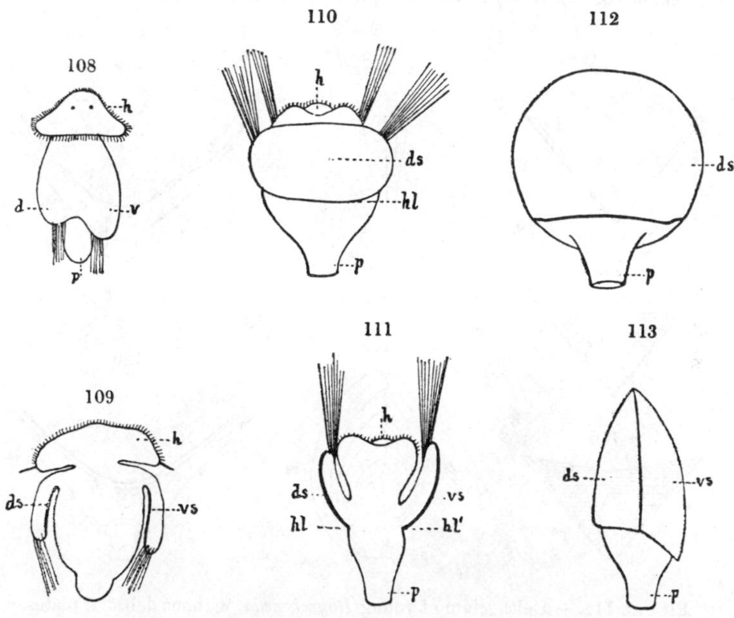

Cistella neapolitana Scacchi.

FIGURE 108. — Lateral view of completed cephalula stage. *h,* head ; *d,* dorsal
lobe of mantle ; *v,* ventral lobe; *p,* pedicle. (After Shipley.)

FIGURE 109. — Dorso-ventral longitudinal section of same : showing posteriorly extended mantle lobes. *h,* head ; *ds* and *vs,* inner surfaces of mantle lobes
which are to form dorsal and ventral valves. (After Shipley.)

FIGURE 110. — Typembryo; dorsal view of larva after transformation. *h,*
head ; *ds,* dorsal valve ; *hl,* hinge-line of dorsal valve ; *p,* pedicle. (After
Kovalevski.)

FIGURE 111. — Dorso-ventral longitudinal section based on preceding ;
showing mantle lobes directed forward, bringing interior shell-secreting surfaces,
ds and *vs* of figure 109, on the exterior. *h,* head ; *ds,* dorsal valve ; *hl,* dorsal
hinge ; *vs,* ventral valve ; *hl',* ventral hinge ; *p,* pedicle.

FIGURE 112. — Dorsal view of early nepionic shell; showing large posterior
opening between valves. (After Kovalevski.)

FIGURE 113. — Profile of same. *ds,* dorsal valve ; *vs,* ventral valve ; *p,*
pedicle.

condition represents the extent of the development of these
parts in *Meristina rectirostris* Hall or *Gwynia capsula* Jef-

freys, which lack deltidial plates in the adult shell. The young of other telotremate species, as *Magellania flavescens* or *Terebratulina septentrionalis*, agree in the same respect.

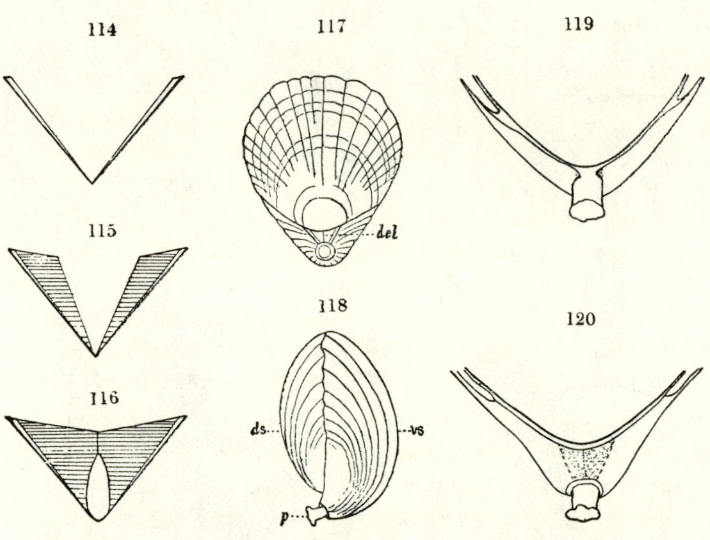

FIGURE 114. — Delthyrium of young *Rhynchonella*, without deltidial plates.

FIGURE 115. — The same at a later stage, with two triangular deltidial plates.

FIGURE 116. — The same after completed growth; showing joining of deltidial plates, and limitation of pedicle-opening to ventral beak.

FIGURE 117. — Dorsal view of *Magellania flavescens*; showing completed deltidial plates, *del.*

FIGURE 118. — The same; profile. *ds*, dorsal valve; *vs*, ventral valve; *p*, pedicle.

FIGURE 119. — Dorsal view of umbonal portion of adult *Terebratulina septentrionalis*, with shell removed by acid; showing slight secondary extension of ventral mantle around pedicle (consequently small deltidial plates are secreted in this species). Mantle areas secreting deltidial plates are shaded.

FIGURE 120. — Dorsal view of umbonal portion of *Magellania flavescens*, with the shell removed by acid; showing the complete envelopment of base of pedicle by secondary expansions from ventral mantle, and consequent production of deltidial plates filling delthyrium except at pedicle-opening. See figure 117.

An examination of the animal at this stage shows that the mantle lobes line only the interior of the valves proper. The exposed edges of the mantle are around the peripheries of the valves and also that portion of the ventral mantle border limiting the deltidial opening and passing along the sides of the pedicle at its base. The ventral mantle gradually extends from each side as two prolongations partially covering the opening and enveloping the proximal portion of the pedicle. As this is an extension of the shell-secreting surface of the mantle, there naturally results the formation of two plates within the deltidial area. Their structure is commonly punctate whenever the valves are punctate.

These outgrowths or extensions of the mantle into the deltidial area finally touch and coalesce until, as in *M. flavescens,* the pedicle emerges through an opening in the ventral mantle, and *pari passu* the deltidial plates unite and limit the pedicle-opening to the beak of the ventral valve. The latter process has been carefully described by Deslongchamps,[6] Clarke, and the writer,[3] and need not be dwelt on here. Figures 119 and 120 of the beaks of *T. septentrionalis* and *M. flavescens* with the shell removed show the relations of the ventral mantle to the pedicle, and the portions which secrete the deltidial plates.

The deltidium and delthyrium are often simulated in the growth of the *dorsal* valve in genera having a high cardinal area in this valve. *Orthis, Leptœna, Clitambonites, Spirifer,* and *Stricklandinia* may be cited as examples. They cannot be properly correlated with similar parts in the ventral valve, for their origin is quite different. Primarily, a deltidial opening is for the extrusion of the pedicle, and this belongs properly to the ventral valve. The dorsal fissure is the space between the diverging teeth sockets, and may be filled by the cardinal process, as in *Leptœna* and *Orthis,* or it may have in addition a convex plate or chilidium covering it, as in *Clitambonites.* In *Spirifer* and *Stricklandinia* the opening remains unclosed.

The true deltidial plates are formed on the side of the pedicle adjacent to the hinge by extensions of the ventral mantle lobe, and begin as two plates. They are likewise expressive of maturity, and are of secondary development, while the deltidium begins as a single plate in the median line, and is eminently a primitive character in the Protremata.

From present knowledge of the group it is difficult to offer an explanation for the presence of an anal opening in the Inarticulata and its absence in the recent Articulata, as the solution of the question depends upon whether the class is to be considered as progressive or degraded. The dorsal beaks of *Amphigenia, Athyris, Cleiothyris, Atrypa*, and *Rhynchonella* are usually notched or perforate. The perforation comes from the union of the crural plates above the floor of the beak leaving a passage through to the apex. A similar opening occurs between the cardinal processes in *Strophomena, Stropheodonta*, and allied genera, and the chilidium may also be furrowed, as in *Leptæna rhomboidalis*. This character is evidently in no way connected with the pedicle-opening, but points to the existence, in the early articulate genera, of an anal opening dorsal to the axial line, as in the recent *Crania*. This dorsal foramen was described and figured by King[13] in 1850, Hall[8] in 1860, and by several authors since, and has commonly been termed a visceral foramen.

Œhlert[21] suggests that it was probably occupied by the terminal portion of the intestine. The persistence of the foramen seems to indicate an anal opening. In reference to this character and the obsolescence of the eyes the class must be viewed as retrogressive since Paleozoic time. Other features, however, are manifestly progressive; namely, the gradual shortening, through time, of the posterior elements of the animal, as the pedicle, visceral portions, and internal shell structures, and the expansion of the anterior parts, as the shell and brachia.

A further advance in specialization is shown in the limitation of the pedicle-opening wholly to the ventral valve in the higher rhynchonelloids, athyroids, spiriferoids, and terebratuloids. The absence of punctæ in all the early radicles and their subsequent development in the derived types may also have a similar bearing.

The features and importance of the protegulum have previously been discussed.[1] It is merely noticed here as the embryonic shell of the completed phylembryonic period, for it is the first stage which can be observed among the fossil species, and is the initial point for the discussions of the relations and affinities of recent and fossil forms. Of the protegulum and later stages, there is abundant material available in nearly every family of brachiopods, ranging through their entire geological history.

Post-embryonic Stages.

In discussing the post-embryonic stages of growth two aspects of development must be clearly differentiated; (*a*) the ontogenetical, and (*b*) the phylogenetical. The ontogeny of a form like *Schizocrania* may be conveniently divided into the nepionic, neanic, and ephebic periods, and such stages may be clearly defined. The ephebic stage of *Schizocrania*, however, is like a neanic stage of *Orbiculoidea*. In other words, *Orbiculoidea*, in its development, passes through a *Schizocrania*-like stage before reaching maturity.* These facts must be viewed from a phylogenetic standpoint. Moreover, in the geological history of a group, certain ephebic characters of early species may become accelerated, and pass into the neanic period of later forms, while other characters remain ephebic. *Discinisca* offers an illustration of this. Its neanic characters agree with *Orbiculoidea* in the form of the valves and in the pedicle-notch, but the circular or elliptical form of the dorsal valve in adult and neanic *Orbiculoidea* appears so early in *Discinisca* that it marks all the nepionic stages. The interpretation of these facts is, of course, very evident, and will be subsequently given in detail. Attention is here called to the statement, that while nepionic, neanic, and ephebic stages represent equal intervals

---

* Attention was called to this fact in a publication preliminary to vol. viii of the *Palæontology of New York*, pp. 131, 132, issued February, 1890. Also, the development of the pedicle-opening in *Orbiculoidea* was fully described.

in the life of each individual, they do not represent conditions of growth, or the possession of characters which always agree, stage for stage, in the species of one family or of different families.

Other distinctions to be made whenever possible are (*a*) whether certain characters (natural or acquired) belong to a species by inheritance, or (*b*) are mere adaptations to special conditions of environment arising at any time in its history. A clear understanding of the first will lead to the true phylogeny of a species or genus, but to reach this the characters of the second category must be excluded. Thus in the series of *Schizocrania*, *Orbiculoidea*, and *Discinisca*, already cited, there is an apparent genetic connection in the facts as stated. The contrary must be the case with shells like *Lingula complanata* Williams and *L. riciniformis* Hall, which initiate a holoperipheral* mode of growth in the ephebic period, for this agreement in the method of concrescence with adult *Orbiculoidea* here appears in the mature stages of this species, and being absent in the early members of the genus cannot therefore be an ancestral character. It is a morphological equivalent, which may or may not be continued in the later species of the series.

Whenever features are present which can be referred to an ancestral origin, their elimination can take place only by the process of acceleration of development. On the other hand, there may be secondary characters of dynamical or homoplastic origin which appear simultaneously or independently in different groups belonging to diverse genetic lines, as the deltidial plates of the Rhynchonellidæ, Terebratulidæ, and Spiriferidæ. Further, many such secondary features may occur anywhere in the geological history of the group, as the high hinge-area of *Orthisina*, *Spirifer*, *Syringothyris*, and *Thecidium*. These statements are in full accord with what Hyatt has determined in the Cephalopoda, and the application of such ideas affords a fertile field of research.

---

* ὅλος whole, and περιφέρεια circumference.

Preliminary to a study of the stages of growth observed in the different orders, a simple characteristic example of each will be taken to show the limitations of the post-embryonic periods.

Nepionic Period. — In brachiopods, as in pelecypods, this period represents the growth of the true shell immediately succeeding the embryonic shell or protegulum, and before the appearance of definite specific characters. In general, the nepionic shells of all groups are marked only by fine concentric lines of growth, and are therefore nearly smooth. Sometimes, however, a few radiating striæ or other ornaments may appear over the nepionic portion, but this is not the prevailing rule. *Obolus pulcher* Matthew shows a cancellated nepionic-stage and is one of the most striking exceptional examples.

Plate XII, figure 1, represents the nepionic stage of *Glottidia albida*, drawn from the beak of a well-preserved adult. The shell at this period had a short straight hinge (originally the hinge of the protegulum), with lines representing anterior and lateral growth, making the outline broadly ovate. It is divided from the succeeding growth of later stages by a strong varix. The form is suggestive of *Obolella*, and as this is the early form of growth of many of the Lingulidæ and allied families, it is here called the *Obolella stage*. It is not known that otherwise the characters agree with those of *Obolella*, but as it is characteristic as well as descriptive the name is used to designate this form of nepionic growth whenever present.

The nepionic stage of *Orbiculoidea minuta* (figure 4) shows a continuance of the straight-hinged condition after the completion of the embryonic shell, with nearly equal incremental lines. As this agrees with the shell of *Paterina* [= *Iphidea*] it is called the *Paterina stage*. The pedicle emerged freely between the cardinal margins of the valves. It will be shown that both this and the *Obolella* stage are represented in the nepionic periods of many genera belonging to the Atremata. They may succeed each other in a single species or one alone may be present. In case both appear, the *Paterina* stage is always the first one to be developed.

The nepionic stage of *Leptœna rhomboidalis* (figure 7, Plate XII) is represented by a shell without radii, having a comparatively large pedicle-opening in the ventral valve and a large deltidium. The hinge is not well defined and the shell is discinoid in form. This term is not used to suggest any special affinities with true discinoid genera, as *Orbiculoidea* or *Discinisca*. The proper name for this stage is not yet apparent to the writer. The external characters as expressed by both valves are manifestly nearer to *Kutorgina* than to any telotremate genus. Until the early forms belonging to the articulate brachiopods, especially to the orthoid and strophomenoid groups, have been thoroughly studied, the interpretation of the nepionic *Leptœna rhomboidalis* may be uncertain. It should be noted, however, that the young of *Chonetes*, *Productus*, *Stropheodonta*, *Orthothetes*, *Leptœna*, *Plectambonites*, and *Strophomena*, all have little or no indication of a straight hinge-line, and that the extension of this member takes place during later neanic and ephebic growth. This in itself is significant, but is more marked when taken with the growth-stages shown by some species of *Strophomena* which have after the protegulum a *Paterina*-like stage, with a straight hinge in the dorsal valve, succeeded by holoperipheral, discinoid, nepionic growth, and finally a renewal of a straight-hinged condition. Thus it has an early straight-hinged form, which is lost during the next stage of growth, and again appears, and is progressively elongated during neanic and ephebic growth.

The nepionic stages of *Terebratulina septentrionalis* (figure 10, Plate XII) represent a decreasing extension of the cardinal line from the protegulum, an open delthyrium, the absence of radii, and the introduction of the shell punctæ. The crura at this stage, as shown by Morse, are short and stout, and the loop is undeveloped.

Neanic Period. — During the progress of this period all

the features which reach their complete growth in the adult organism are introduced and progressively developed. Usually they appear in succession, and gradually assume mature conditions. Thus in many species with radiate plications or striæ, a few radii appear in early neanic growth, and are added to until the full number is present. Species with deltidial plates develop them in this period. The early stages may offer many points for comparison with the adult, but later stages usually differ little except in size. Figures 2, 5, 8, and 11, Plate XII, represent a neanic stage in each of the four species taken as examples. Others from the same species could be given, but these suffice to show that one or more characteristic adult features have made their appearance.

Ephebic Period. — The period of complete normal growth, or the maximum of individual perfection. This corresponds to the adult, or mature organism, and is so well understood that no further explanation is necessary. For the sake of completing the series, the ephebic shells of the species given are represented in figures 3, 6, 9, and 12, Plate XII.

Gerontic Period. — The variations due to old age may be numerous and complex. As shown by Clarke and the writer,[3] the valves generally become thickened, and, as a consequence, the margins are truncate or varicose, the vertical diameter of the shell is increased, the beaks involuted, and the margins of the valves often lose the ornamentation characteristic of the species. The deltidial plates or deltidium may be resorbed as well as the beaks of the valves. Usually the ephebic characters disappear in inverse order to their introduction. Thus in a normal adult brachiopod having a plicate shell and deltidial plates, which characters were introduced during the neanic period, the expression of old age will be found in the absorption of the deltidial plates and in the obsolescence of the plications. Large specimens of *Terebratella transversa* Sowerby often furnish examples of this condition.

The gerontic development of *Bilobites*[2] consists in the obsolescence, in *B. varicus* Conrad, of the bilobed form of the shell, thus reverting to an early neanic condition equally characteristic of *B. bilobus* and *B. Verneuilianus.*

Another aspect of growth and decline is manifest when the size of individuals and the chronological history of groups are taken into consideration. Each genus and family began with small representatives, and rapidly developed the more radical varieties of structure. Then came the culmination and final reduction in size, with abundance of gerontic and pathologic forms. The oldest known shell with calcareous spires, *Zygospira*, is a comparatively minute form. Nearly all the types of the sub-order to which this genus belongs (Helicopegmata) appear in the Upper Silurian. Species presenting the maximum size belong to the Devonian and Carboniferous. Before the extinction of the sub-order in the Trias, the individuals are small, and such abnormal genera as *Thecospira, Koninckina,* and *Amphiclina* abound. *Productus* begins with small species (*Productella*) in the Lower Devonian, and in the Carboniferous attains the largest dimensions of any known brachiopod (*P. giganteus*). During the Permian the species have dwindled in size, and the gerontic *Strophalosia* and *Aulosteges* are the chief representatives.

The culmination of gerontic growth results in the reversion of the animal to its own nepionic period, and is called the *paragerontic stage.* As this is an extreme condition, it can be found only in certain genera and species which have been developed by a process of accelerated gerontic heredity. If *Gwynia* * is accepted as a valid genus, it belongs to a pronounced paragerontic type. The shell has a small internal plate on each side of the dorsal umbo, evidently the bases of crural plates. King,[14] the author of the genus, states that

---

* Some authors have been disposed to consider this form as the young of a species not yet determined. It has also been referred to *Macandrevia cranium, Cistella cistellula,* and *C. neapolitana.* This question cannot be at present determined, although some characters of the shell indicate a mature organism.

the labial appendages are attached directly to the shell, and not to a loop, as in other genera of the family. *Cistella* may be taken as a representative of paragerontic development among the terebratuloids. The species are smooth or pauciplicate, and small; deltidial plates obsolescent, loop more or less undeveloped. In *C. neapolitana* the lamellæ of the loop are nearly obsolete and are free only near the crura, while the anterior portions are confluent with the valve (Shipley). A slight progression of these reversions would naturally result in a degenerate form like *Gwynia*, which is without a calcareous loop; with no surface ornamentation; deltidial plates absent; punctæ few and large, all of which features are strictly nepionic. Besides *Cistella* and *Gwynia*, other loop-bearing genera present paragerontic features of importance in a natural classification. These consist mainly in their small size; the absence of surface ornaments; the obsolescence of deltidial plates, and the loss of a complete loop supporting the arms. In the Terebratulidæ *Kraussina* and *Platydia* may be mentioned as belonging to gerontic types with a paragerontic tendency. Likewise, in other groups, *Atretia* in the Rhynchonellidæ, and *Strophalosia* and *Aulosteges* in the Productidæ, are examples of paragerontic types.

Cistella and *Gwynia* among the genera of brachiopods, therefore, bear the same relation to the terebratuloids that *Baculites* among the cephalopods bears to the ammonoids.

Synopsis.

Protembryo. — Ovum and segmented stages before formation of blastula cavity.

Mesembryo. — Blastosphere.

Metembryo. — Gastrula.

Neoembryo. — Trochosphere and cephalula, with posteriorly directed mantle lobes, and bundles of setæ from body segment.

Typembryo. — Larva with mantle lobes folded anteriorly over head segment.

Phylembryo. — Bachiopod covered by protegulum, tentacles of arms developed, bundles of setæ dehisced, definition of stomach and œsophagus, direct transformation of larval muscles into those corresponding to muscles of adult animal.

Deltidium. — A single plate developed at an early period by the body and pedicle of animal posterior to dorsal hinge, and later ankylosed to ventral valve.

Deltidial Plates. — A neanic and adult feature produced by the extensions of the ventral mantle lobe into the delthyrium.

Brachiopoda. — Retrogressive in loss of anal opening and eyes, progressive in concentration of posterior elements, expansion of anterior elements, and limitation of pedicle-opening to one valve.

Nepionic Period. — Young shells before the appearance of distinctive specific characters.

Neanic Period. — Progressive development of the specific features which reach their complete growth in the adult.

Ephebic Period. — Normal adult condition.

Gerontic Period. — Special manifestations of old age in ontogeny and in phylogeny.

Paragerontic Types. — Extremes of geratology represented by *Cistella*, *Gwynia*, and *Atretia*.

References.

1. Beecher, C. E., 1891. — Development of the Brachiopoda. Part I. Introduction. *Amer. Jour. Sci.* (3), vol. xli, April.

2. —— 1891. — Development of Bilobites. *Amer. Jour. Sci.* (3), vol. xlii, July.

3. —— and Clarke, J. M., 1889. — The Development of some Silurian Brachiopoda. *Mem. N. Y. State Mus.*, vol. i. No. 1.

4. Brooks, W. K., 1879. — The Development of Lingula and the Systematic Position of the Brachiopoda. *Johns Hopkins Univ., Chesapeake Zoöl. Lab., Sci. Results*, Session of 1878.

5. Davidson, T., 1851–1885. — A Monograph of the British Fossil Brachiopoda. *Pal. Soc.*, London.

6. Deslongchamps, E., 1862. — Note sur le développement du deltidium chez les brachiopodes articulés. *Bull. Soc. Géol. France* (2), t. xix.

7. Fewkes, J. W., 1885. — On the Larval Forms of Spirorbis borealis, Daudin. *American Naturalist,* March.

8. Hall, James, 1860. — *Palæontology of New York,* vol. iii.

9. Hyatt, A., 1888. — Values in Classification of the Stages of Growth and Decline, with Propositions for a New Nomenclature. *Proc. Boston Soc. Nat. Hist.,* vol. xxiii, March.

10. —— 1889. — Genesis of the Arietidæ. *Mem. Mus. Comp. Zoöl.,* vol. xvi, No. 3.

11. Jackson, R. T., 1890. — Phylogeny of the Pelecypoda. The Aviculidæ and their Allies. *Mem. Boston Soc. Nat. Hist.,* vol. iv, No. viii.

12. Joubin, L., 1886. — Recherches sur l'Anatomie des Brachiopodes Inarticulés. *Archiv Zool. Expérimentale* (2), t. iv.

13. King, W., 1850. — A Monograph of the Permian Fossils of England. *Pal. Soc.,* London.

14. King, W., 1859. — On Gwynia, Dielasma, and Macandrevia, three new genera of Palliobranchiate Mollusca, one of which has been dredged in Belfast Lough. *Proc. Dublin Univ., Zoöl. Bot. Assoc.,* vol. i.

15. Kovalevski, A. O., 1874. — Observations on the Development of Brachiopoda. *Proc. Imp. Soc. Amateur Naturalists,* etc., held at the University of Moscow, 11th year, vol. xiv.

16. Lacaze-Duthiers, H., 1861. — Histoire naturelle des Brachiopodes vivants de la Mediterranée. *Ann. Sci. Nat. Zool.,* t. xv.

17. Morse, E. S., 1873. — On the Early Stages of Terebratulina septentrionalis (Couthouy). *Mem. Boston Soc. Nat. Hist.,* vol. ii.

18. —— 1873. — Embryology of Terebratulina. *Mem. Boston Soc. Nat. Hist.,* vol. ii.

19. —— 1873. — On the Systematic Position of the Brachiopoda. *Proc. Boston Soc. Nat. Hist.,* vol. xv.

20. Müller, F., 1860. — Beschreibung einer Brachiopodenlarve. *Archiv Anat. Physiol.,* Jahrg. 1860.

21. Œhlert, D. P., 1887. — Brachiopodes. *Manuel de Conchyliologie,* Paul Fischer. Appendice.

22. Shipley, A. E., 1883. — On the Structure and Development of Argiope. *Mittheil. Zool. Station Neapel,* Bd. IV.

23. Walcott, C. D., 1888. — A Fossil Lingula preserving the Cast of the Peduncle. *Proc. U. S. Nat. Mus.*

PART III.   MORPHOLOGY OF THE BRACHIA[*]

THE diagnostic value of the brachidium, or calcareous arm supports, of brachiopods has long been recognized, and forms one of the chief characters for generic and family sub-division among the Terebratulacea and Spiriferacea.   This character fails in all other brachiopods, which have simply fleshy arms, unsupported by calcareous skeletons.   There is, however, generally the most obvious analogy and intimate relationship between the arms themselves and the brachidium, so that whenever either structure can be ascertained it furnishes important data aiding in the determination of the systematic position of any genus within a family or order.

The growth of the arms, or lophophore, in recent genera may be divided into distinct stages, which often have a direct correlation with other important features of the shell.   In many cases it is also possible to infer the form and arrangement of the brachia in fossil genera from markings on the interior of the valves and from the calcareous arm supports, and thus to obtain the chronogenetic as well as the morphogenetic history of these organs.

The most detailed accounts of arm development are given by Brooks [5][†] for *Glottidia*, by Morse [11] for *Terebratulina*, and by Kovalevski [10] for *Cistella* and *Thecidea*.   These results, combined with original observations by the writer [1,2] and occasional descriptions of arm structure by Davidson [7] and other authors, are sufficient to include and to interpret properly all the leading varieties of structure.

As shown by Brooks [5] the tentacles, or cirri, in *Glottidia* originate on the dorsal side of the oral disk.   They grow in pairs, one on each side of a central lobe.   New tentacles are added between the first pair formed and the median lobe.

---

[*] *Bulletin* 87, *U. S. Geol. Surv.*, Chapter IV, 105–112, 1897.

[†] The references to the literature will be found at the end of this chapter.

Thus the cirri farthest removed from the median lobe are the oldest. Tentacles are added rapidly until the first arc is extended to a semi-circle, and then progressively the whole disk becomes surrounded by a circle of these organs. The further introduction of cirri can take place only by the enlargement of the oral disk or through the deformation of the circle by lobes, loops, or extensions. In *Glottidia, Lingula, Discinisca, Crania,* and *Rhynchonella* the two points of tentacular increase, originally together and on opposite sides of a median lobe, or tentacle, gradually separate, and the further multiplication of tentacles results in strap-shaped extensions on each side, which finally assume a coiled form, due to the limited space in which they grow. Therefore the arms in adult individuals of these genera have a single cirrated edge, extending from their free extremities to the sides of the oral disk, and, continuing posteriorly, unite on the ventral side of the disk behind the mouth. Each cirrated edge in the adult lophophore apparently has two approximate rows of alternating cirri (Hancock[9]), but as they were originally a single row in early stages, this appearance is evidently the result of a crowding of the cirri or a crumpling of the edge.

Kovalevski[10] has shown that in *Cistella* the tentacles also originate in pairs on each side of the dorso-median line, without a central tentacle or lobe. The same mode of increase has been shown by the writer[2] to be present in *Magellania* and *Terebratalia*. In young stages of *Cistella, Terebratulina, Magellania,* and other terebratuloid genera, as well as in *Thecidea,* after the circlet of tentacles is complete the two points at which new ones are added do not separate, but remain close together throughout the life of the animal. In this case the cirrated margin is lengthened by means of lobation and looping, and often by the final growth of a single, median, coiled arm, cirrated on both margins. *Gwynia* illustrates the completed circle of tentacles about the mouth. Adult *Cistella* shows an advance in having the anterior margin of the lophophore introverted, making it bilobed. *Megathyris* is slightly more complicated by two additional lobes. This simple method of increase is further elaborated in the Thecidiidæ. In the higher genera, especially among the Terebratulidæ, the maximum is reached by means of a median, unpaired, coiled arm, as in *Magellania* and *Terebratulina*.

The development of the different types and varieties of arm structure is presented in the accompanying figures (121–125), which are necessarily somewhat diagrammatic in order to show the features clearly, but the essential structure can be readily verified from consultation of the works cited or from a study of actual specimens. In the case of fossil forms, such as *Dielasma*, the Atrypidæ, and Athyridæ, the brachial supports have sufficient analogy with the arm structures of *Terebratulina* and *Rhynchonella* to warrant their interpretation as given. Also, the spiral impressions on the valves of *Davidsonia*, and those occasionally present in *Leptæna* and *Productus*, clearly point to the possession of coiled arms by these genera.

Classification of Brachial Structures.

From what has already been shown it is seen that the various types of lophophore admit of a simple classification into stages and groups. It is proposed to give to these distinctive names, which may be used with facility in making comparisons and correlations. They may be found useful, also, in designating the kind of brachial complexity attained in any genus the arm structure of which can be determined, thus helping to fix its place in a genetic scale. It should be emphasized, however, that the form and complexity of the cirrated margin of the lophophore can have a taxonomic value only within comparatively narrow limits. This at once becomes evident when the arms of *Lingula*, *Discinisca*, *Crania*, *Rhynchonella*, and all the Spiriferacea are considered. Each has spiral arms, which were probably developed through similar changes of form, and yet each is genetically distinct, as shown by all the other leading characters.   But when this classification of arm structures is applied within a family or genus, or even when made the basis of comparison among some closely related families, it is sometimes possible to reach very satisfactory conclusions relating to the systematic position of various forms.

Leiolophus Stage.

It is hardly necessary to direct attention to the embryonic brachial structure before the growth of any of the tentacles, or cirri, on the edge of the lophophore, while the animal is in the typembryonic stage.   For the sake of designating all the stages, this may be called the *leiolophus stage*, though it has no special significance beyond indicating the beginning of the lophophore.

Taxolophus Stage.

The first stage in which a true brachial structure is manifest is an early larval form, often the protegulum stage, when the tentacular portion of the lophophore is a simple arc or crescent.   This may be called the *taxolophus*.   The tentacles are few in number, and increase takes place on each side of the median line, dorsally, in front of the mouth. In figures 121, *a*, *e*, 122, *a*, *f*, 124, *a*, this character is clearly shown.   The tentacles at the ends of the arc are the oldest, and new ones are being formed in the middle portion. In *Thecidea*, *Cistella*, and *Magellania* the tentacles of the taxolophus are centripetal, due to the edge of the lophophore being near the margin of the shell; while in *Terebratulina*, *Discinisca*, and *Lingula* they are centrifugal, due to the smaller and central lophophore.

So far as known, there is no adult living form which has the taxolophian brachial structure.   It may have been present in adult *Iphidea* of the Cambrian.

Trocholophus Stage.

By the continual addition of new cirri and the pushing back of the old ones, the fringed margin of the lophophore passes from a crescentic to a circular form, thus making a complete ring about the mouth. This may be termed the *trocholophus stage*. It appears in the late larval and early adolescent stages of *Thecidea* (figure 121, *b*), *Cistella* (figure 121, *f*), *Magellania* and *Terabratalia* (figure 122, *b*), *Terebratulina* (figure 122, *g*), *Glottidia* (figure 124, *b*), and *Discinisca*, and, like the former stages, is undoubtedly common to all brachiopods, except, perhaps, *Iphidea.*

Gwynia is an adult living representative of this stage, and never develops any higher type of brachial structure. *Dyscolia* also belongs here, since it has a discoid lophophore surrounded by a marginal fringe of tentacles (Fischer and Œhlert[8]). It is possibly a little more advanced than *Gwynia*, as it has a slight median anterior notch, suggesting the beginning of the bilobed structure of the next higher type.

The absence of septum, hinge-plate, and dental plates are other primitive characters belonging to *Dyscolia.*

Schizolophus Stage.

After the completion of the trocholophus stage in all brachiopods, except such simple forms as *Gwynia* and *Dyscolia*, no further increase in the cirrated edge of the lophophore can occur without some deformation of the circle. This is first accomplished by an introversion of the anterior median edge, thus dividing the lophophore into two lobes, and suggesting the name *schizolophus* for this type. (See figures 121, *c*, *g*, 122, *c*, *h*, 124, *c*.)

Several brachiopods retain the schizolophian brachia as an adult character. Of these, *Cistella* is perhaps the best example, as it agrees exactly with an early stage of arm structure among the Terebratellidæ, which has been called the cistelliform stage (figure 122, *c*). *Terebratulina* (figure

122, *h*), *Glottidia* (figure 124, *c*), and other higher forms also have corresponding schizolophian stages, but are without the median septum. *Lacazella mediterranea* presents a similar larval structure, and in *L. Barretti* it is retained to maturity. The fossil genera *Davidsonella* and *Thecidella* of the Thecidiidæ, and *Zellania* of the Terebratellidæ, never developed beyond the schizolophus-stage, and they must therefore be considered as quite primitive genera in their respective families.

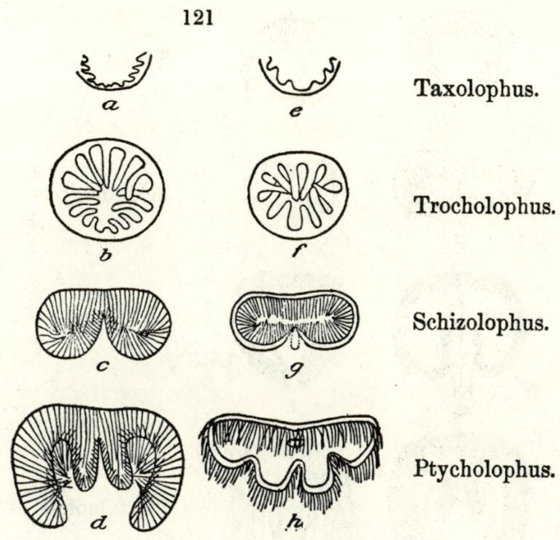

FIGURE 121.— Stages of growth of the lophophore in *Thecidea*, *Cistella*, and *Megathyris*. *a, b, c, d*, stages in the growth of the lophophore in *Thecidea* (*Lacazella*) *mediterranea*. Enlarged. (*a-c*, after Kovalevski; *d*, after Lacaze-Duthiers.) *e, f*, early stages of lophophore of *Cistella neapolitana*. Enlarged. (After Kovalevski.) *g*, adult lophophore of *Cistella* (*C. cistellula*). Enlarged. (After Davidson.) *h*, labial appendages of *Megathyris decollata*. Enlarged. (After Davidson.)

From this point the further development and complication of arm structure proceeds in three distinct diverging lines, producing the three characteristic types of brachia of all the higher brachiopods, as exemplified in *Thecidea*, *Terebratulina*, and *Rhynchonella*.

Ptycholophus Stage.

The simplest of the types of brachia just cited is developed out of the schizolophus by the additional lobation, or looping, of the primary lobes, making a structure which may be

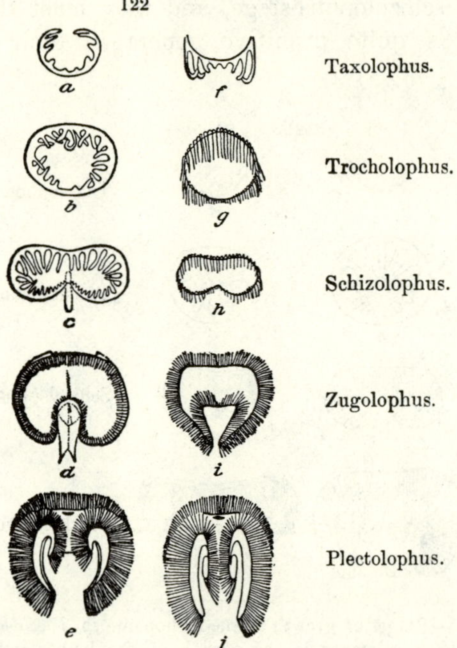

FIGURE 122. — Stages of growth of the lophophore in the Terebratellidæ and Terebratulidæ. *a, b, c, d, e,* five stages in the development of the lophophore in the Terebratellidæ. *a–d, Terebratalia obsoleta.* Enlarged. (After Beecher.[2]) *e, Magellania kerguelenensis.* Natural size. (After Davidson.[7]) *f, g, h, i, j,* development of lophophore in the Terebratulidæ. *f–i,* early stages in *Terebratulina septentrionalis.* Enlarged. (After Morse.[11]) *j,* adult *Terebratulina cancellata.* (After Davidson.[7])

called the *ptycholophus.* *Megathyris* and *Lacazella mediterranea* both have four lobes (figure 2, *d, h*); *Thecidea radiata* has six; *T. vermicularis* and *Eudesella mayale,* eight; *E. digitata,* ten; *Pterophloios* and *Oldhamina,* about twenty. Lobation in some (*Thecidea*) is produced by the forking or branching of the median septum; in others (*Pterophloios*) the septum remains simple, while the lateral borders of the lophophore are lobed.

Zugolophus and Plectolophus Stages.

All the higher Terebratulacea reach the final growth of the lophophore through an intermediate stage which from its form may be called the *zugolophus* (figure 122, *d, i*). *Eucalathis* and *Platidia* (? *Tropidoleptus*) are apparently adult representatives of this stage, while *Kraussina* and probably *Bouchardia* are slightly more advanced by the growth of a short median, coiled arm, and lead to the next higher, or *plectolophus*, stage, in which there is a well-developed spiral arm with a fringe of cirri on each edge (figure 122, *e, j*).

A long loop pointed in front, like *Rensselæria* and *Centronella*, could not have supported a median arm, as the pallial cavity is thus fully occupied, and the development of the brachidium in the Terebratellidæ shows that the central space between the branches of the loop is to accommodate such an organ. The same is doubtless true of *Dielasma*, which first has a *Centronella*-like loop, and through the subsequent resorption of the anterior portion the ascending branches are formed and space allowed for the median arm (figure 123, *a–d*). In a spire-bearing genus like *Zygospira* this is more obvious, for here the transverse process or jugum is clearly the result of the growth and resorption of the centronelliform loop to admit the spiralia.

123

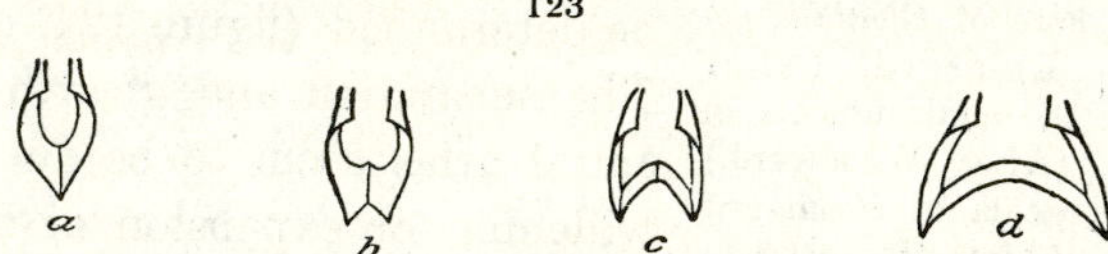

FIGURE 123. — Metamorphoses of the brachidium in *Dielasma turgidum*. Enlarged. (After Beecher and Schuchert.)

The calcareous loop in *Terebratulina* and *Liothyrina* is only a posterior basal support, and does not repeat the outline of the cirrated margin of the lophophore, exclusive of the arm. Therefore it is impossible in these and closely allied genera to infer the stage of development of the lophophore from the loop alone. *Dyscolia* is an excellent example, since the loop is the same as in *Terebratulina*; but the lophophores are quite distinct in each, the former being of the trocholophus type and the latter belonging to the plectolophus.

124

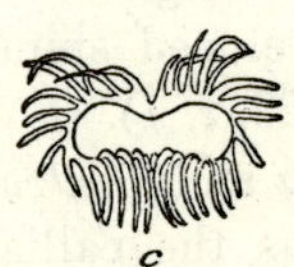

Taxolophus.

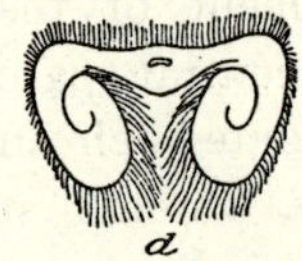

Trocholophus.

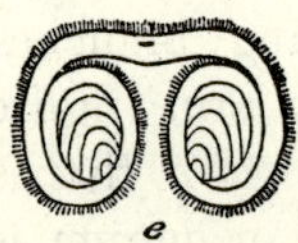

Schizolophus.

Spirolophus.

FIGURE 124. — Early stages of lophophore of *Glottidia*, and adult brachia in *Lingula* and *Hemithyris*. *a, b, c*, early stages of lophophore of *Glottidia pyramidata*. Enlarged. (After Brooks.) *d*, adult brachia in *Lingula*. (After Woodward.) *e*, adult brachia in *Hemithyris psittacea*. (After Hancock.)

Spirolophus Stage.

The last type to be noticed is the one in which there are two separate coiled arms, each with a row of cirri on one edge only (figure 124, *d, e*). It embraces the greater part of the families of brachiopods in the orders Telotremata and Protremata, and includes all the living species in the orders Atremata and Neotremata.

In the early stages of development of the spiral lophophore there is an agreement with the early stages of the families already noticed, and the taxolophus, trocholophus, and schizolophus stages may be determined (figure 124, *a, b, c*). The separation and growth of the spiral arms seem to be due to the widening or expansion of the median lobe or tentacle, on each side of which is the formative tissue for new cirri. This is very apparent in the young *Discinisca* described by Müller,[12] and the *Glottidia* described by Brooks.[5]

The brachidium in *Zygospira* passes through a series of

changes which have been described in detail elsewhere.[4]
These metamorphoses are of great assistance in understanding the development and comparative morphology of this
feature in other groups of the Spiriferacea. The earliest
stage observed (figure 125, *a*) has the form of a simple terebratuloid loop, which, from its resemblance to *Centronella*,
was called the centronelliform stage. Since approximately
this form of brachidium is also characteristic of the young
of recent terebratuloids, it may be taken in *Zygospira* as
indicative of the trocholophus stage of brachial development.
With this as a starting-point for comparison, the further
correlation of the succeeding stages is very simple.

The first resorption of the end of the loop in *Zygospira*
produced a schizolophus condition, and further resorption
carried the brachidium to a stage closely resembling *Dielasma*
(figure 125, *b*). The dielasmatiform stage has already been
explained as due to the requirements of space for the growth
of the coiled brachia. Next, the initial calcification of
the spiral arms resulted in the extension of the descending
branches beyond the jugum (figure 125, *c*), and, lastly, complete calcification manifests the spirolophus structure and
produced the characteristic brachidium of the Spiriferacea.

The Atrypidæ and the Athyridæ seem to stand to each
other in the same relation as the Terebratellidæ and Terebratulidæ. In the first the descending branches are widely
separated and follow the edges of the valves; in the second
the descending branches are close together. This difference
in the Spiriferacea produces the converging cones of the
Atrypidæ (figure 125, *d*) and the diverging cones of the
Athyridæ, Spiriferidæ (figure 125, *e*), etc.

It seems doubtful whether the fleshy portions of the brachia
in the Meristellidæ and Athyridæ possessed additional characters expressing the complexity and elaboration reached by
the jugal processes, even when the lamellæ were duplicated,
as in *Koninckina* and *Kayseria*.

From the foregoing descriptions and illustrations it appears
that the mode of growth of the cirrated lophophore, or

brachia, is alike in the larval stages of all brachiopods. They first develop tentacles in pairs on each side of the median line in front of the mouth (taxolophus stage). New tentacles are continually added at the same points, until, by pushing back the older ones, they form a complete circle about the mouth (trocholophus stage), later becoming introverted in front (schizolophus stage). From this common and simple structure all the higher types of brachial complication are developed through one of two methods: (1) The growing

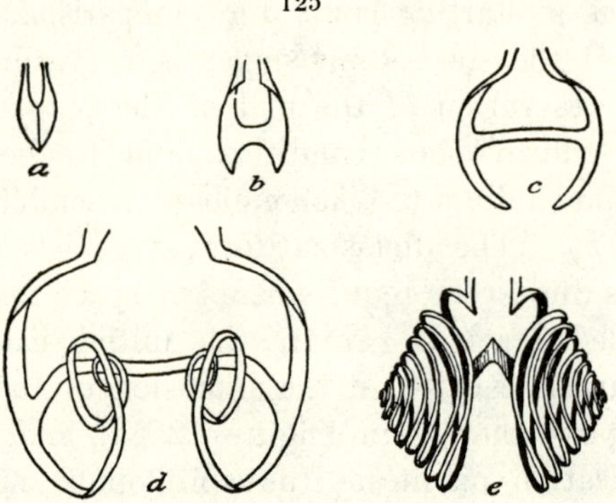

FIGURE 125. — Metamorphoses of brachidium of *Zygospira* and adult brachidium of *Rhynchospira*. *a, b, c, d*, metamorphoses of brachidium of *Zygospira recurvirostris*. Enlarged. (After Beecher and Schuchert.) *e*, brachidium of *Homœospira evax*. (After Beecher and Clarke.)

points of the lophophore, or points at which new tentacles are formed, remain in juxtaposition; or (2) they separate. Complexity in the first is produced (*a*) by lobation, as in *Megathyris, Eudesella, Pterophloios Thecidea*, etc. (ptycholophus type), and (*b*) by looping (zugolophus) and the growth of a median, unpaired coiled arm (plectolophus), as in *Magellania, Terebratulina*, etc.; in the second (*c*) by the growth of two, separate, coiled extensions or arms, one on each side of the median line (spirolophus), as in *Lingula, Crania, Discinisca, Rhynchonella, Leptœna, Davidsonia, Spirifer, Athyris, Atrypa*, etc.

References.

1. Beecher, C. E., 1893. — Revision of the families of loop-bearing Brachiopoda. *Trans. Conn. Acad. Sci.*, vol. ix.
2. —— 1893. — The development of *Terebratalia obsoleta* Dall. *Trans. Conn. Acad. Sci.*, vol. ix.
3. —— and J. M. Clarke, 1889. — The development of some Silurian Brachiopoda. *Mem. N. Y. State Mus.*, vol. i, No. 1.
4. —— and Charles Schuchert, 1893. — Development of the brachial supports in Dielasma and Zygospira. *Proc. Biol. Soc. Washington*, vol. viii.
5. Brooks, W. K., 1879. — The development of Lingula and the systematic position of the Brachiopoda. *Johns Hopkins Univ.*, *Chesapeake Zoöl. Lab.*
6. Davidson, T., 1851–1885. — A monograph of the British fossil Brachiopoda. *Pal. Soc.*, London.
7. —— 1886–1888. — A monograph of recent Brachiopoda. *Trans. Linn. Soc.*, London, vol. iv.
8. Fischer, P., and D.-P. Œhlert, 1892. — Résultats des campagnes scientifiques accomplies sur son yacht par Albert 1er, Prince Souverain de Monaco. Fascicule III. Brachiopodes de l'Atlantique Nord.
9. Hancock, A., 1858. — On the organization of the Brachiopoda. *Phil. Trans.*, vol. cxlviii.
10. Kovalevski, A. O., 1874. — Observations on the development of Brachiopoda. *Proc. Imp. Soc. Amateur Naturalists*, etc., held at the University of Moscow, 11th year, vol. xiv.
11. Morse, E. S., 1873. — On the early stages of *Terebratulina septentrionalis* (Couthouy). *Mem. Boston Soc. Nat. Hist.*, vol. ii.
12. Müller, F., 1860. — Beschreibung einer Brachiopodenlarve. *Archiv nat. Physiol.*, Jahrg. 1860.

2. SOME CORRELATIONS OF ONTOGENY AND PHYLOGENY IN THE BRACHIOPODA*

(Plate XIII)

The parallelism between the ontogeny and phylogeny in the Brachiopoda has been worked out in numerous instances.† To illustrate these, some more or less familiar genera may be taken as characteristic examples.

Lingula has been shown by Hall and Clarke (*Pal. N. Y.*, vol. viii, 1892) to have had its inception in the Ordovician. In the ontogeny of both recent and fossil forms, the first shelled stage has a straight hinge-line, nearly equal in length to the width of the shell. This stage may be correlated with the more ancient genus *Paterina* [= *Iphidea*], from the lowest Cambrian. Subsequent growth produces a form resembling *Obolella*, a Cambrian and Lower Silurian genus. Then the linguloid type of structure appears at an adolescent period, and is completed at maturity. Thus *Lingula* has ontogenetic stages corresponding to (1) *Paterina* [= *Iphidea*], (2) *Obolella*, and (3) *Lingula*, of which the first two occur as

---

* *American Naturalist*, XXVII, 599–604, pl. xv, 1893.

† C. E. Beecher. Development of the Brachiopoda. Part I. Introduction. *Amer. Jour. Sci.*, XLI, April, 1891.

—— Development of the Brachiopoda. Part II. Classification of the Stages of Growth and Decline. *Amer. Jour. Sci.*, XLIV, August, 1892.

—— Development of Bilobites. *Amer. Jour. Sci.*, XLII, July, 1891.

—— Revision of the Families of Loop-bearing Brachiopoda. *Trans. Conn. Acad. Sci.*, IX, May, 1893.

Deslongchamps, E. Études critiques sur des Brachiopodes nouveaux ou peu connus, 1884.

Fischer and Œhlert. Brachiopodes: Mission Scientifique du Cap Horn, 1882–1883. *Bull. Soc. Hist. Nat. d'Autun*, V, 1892.

adult forms in geological formations older than any known *Lingula.*

Paterina [= *Iphidea*] represents the prototype of the brachiopods. It shows no separate stages of growth in the shell, is found in the oldest fossiliferous rocks, and corresponds to the embryonic shelled condition (protegulum) of the class.

The genus *Orbiculoidea* of the Discinidæ first appears in the Ordovician and continues through the Mesozoic. The early stages in the ontogeny of an individual are, as in *Lingula*, first a *Paterina* stage, followed by an *Obolella* stage. Then, from the mechanical conditions of growth, a *Schizocrania*-like stage follows, and complete growth results in *Orbiculoidea.*

The elongate form of the shell in *Lingula* as well as in many other genera is determined by the length of the pedicle and freedom of motion. The discinoid, or discoid, form of *Orbiculoidea* and *Discinisca* among the brachiopods, and *Anomia* among pelecypods, is determined by the horizontal position of the valves, which are attached to an object of support by a more or less flexible, very short organ, — a pedicle or byssus, without calcareous cementation. This mode of growth is characteristic of all the discinoid genera, but, as already shown, the early stages of Paleozoic *Orbiculoidea* have straight hinge-lines and marginal beaks, and in the adult stages of the shell the beaks are usually sub-central and the growth holoperipheral. This adult discinoid form, which originated and was acquired through the conditions of fixation of the animals, has been accelerated in the recent *Discinisca* so that it appears in a free-swimming larval stage. Thus, a character acquired in adolescent and adult stages of Paleozoic species, through the mechanical conditions of growth, appears by acceleration in larval stages of later forms before the assumption of the condition of fixation which first produced this character.

The two chief sub-families of the Terebratellidæ undergo complicated series of metamorphoses in their brachial structure. Generic characters in this family are usually based upon the form and disposition of the brachia and their supports. The highest genera in one sub-family, which is austral in distribution, pass through stages correlated with the adult structure in the genera *Gwynia*, *Cistella*, *Bouchardia*, *Megerlina*, *Magas*, *Magasella*, and *Terebratella*, and reach their final development in *Magellania* and *Neothyris*. The higher genera in another sub-family, boreal in distribution, pass through metamorphoses correlated with the adult structures of *Gwynia*, *Cistella*, *Platidia*, *Ismenia*, *Mühlfeldtia*, *Terebratalia*, and *Dallina*. The first two stages in both sub-families are related in the same manner to *Gwynia* and *Cistella*. The subsequent stages are different except the last two, so that the *Magellania* structure is similar in all respects to the *Dallina* structure, and *Terebratella* is like *Terebratalia*. Therefore *Magellania* and *Terebratella* are respectively the exact morphological equivalents to, or are in exact parallelism with, *Dallina* and *Terebratalia*.

The stages of growth of the genera belonging to the two sub-families Dallininæ and Magellaniinæ are further correlated in the tables on page 303.

The simplest genus, *Gwynia*, as far as known, passes through no brachial metamorphoses, and has the same structure throughout the adolescent period, up to and including the mature condition. In the ontogeny of *Cistella* the *gwyniform* stage, through acceleration, has become a larval condition. In *Platidia* the *cistelliform* structure is accelerated to the immature period, and in *Ismenia* (representing an *ismeniform* type of structure in the higher genera), the *gwyniform* and *cistelliform* stages are larval, and the *platidiform* represents an adolescent condition. Similar comparisons may be made in the other genera. Progressively through each series, the adult structure of any genus forms the last immature stage of the next higher, until the highest member in its ontogeny represents serially, in its stages of growth, all the adult structures, with the larval and immature stages of the simpler genera. It is evident that in the identification of species belonging to the Terebratellidæ, whether recent or fossil, the strict specific characters must be given first consideration. Species, therefore, must be based upon surface ornaments, form and color, within certain limits, and genera only upon structural features developed through a definite series of changes, the results of which are permanent in individuals evidently fully adult.

In each line of progression in the Terebratellidæ, the acceleration of the period of reproduction, by the influence of environment, threw off genera which did not go through the complete series of metamorphoses, but are otherwise fully adult and even may show reversional tendencies due to old age; so that nearly every stage passed through by the higher genera has a fixed representative in a lower genus. Moreover the lower genera are not merely equivalent to, or in exact parallelism with, the early stages of the higher, but they express a permanent type of structure, as far as these genera are concerned, and after reaching maturity do not show a tendency to attain higher phases of development, but thicken the shell and cardinal process, absorb the deltidial plates, and exhibit all the evidences of senility.

3. REVISION OF THE FAMILIES OF LOOP-BEARING BRACHIOPODA[*]

(Plates XIV and XXIV)

The recent publications of Fischer and Œhlert,[8,9,10,†] combined with previous observations by Friele[11] and Deslongchamps,[7] furnish material which suggests a natural grouping of the terebratuloids. The present knowledge is incomplete in some details, especially as regards the fossil genera, yet enough is available to simplify the arrangement of the leading terebratuloid types, and to show their common relationships. By far the best classifications have been those proposed by Dall[3] in 1870, and by Deslongchamps[7] in 1884. Only in the light of recent discoveries is it possible to offer a new arrangement of the genera.

The sub-order Ancylobrachia, proposed by Gray[12] in 1848, includes, with some emendations, all the genera currently known as terebratuloids. Taking Gray's name for the entire group, since it has priority over Kampylopegmata, Waagen,[16] 1883, it is found to comprise two distinct types of brachial structure, each with a separate genetic history. It is here proposed to recognize these two types as of family importance, according to the interpretation of family characters given by Agassiz.[1]

The Terebratulidæ.

In the first family, the Terebratulidæ, the loop is always free and may be long or short. It is developed by the growth

[*] *Trans. Conn. Acad. Sci.*, IX, 376–391, 395–398, pls. i, ii, 1893.

[†] The works referred to by numbers are cited in full in the list appended. An excellent summary and review of Fischer and Œhlert's papers, [8,9,10] by Miss Agues Crane,[2] appeared in the January number of *Natural Science*, 1893.

of two lamellæ, or descending branches, from the points of the crura, uniting in the median line. The central portion may be narrow or medially expanded. In some genera, recurved ascending branches are produced by the partial resorption of the broad band or plate forming the connection between the descending branches. The cirri in early stages of the animal are centrifugal or directed outward. The growth of the loop in *Terebratulina* has been illustrated by Morse.[15]

Terebratula (Liothyrina) and *Terebratulina* may be selected as best representing the Terebratulidæ; for *Dyscolia*, *Agulhasia*, and *Eucalathis* do not represent the highest development of the family type, but must be regarded as degraded forms. Among fossil genera *Cryptonella*, *Megalanteris*, *Dielasma*, *Centronella*, *Rensselæria*, *Stringocephalus*, and some others, probably belong here. The following sub-families can be recognized: (1) the Centronellinæ, (2) Stringocephalinæ, (3) Terebratulinæ, and (4) Dyscoliinæ. The adult arm structure in *Dyscolia* is homologous with early larval features in *Terebratulina;* also the cirri are centrifugal or directed outward, as in early stages of *Terebratulina*, and not centripetal as in larval *Magellania*.

The Terebratellidæ.

The loop in the second family, for which the name Terebratellidæ is retained, undergoes a series of metamorphoses while attached to a dorsal septum during the larval and immature stages of the animal, and in the higher forms results in a loop of secondary growth much like the primary loop of some of the early genera of the Terebratulidæ. The cirri in larval stages of the animal are centripetal or directed inwardly.

In one division of the Terebratellidæ the stages of growth may be correlated with the adult loops in the genera *Gwynia*, *Cistella*, *Platidia*, *Ismenia*, *Mühlfeldtia*, *Terebratalia*,* and *Dallina;*† while in another division a quite different series

---

* Type *Terebratula transversa* G. B. Sowerby.
† Type *Terebratula septigera* Lovén.

of transformations takes place.   These have been termed, by Fischer and Œhlert,[8] the *præmagadiform, magadiform, magaselliform, terebratelliform,* and *magellaniform* stages, from their resemblance to the loops of the genera suggesting these names.   The *præmagadiform* stage is here divided into the *bouchardiform* and *megerliniform* stages.   To these may be added the earlier larval stages resembling *Gwynia* and *Cistella,* as in the previous group, and showing a parallel development in the first two stages.

These two groups of the Terebratellidæ usually have been considered as part of the family Terebratulidæ, although King,[14] in 1850, proposed the name Terebratellidæ to include *Terebratella, Mühlfeldtia,* and *Ismenia,* on account of the attachment of the loop to the septum of the dorsal valve in these genera.   Friele[11] and Deslongchamps[7] next showed that *Macandrevia cranium* and *Dallina septigera* passed through a series of changes in which the loop was united to a septum in all but the last stage.   This completed loop in *Macandrevia,* composed of two descending and ascending lamellæ, was believed to be homologous with the loop of *Terebratulina* and *Liothyrina,* and the family proposed by King fell into disuse. It can now be shown, however, that the loop of *Macandrevia* is made up of a primary portion corresponding to the entire loop of *Liothyrina,* and a secondary part which has no equivalent in the calcified lamellæ of *Liothyrina* or *Terebratulina,* but in them is represented in the fleshy portion of the arms, as previously recognized by Hancock.[13]

The loop in *Terebratulina* is equivalent to the descending lamellæ in *Terebratella,* from the crural points down to and including the bands connecting with the septum.   In *Magellania* and *Macandrevia* the connecting bands of *Terebratella* are represented, except in old specimens, by slight projections from the descending branches, and in these genera, therefore, the primary loop is incomplete.*   The true rela-

---

* The prongs or points below the ends of the crura on the primary lamellæ in *Spirifer* also represent portions of a loop.   More close analogy is seen in later forms of *Atrypa* having a disunited loop.

tions and homologies of these parts can best be shown in a
series of figures.

Plate XIV, figures C1, D1, represent the loop in a young
Macandrevia cranium in the so-called *platidiform* stage, showing a complete primary loop and the beginning of a secondary
loop in the middle, on top of the septum.   A later stage of
the same species (Plate XIV, figure G1) has the structure
of *Terebratalia*.   The descending lamellæ and the median
V-shaped plate correspond to the primary loop, while the
secondary loop or posteriorly recurved portion has greatly
increased in size.   A later stage, nearly complete (Plate
XXIV, figure 1), shows two points (*p*) on the descending
lamellæ, which are remnants of the connecting band in previous stages.   The parts homologous with the loop of the
first stage and with the loop of *Terebratulina* are shaded.
Greater emphasis is expressed by figures 2, 3, Plate XXIV,
where the cirrated brachia and calcareous supports are both
represented in the genera *Terebratulina* and *Magellania*.   It
is readily seen that the arm structure is the same in both, but
that the calcareous loops which are darkly shaded are very
different in form.

The family Terebratellidæ should therefore be reinstated
on the evidence here given.   The development of *Terebratella* may be reviewed for the leading characteristics of one
division of the family.   The type is *Terebratella chiliensis*
Broderip, sp. = *T. dorsata* Gmelin, sp., from the Straits of
Magellan.   Fischer and Œhlert[8] have described in detail the
development of the loop in this form.   Their researches also
include *Magellania venosa* Solander, sp., which was found to
pass through all the stages of *Terebratella dorsata*, and after
losing the processes connecting the primary lamellæ with the
septum finally results in adult *Magellania*.

Magellaniinæ.

The first stage described by these authors (Plate XIV,
figure B) showed only a septum anterior to the middle of the

dorsal valve.* The next stage was called the *præmagadiform* stage (Plate XIV, figures C*a*, D*a*), but it may well be divided into two stages, which correspond in structure to adult *Bouchardia* and *Megerlina*. The *bouchardiform* stage (figure C*a*) has a high quadrangular septum in the dorsal valve, and on the posterior distal angle there is a small circle, or calcareous ring. The crura are present, but the primary lamellæ have not yet appeared. In the next stage, the *megerliniform* (figures D*a*, D*a'*), the ring has increased in size, and below, on the septum, have appeared two projections or points, which are the beginnings of the descending primary branches.

The subsequent, or *magadiform*, stage (Plate XIV, figure E*a*) shows the completion of the descending branches to form the primary loop, and also the enlargement of the secondary loop or ring. During further growth the primary and secondary loops approach each other on the septum, then coalesce and make the *magaselliform* condition represented in figure F*a*.

The ventrally projecting, free portion of the septum next is absorbed, and the branches of the loop become attenuated, but still the descending branches remain connected with the septum, and thus the *terebratelliform* stage is completed (Plate XIV, figure G*a*).

Magellania venosa, after passing through all the stages described, including the *terebratelliform*, loses the connecting bands, and develops into the final *magellaniform* type of structure (Plate XIV, figure H*a*). Moreover, *Magellania lenticularis*, *M. flavescens*, *Terebratella cruenta*, and *T. rubicunda*, as far as observed, correspond closely in their development with the morphogeny of *M. venosa*.

A fact of importance noticed by Fischer and Œhlert[8] is that these species are confined to the southern hemisphere. The other austral types of terebratuloids, exclusive of the genera of Terebratulidæ, as here restricted, are *Magasella* (*M. Cumingi*), *Kraussina*, *Megerlina*, and *Bouchardia*. In

---

* An earlier *gwyniform* stage has been observed by the writer in a young example of *Magellania flavescens*.

their brachial supports these all approximate early stages of the higher genera *Magellania* and *Terebratella*. They must be regarded as arrested and degraded forms.

The brachial supports in *Kraussina* and *Bouchardia* are merely portions of the ascending branches, or secondary loop, on the septum, without any traces of the descending branches, or primary lamellæ. These genera may be compared with the *bouchardiform* stage of *Terebratella dorsata*. One grade higher is exhibited in *Megerlina* (type *M. Lamarckiana* Davidson) in which there is added to the *Kraussina* structure two processes apparently homologous with the points belonging to the descending branches appearing on the septum in the *megerliniform* stage of *T. dorsata*. These atavistic genera are all austral in their distribution, but not strictly polar, occurring as they do off the coasts of South Africa, Brazil, Australia, St. Paul's Island, etc.

In reviewing this group of genera, it is seen that the highest member of the series is *Magellania*, which reaches its maximum development in size and number of species in antarctic seas. The next genus below, *Terebratella*, ranges still further toward the equator, while the atavistic types *Kraussina*, *Megerlina*, and *Bouchardia* do not occur in polar regions, but are nevertheless austral in their distribution.

Dallininæ.

The northern hemisphere furnishes a series of genera and species, which, passing through a different and distinct series of loop metamorphoses, attains in the higher members the same result as those of the southern fauna, constituting a case of exact parallel development. Thus the northern *Macandrevia cranium*, *Dallina septigera*, *D. Raphaelis*, *D. Grayi*, *Terebratalia transversa*, *T. coreanica*, *T. spitzbergensis*, and *T. frontalis* are very similar in the adult characters of the loop to the southern *Magellania venosa*, *M. kerguelenensis*, *M. Wyvillii*, *M. flavescens*, *M. lenticularis*, *Terebratella dorsata*, *T. cruenta*, and *T. rubicunda*. It is only when their development is examined that a difference is manifest.

By observing the stages of development in the austral and boreal terebratellids, it is seen that both start from a common larval stage, and divergence into two lines begins in the first adolescent stages, so that the series of metamorphoses in each is quite distinct nearly to the end. This in itself might not require that the austral and boreal species should be referred to different genera and placed in different sub-families; but when it is found that all the other southern genera of the Terebratellidæ represent arrested and degraded stages in the development of a southern *Terebratella* or *Magellania*, and that the northern genera represent similar stages in the development of a northern high type, such a separation necessarily follows.* Moreover, these stages have a more profound significance, as several of them in both regions represent established genera now extinct.

A feature which may be of service in distinguishing adult recent shells is, that the Dallininæ have small cardinal processes, and the interior of the dorsal beak is usually grooved to the apex, while in Magellaniinæ there is a well-developed projecting cardinal process often filling the cavity of the beak. The lower genera can be readily determined by the characters of the loop and by the median septum, which is generally low in the Dallininæ and projecting above the loop in the Magellaniinæ. With these considerations in mind, the metamorphoses and relations of the northern Terebratellidæ may be described.

There are two finished types of northern genera, which are

---

* *Platidia* seems to be an exception in the distribution of the northern genera, as it has been recorded from Marion Island, in the southern Indian Ocean. The northern forms referred to *Magasella* are without the characteristic high septum of *M. Cumingi*, and appear to be stages of development of a higher northern form.

In Fischer's "Manuel de Conchyliologie," p. 1246, Œhlert, in discussing the geographical distribution of brachiopods, says : " Parmi les Brachiopods il en est, dont la distribution est en rapport avec la température régionale ; c'est ainsi qu'un certain nombre d'espèces sont particulières aux mers qui avoisinent les pôles, chaque hémisphère ayant ses formes spéciales qui lui appartiennent en propre, à l'exception de *Terebratulina caput-serpentis*, var. *septentrionalis*, qui se trouve à la fois dans l'hémisphère austral et dans l'hémisphère boréal."

taken as characteristic examples. One is the *Macandrevia cranium* Müller, and the other has been called *Magellania* (*Waldheimia*) *septigera* Lovén. In the light of the geographic, genetic, and ontogenetic facts, the application of the law of morphogenesis necessitates a new generic name for the second. *Magellania* cannot be retained, as the type is *M. venosa* from Tierra del Fuego, and therefore belongs to the southern line having a different series of metamorphoses. Neither can it be referred to *Macandrevia*, on account of its well-developed septum at maturity; nor to *Endesia* (type *E. cardium* Lamarck, from the Jurassic), since that genus has strong dental plates in the ventral valve, dividing the cavity of the beak into three chambers. Waagen [16] shows that these features — septal and dental plates — are entitled, in the terebratuloids, to rank as generic characters.

The name *Dallina*, nov. gen., is therefore proposed, to include shells of the type of *Terebratula septigera* Lovén; as *Dallina Raphaelis* Dall, sp., *D. Grayi* Davidson, sp., and *D. floridana* Pourtales, sp. The genus is given in honor of William H. Dall, whose name has long been intimately associated with the best work on recent Brachiopoda.

There still remain the northern species heretofore referred to *Terebratella*, which differ from true *Terebratella* (type *T. dorsata*) in the same manner and degree as *Dallina* from *Magellania*. These also require a special designation, and the name *Terebratalia*, nov. gen., is proposed, based on *Terebratula transversa* G. B. Sowerby, as the type.

The earliest stages of development in the *Dallina* and *Terebratalia* branch of the Terebratellidæ have been observed by the writer in *T. transversa* Sowerby and *T. obsoleta* Dall. [4] * They represent first a shell without a septum in the dorsal valve, and without calcified supports to the brachia (Plate XIV, figure A). The structure just before the appearance

---

* Originally described as *Terebratella occidentalis*, var. *obsoleta*, by Dall, but now considered by him as a distinct species. The complete development of the brachial supports in this species is shown in paper No. 8 of this series.

of the septum is the same as that described in *Gwynia* by King. The brachia form a slender fleshy ellipse or circle, resting in front on the floor of the interior of the dorsal valve, with the tentacles or cirri centripetal or directed inward, as in an early stage of *Cistella*. After this *gwyniform* stage the growth of the septum inflects the circlet of tentacles, producing a condition identical with that in adult *Cistella* (Plate XIV, figure B). It is therefore called the *cistelliform* stage.

The succeeding transformations in *Dallina septigera* and *Macandrevia cranium* have been fully described by Friele.[11] These species, with *Terebratalia obsoleta* Dall, sp., make three typical northern forms whose development has been observed. They agree in every essential detail, and may be described in general terms. The first stage described by Friele (Plate XIV, figures C1, D1), showed the growth of the descending lamellæ, their union with the septum, and the appearance of a small ring on the top of the septum, which is the beginning of the ascending branches, or secondary loop. This condition was correlated with the genus *Platidia*, by Deslongchamps,[7] and was called the *platidiform* stage. It has also been called the *centronelliform* stage by Fischer and Œhlert,[8] but, as *Centronella* is not known to have a septum supporting the loop, the name is not adopted here.

The lower anterior part of the secondary loop begins to divide very early (Plate XIV, figure D1), and, at the same time, the ends of the descending branches broaden and approach the top of the septum, being thus in juxtaposition to the ascending branches (as in figure E1), called the *ismeniform* stage. Lacunæ are then produced by resorption in the broad plates forming the ascending branches, and the structure of the supports at this time (figure F1), resembles that in adult *Mühlfeldtia sanguinea* and *M. truncata* (figure F3), in which the secondary loop is still attached to the septum. This stage is here termed the *mühlfeldtiform* stage. A further broadening of the loop and completion of the structures already outlined, with the recession of the secondary connecting bands from the septum, result in *Laqueus* (figures G4, G5), which has connecting bands from the ascending to the descending lamellæ and from the latter to the septum.*

The absorption of the connecting bands from the ascending branches completes the *Terebratalia* stage (Plate XIV, figures G1, G2), in *Macandrevia* and *Dallina*, and is the adult condition in *Terebratalia transversa*, *T. obsoleta*, *T. frontalis*, and *T. coreanica* (figure G3).†

Finally, the resorption of the connecting bands from the descending branches produces the *Dallina* structure, and the further resorption of the septum terminates the series in *Macandrevia* (Plate XIV, figure H1).

Comparisons and Homologies.

Thus the genera of the Terebratellidæ begin their larval development in a form like *Gwynia*, having no calcified brachial supports, and with a simple circle of centripetally directed tentacles. Then by the growth of a septum in the middle of the dorsal valve, a *cistelliform* stage is reached. From this point divergence begins, and there is one series of transformations resulting in *Macandrevia*, and another terminating in *Magellania*, the mature loops in both groups being practically alike. *Macandrevia* and *Dallina* are morphically equivalent to *Magellania*, and *Terebratalia* is also in exact parallelism with *Terebratella*.

A more graphic presentation of the development and relations of the genera is shown on Plate XIV, in which the stages of growth of *Magellania* and *Macandrevia* are represented on two ontogenetic lines. Outside are placed other species and genera, with their known stages of growth so arranged

---

* The name *Megerlina Jeffreysi* was given to a stage of *Laqueus californica* from its having a structure like *Megerlina* (= *Mühlfeldtia*) *truncata*, thus indicating clearly the close relationship of these genera.

† *T. spitzbergensis* and *T. Mariæ*, from the unfinished appearance of their brachial supports, possibly will be found to belong to a higher member of the series; for example, *Dallina*.

that their equivalent stages fall into parallel lines with those of *Magellania* and *Macandrevia.**

The line begins in an early larval stage (Plate XIV, figure A), in which there is a simple circlet of tentacles without a calcified loop, a structure comparable in every respect with *Gwynia* (figure A*a*).

The next stage (Plate XIV, figure B) shows the growth of a septum inflecting the line of tentacles, and producing an arrangement of parts similar to *Cistella*, although the loop is not calcified. The completion of this structure results in *Cistella* (figure B1), and specimens having the characters presented by figures B, B1, are referred to the *cistelliform* stage. A fossil representative of this type is *Zellania*, from the Jurassic (figure B*a*).

Megathyris (figure B2) offers nearly the same structure as *Cistella*, but the growth of two lateral septa or projections has produced two additional inflections in the loop. This completes the line of development in the Megathyrinæ.

Next, considering the Dallininæ in their ontogeny and morphology, it is found that after passing through the *gwyniform* and *cistelliform* stages (Plate XIV, figures A, B) a form like *Platidia* is reached (figure C1), of which *Platidia* is the living adult representative (figure C4). The *platidiform* stage is shown in *Macandrevia cranium* (figures C1, D1); *Dallina septigera* (figure D2); *D. floridana* (figure C3); and *Mühlfeldtia sanguinea* (figures C2, D3).

The growth of the secondary loop on the septum and the subsequent partial resorption produces a structure (1) like that in the fossil genus *Ismenia*, and (2) one identical with that of adult *Mühlfeldtia* (*M. truncata* and *M. sanguinea*). In Plate XIV, figures E1–E5 show the *ismeniform* stage of *Macandrevia*, *Dallina*, and *Mühlfeldtia*, as well as the final condition of *Ismenia*,† and figures F1–F4 represent the *mühlfeldti-*

* The illustrations on Plate XIV are taken from Davidson,[5,6] Fischer and Œhlert,[8] Deslongchamps,[7] and Friele,[11] with original drawings by the writer.

† Figures F5 and F6, Plate XIV, are from Davidson.[5] They are Jurassic species, and were referred to *Terebratella* (*T. furcata* Sowerby, figure F5, and

form stage of *Macandrevia, Dallina, Laqueus,* and the adult structure of *Mühlfeldtia sanguinea.* After this, the union of the primary and secondary loops and their removal from the septum to which they remain attached only by connecting processes form a structure like that in *Laqueus* (figures G4, G5), and the resorption of the connecting bands from the ascending branches of the loop completes the *terebrataliform* stage of *Macandrevia* and *Dallina,* as shown in Plate XIV, figures G1, G2. *Terebratalia* is the present fixed genus of this type of structure (figure G3), and *Trigonosemus* (figure G5), is a Cretaceous representative. Finally, by the resorption of the bands of the *terebrataliform* stage, the structure of the highest genera, *Macandrevia* and *Dallina,* is reached (figures H1–H5).

The first stage after the *cistelliform* in the Magellaniinæ, the austral branch of the Terebratellidæ, is represented for *Terebratella dorsata,* in Plate XIV, figure Ca. *Kraussina* (figures C*b,* C*c*) has a simple fork or V-shaped process on the septum, which apparently represents an incomplete secondary loop. The relations of *Bouchardia* (figure C*d*) to this *bouchardiform* stage of *Terebratella* are more evident. After this stage the beginnings of the primary loop, or descending branches, appear as two projections on each side of the septum (figure D*a'*). *Megerlina* (figure D*b*) shows this advance over *Kraussina.*

The completion of the descending branches in the next, or *magadiform,* stage is represented for *Terebratella dorsata,* in Plate XIV, figure E*a ;* *T. cruenta,* figure E*c ;* *T. rubicunda,* figure E*d ; Neothyris lenticularis,* figure E*b ; Magasella Cumingi,* figure E*f.* The Cretaceous equivalent, *Magas,* is shown in figure E*e.* In all these forms the septum projects above the descending lamellæ nearly to the ventral valve.

T. Buckmani Moore, figure F6). A strict interpretation of that genus based upon *T. dorsata,* the type, excludes these species, which agree with the definition of *Ismenia* in that the ascending and descending branches are attached directly to the septum. They may be, however, stages of growth of higher forms.

After the *magadiform* stage the descending and ascending branches approach and unite, and at the same time there is a narrowing of the latter (Plate XIV, figures F*a*-F*e*). *Magasella Cumingi* Davidson seems to be the only permanent adult representative of this structure which has yet been found.

The further narrowing of the lamellæ, broadening of the loop, and absorption of the free portion of the septum, result in the *terebratelliform* structure (Plate XIV, figures G*a*–G*d*), comparable directly with figures G1–G6 of the Dallininæ or boreal genera. Also, as in the Dallininæ, the disappearance of the connecting bands completes the *magellaniform* stage, and terminates the series (figures H*a*–H*e*).

The stages of growth of the genera belonging to the three sub-families of the Terebratellidæ, — the Megathyrinæ, Dallininæ, and Magellaniinæ, are further correlated in the accompanying tables. It must be understood, of course, that the larval and immature stages have not been observed in all the genera, but from the known ontogeny of several of the lower and higher forms, and from evident homologies of structure, such stages may be inferred.

Morphogeny from Gwynia to Megathyris.

Periods.	Stages.	Stages.	Stages.
Larval	gwyniform?	gwyniform	gwyniform
Adolescent	gwyniform	cistelliform	cistelliform
Mature	*Gwynia*	*Cistella*	*Megathyris*

The simplest genus, *Gwynia*, as far as known, passes through no metamorphoses, and has the same structure throughout the adolescent period, up to and including the mature condition. In the ontogeny of *Cistella* the *gwyniform* stage through acceleration has become a larval condition. In *Platidia* the *cistelliform* structure is accelerated to the immature period, and in *Ismenia* (representing an *ismeniform* type of structure in the higher genera), the *gwyni-*

Morphogeny from Gwynia to Dallina.

Periods.	Stages.	Stages.	Stages.	Stages.	Stages.	Stages.	Stages.
Larval	gwyniform?	gwyniform	gwyniform	gwyniform cistelliform	gwyniform cistelliform	gwyniform cistelliform	gwyniform cistelliform
Adolescent	gwyniform	cistelliform	cistelliform	platidiform	platidiform ismeniform	platidiform ismeniform mühlfeldtiform	platidiform ismeniform mühlfeldtiform terebrataliform
Mature	*Gwynia*	*Cistella*	*Platidia*	*Ismenia*	*Mühlfeldtia*	*Terebratalia*	*Dallina*

Morphogeny from Gwynia to Magellania.

Periods.	Stages.	Stages.	Stages.	Stages.	Stages.	Stages.	Stages.	Stages.
Larval	gwyniform?	gwyniform	gwyniform	gwyniform cistelliform	gwyniform cistelliform	gwyniform cistelliform	gwyniform cistelliform	gwyniform cistelliform
Adolescent	gwyniform	cistelliform	cistelliform	bouchardiform	bouchardiform megerliniform	bouchardiform megerliniform magadiform	bouchardiform megerliniform magadiform magaselliform	bouchardiform megerliniform magadiform magaselliform terebratelliform
Mature	*Gwynia*	*Cistella*	*Bouchardia*	*Megerlina*	*Magas*	*Magasella*	*Terebratella*	*Magellania*

form and *cistelliform* stages are larval, and the *platidiform*
represents an adolescent condition. Similar comparisons may
be made in the other genera. Progressively through each
series the adult structure of any genus forms the last immature stage of the next higher, until the highest member in
its ontogeny represents serially, in its stages of growth, all
the adult structures, with the larval and immature stages of
the simpler genera.

Conclusions.

It is impossible, with present knowledge, to go deeply into
the chronological history of the genera of the Terebratellidæ.
They certainly appeared before Jurassic time, because they
were then well represented by several characteristic genera;
namely, *Kingena, Ismenia, Zellania,* and *Megathyris,* although some of them may represent incomplete develop-
ments of higher forms. Also, as among recent species,
several separate generic and specific names may have been
given to stages of growth of a few species. It is evident that,
in the identification of specimens in this family, whether recent or fossil, the strict specific characters must be given first
consideration. Species, therefore, must be based upon surface
ornaments, form, and color, within certain limits, and genera
only upon structural features developed through a definite
series of changes, the results of which are permanent in individuals evidently fully adult.

The austral distribution of the Magellaniinæ and the boreal
distribution of the Dallininæ have been emphasized already,
but, as has been stated, this difference in geographical position in itself does not necessarily constitute a basis for sub-
family separation. The facts of development, however, and
the distribution of the genera show that the radical stock
was dispersed probably in Mesozoic time, and since then
evolution has gone forward through different lines of progression, and has terminated in similar types of structure.

In each line of progression the acceleration of the period

of reproduction, by the influence of environment, threw off genera which did not go through the complete series of metamorphoses, but are otherwise fully adult, and even may show reversional tendencies due to old age; so that nearly every stage passed through by the higher genera has a fixed representative in a lower genus. Moreover, the lower genera are not merely equivalent to or in exact parallelism with the early stages of the higher, but they express a permanent type of structure, as far as these genera are concerned, and after reaching maturity do not show a tendency to attain higher phases of development, but thicken the shell and cardinal process, absorb the deltidial plates, and exhibit all the evidences of senility and reversion presented during the old age of the higher genera. *Kraussina* shows a partial loss of deltidial plates and a thickened septum, with two strong prongs representing the ascending branches, which in other genera are very delicate. *Bouchardia, Magasella* (*M. Cumingi*), and *Agulhasia* have excessively thickened hinge-plates and muscular fulcra; the deltidial plates are obsolete, but the pedicle is enclosed by the growth and overlapping of the edges of the delthyrium. *Megathyris* and *Cistella* deposit lime on the interior of the valves, often in the form of nodes and ridges, while the pedicle in its growth encroaches upon the beaks of both valves to the final elimination of the deltidial plates. *Platidia* represents the extreme of this encroachment. Thus there are incomplete types of brachial development in these genera, accompanied by positive evidences of senility and retrogression in other shell characters.

Classification.

In the light of the previous discussion, the classification of the families and genera of Ancylobrachia, with the exception of the fossil forms, is comparatively simple. Among the extinct genera of the Terebratulidæ, or the family in which the loop is a development of the descending branches, *Centronella* and *Rensselæria* belong to one sub-family; *Stringocephalus* to another; and *Megalanteris, Cryptonella, Dielasma, Dictyothyris,* and others, will come in the Terebratulinæ.

As for the long-looped Mesozoic forms, it is evident that many of them underwent metamorphoses in their development while attached to a dorsal septum. The presence with them of such genera as *Kingena*, *Ismenia*, and *Mühlfeldtia*, points to their intimate relations with the boreal stock of the Terebratellidæ, and therefore indicates that the Mesozoic species with long recurved loops will in all probability be found to agree with *Macandrevia*, *Eudesia*, *Dallina*, *Trigonosemus*, *Lyra*, *Terebratalia*, *Laqueus*, *Kingena*, *Ismenia*, and *Mühlfeldtia*. *Magas* is exceptional, as it belongs to the austral group of genera, and suggests the likelihood that some of the higher species and genera associated with it geologically may prove to belong to the same branch.

The septum in the Terebratellidæ appears before the branches of the loop, and is an important character in the series of metamorphoses until the last stage. At first it is merely a support to the introverted growing cirrated portion of the lophophore. This elongation of the cirrated margin next results in an arch on the septum in the *platidiform* and *ismeniform* stages. Further elongation produces the median coiled arm, and the structures impeding its growth become resorbed, so that the transverse space between the lamellæ widens, the connecting bands and high septum recede, and finally even these disappear in *Macandrevia* and *Magellania*.

The following table represents the genera arranged according to the views expressed in this paper: —

Family TEREBRATULIDÆ Gray.

Loop free, developed by the growth and modification of two primary lamellæ. Cirri directed outward in larval stages.

CENTRONELLINÆ Waagen.

Loop composed of two descending lamellæ uniting in the median line, forming a broad arched plate.

Centronella Billings.	*Juvavella* Bittner.
Rensselœria Hall.	*Nucleatula* Zugmayer.
Newberria Hall.	

Stringocephalinæ Dall.

Loop long, following the margin of the dorsal valve, not recurved in front. Probably no median coiled arm.

Stringocephalus Defrance.

Terebratulinæ Dall.

Loop usually short. Ascending branches, when present, formed by partial resorption of a primary *centronelliform* loop. A median unpaired coiled arm exists in recent genera.

Cryptonella Hall.	*Terebratulina* d'Orbigny.
Megalanteris Suess.	*Dictyothyris* Douville.
Dielasma King.	*Pygope* Link.
Dielasmina Waagen.	*Cœnothyris* Douville.
Liothyrina Œhlert.	*Glossothyris* Douville.

Dyscoliinæ (= Dyscoliidæ Fischer and Œhlert *emend.*).

Loop short, continuous with the cirrated edge of the lopho-phore.   No coiled median arm.

Dyscolia Fischer and Œhlert.	*? Agulhasia* King.
Eucalathis Fischer and Œhlert.	

Family Terebratellidæ King *emend.*

Loop in the higher genera composed of two primary and two secondary lamellæ, development indirect, passing through a series of distinct metamorphoses while attached to a dorsal septum. Cirri directed inwardly during larval stages, at which time the structure is comparable to that of *Gwynia* and *Cistella*.

Dallininæ, n. sub-fam.

Loop composed of descending and ascending lamellæ, pass-ing in the highest genera through metamorphoses comparable to the adult structure of *Platidia, Ismenia, Mühlfeldtia,*

Terebratalia, and *Dallina.* The lower genera, therefore, do not progress to the final stages.

Macandrevia King.	*Laqueus* Dall.
Dallina Beecher.	*Mühlfeldtia* Bayle.
Eudesia King.	*Kingena* Davidson.
Terebratalia Beecher.	*Ismenia* King.
Trigonosemus Koenig.	*Platidia* Costa.
Lyra Cumberland.	

MAGELLANIINÆ, n. sub-fam.

Loop composed of descending and ascending branches, passing in the higher genera through metamorphoses comparable to the adult structure of *Bouchardia, Magas, Magasella, Terebratella,* and *Magellania.* The lower genera become adult before reaching the terminal stages.

Magellania Bayle.	*Megerlina* Deslongchamps.
Terebratella d'Orbigny.	*Bouchardia* Davidson.
Magasella Dall.	*Kraussina* Davidson.
Magas Sowerby.	

MEGATHYRINÆ Dall.

Loop composed of descending branches only, passing in the highest genus through stages correlative with *Gwynia, Cistella,* and *Megathyris.* The lower genera do not complete the series.

Megathyris d'Orbigny.	*Zellania* Moore.
Cistella Gray.	*Gwynia* King.

References.

1. Agassiz, L., 1873. — Methods of Study in Natural History, eighth edition.
2. Crane, Agnes, 1893. — The Distribution and Generic Evolution of some Recent Brachiopoda. *Natural Science,* vol. ii.
3. Dall, W. H., 1870. — A Revision of the Terebratulidæ and Lingulidæ with remarks on and descriptions of some recent forms. *Amer. Jour. Conchology,* vol. vi, pt. ii.

4. Dall, W. H., 1891. — On some new or interesting West American Shells obtained from the dredgings of the U. S. Fish Commission Steamer *Albatross* in 1888, and from other sources. *Proc. U. S. Nat. Mus.*, vol. xiv.

5. Davidson, Thomas, 1876. — Monograph of the British Fossil Brachiopoda, Pt. II. No. 1. *Pal. Soc.*, London, vol. xxx.

6. —— 1886–88. — A monograph of Recent Brachiopoda. *Trans. Linn. Soc.*, London, vol. iv.

7. Deslongchamps, E., 1884. — Études critiques sur des Brachiopodes nouveaux ou peu connus.

8. Fischer, P., and Œhlert, D.-P., 1892. — Brachiopodes: Mission Scientifique du Cap Horn, 1882–1883. *Bull. Soc. Hist. Nat. d'Autun*, vol. v, with 5 plates.

9. —— 1892. — Résultats des campagnes scientifiques accomplies sur son yacht par *Albert* 1er, Prince Souverain de Monaco. Fascicule III. Brachiopodes de l'Atlantique Nord, 2 plates.

10. —— 1892. — Sur l'évolution de l'appareil brachial de quelques Brachiopodes. *Comptes Rendus*, Nov., 1892.

11. Friele, H., 1877. — [On the Development of Waldheimia.] *Arch. Math. Nat.*, Bd. xxiii.

12. Gray, J. E., 1848. — On the Arrangement of the Brachiopoda. *Ann. Mag. Nat. Hist.* (2), vol. ii.

13. Hancock, A., 1858. — On the Organization of the Brachiopoda. *Phil. Trans.*, vol. cxlviii.

14. King, W., 1850. — A Monograph of the Permian Fossils of England. *Pal. Soc.*, London.

15. Morse, E. S., 1873. — On the Early Stages of Terebratulina septentrionalis (Couthouy). *Mem. Boston Soc. Nat. Hist.*, vol. ii.

16. Waagen, W., 1879–85. — *Palæontologia Indica* (13), vol. i. Salt Range Fossils.

4. DEVELOPMENT OF SOME SILURIAN BRACHIOPODA *

(Plates XV–XXII)

Introduction

The fossil faunas of rock systems rarely furnish material for tracing the individual development of any of the contained species. Much will doubtless be done toward ascertaining such development when large collections from suitable localities have been studied with this object in view, and when the number of new species discovered and described each year approaches a minimum. A comparatively full and satisfactory account of the development of the individual organism in several species of trilobites is given in the works of Barrande, Walcott, Ford, and Matthew; Hyatt, Branco, Mojsisovics, and others have demonstrated the developmental characters of many of the fossil cephalopods, and Verworn has elicited similar facts from certain extinct species of Ostracoda. Further than this but little has been attempted, although the field is a most extensive, important, and inviting one.

As a general rule, the treatment of fossil organisms has rested mainly with geologists having more or less of a zoölogical training, and the principal aim has been to present the faunal aspects of each horizon for the purpose of chronological identification. This process has frequently become so involved with the imperfect description of species, that the systematic zoölogist or paleontologist is unable to make any

---

* Beecher and Clarke. *Mem. N. Y. State Mus.*, I, 1–95, pls. i–viii, 1889. The order of the names of the authors of this paper is without significance. The work was equally divided and jointly reviewed.

use of a large proportion of the species as a means of studying their taxonomic relations or their structural affinities with each other and with recent forms.

Each revision of a group of fossil animals has resulted in the establishment of numerous specific and generic synonyms. Many of these are owing, of necessity, to the imperfection of the material, and many names which are finally relegated as synonyms have been created under a misconception of the full significance of age, sex, habitat, and condition of preservation. Additional confusion often results from the inclusion, in a generic or specific description, of characters which pertain not alone to a normal individual, but interspersed with certain normal adult features are those belonging to various stages of morphological development and peculiarities arising from accident, disease, and impoverished conditions.

In the case of rare species, or of meagre material belonging to common forms, it is to be noticed that assertions regarding specific and generic characters are usually very positive; while, with an abundance of specimens representing many stages of growth and the extremes of individual variation, the descriptions are qualified, the latitude of genera and species is extended, and the points of relationship with allied forms are multiplied, thus binding a group of organisms into comparative uniformity, without anomalous differences such as often occur where the dividing lines are rigidly drawn.

During the years 1878–79 the collection of fossils made from the Niagara group at Waldron, Indiana, for the New York State Museum, was studied and arranged by one of the writers. This is probably the largest collection yet brought together from that celebrated locality, and some conception of its size may be obtained from the fact that, when received, it weighed about seven tons. At the time mentioned all the mature specimens were selected and specifically separated. Many immature forms were also reserved and used in arranging the series prepared for exhibition in the State Museum. It was designed to represent in the arrangement each species by a series of specimens showing the gradations of size and form from mature individuals down to as young and small specimens as could be found. Abnormal examples, also, were reserved and grouped with them. It was the intention of the writers to accompany this memoir with photographic illustrations of these series, representing each species here discussed; but it has not been found wholly feasible, and the illustrations are largely restricted to the presentation of the immature and adult conditions of growth, with the exception of the several series which are given on Plate XXII.

The product obtained from washing the slabs was preserved and passed through sieves to assort the material into different grades of fineness. It was found that these washings contained a great number of partially developed shells, and it is from them that the extremely young brachiopods treated of in the present paper have been derived. The writers have carefully examined all the residue of these washings, and have picked out about fifty thousand specimens, most of which are less than five millimetres in length, and many have a length of not more than one millimetre. After all the imperfect and badly preserved individuals were rejected, there still remained more than fifteen thousand inchoate individuals.

The sediments at Waldron consist of fine calcareous shales weathering into clays. A stratum of Niagara limestone overlies the shales at this locality, but none of the fossils derived from this limestone have been used in the preparation of the present paper, and so far as known, it has a comparatively different fauna and does not furnish such material as is here described. The calcareous matter in the shales consists almost entirely of fossils and fragments of fossils, principally branches of corals and bryozoa, segments of crinoid columns, and broken crinoid plates. The Brachiopoda are all calcareous, and the original shell structure is more or less preserved, depending upon the absence or presence of pyrite.

The occurrence in such great numbers of immature shells in these deposits may be explained by the luxuriant fauna which flourished in this Niagara basin, by the quiet seas of this region, and by the rapid sedimentation of the shales. The richness of the material is shown by the great profusion of specimens representing the sponges, corals, crinoids, Bryozoa, brachiopods, gastropods, annelids, and crustaceans, comprising altogether about one hundred and fifty species. The lamellibranchs and cephalopods were also doubtless abundant; but the conditions existing for the preservation of their remains were not favorable, probably on account of the composition of their shells, and but sixteen species have been noted. That the fauna was protected from excessive storms and the action of sea currents, is evinced by the usual perfection of the fossils. Some of the crinoids are unbroken and remain attached by their roots, retaining their arms in place; also large colonies of delicate branching corals and Bryozoa still preserve their unity. The specimens were rapidly buried in the soft calcareous mud, and show none of the eroding or disintegrating action of the water, such as would have been produced had they lain for any considerable period unprotected on the sea-bottom. It is true that many specimens are incrusted with Bryozoa, annelids, Cranias, and other fixed and incrusting forms, but the majority of these seem to have flourished during the life of their hosts.

Besides the embryonic Brachiopoda occurring in these shales, there are other classes represented by immature forms, notably the Gastropoda and Crinoidea. These, with the Brachiopoda, embrace almost all the young forms found. The small gastropods are of little interest, on account of the limited number of species, and because they undergo no important modification in their subsequent growth, and merely represent the apical portion of mature individuals. Among the crinoids the modifications of form and structure from the embryo state to maturity are more profound and essential, although the material is not sufficiently complete to furnish any very important results.

It is necessary to state that nearly all these observations on the development of the Brachiopoda are based upon the study of the material derived from a single locality, and some of the minor deductions may not apply, in every case, to the individuals of the same species found in other regions. The writers have also refrained, except when essential to the proper exposition of a species, from entering into details of synonymy or generic controversy as to the correct reference of the species. This course is considered advisable, from a desire not to introduce any discussions alien to the descriptions of the developmental changes in these organisms. Aside from this it is believed that a number of important facts are here added to the knowledge of the Brachiopoda, and that many of them will be found to be of general application. The investigation has also resulted in elucidating several obscure and anomalous features of the shell and of the cardinal area, which appear in their proper place in the description of the species and in the general summary.

The following list includes all the species of Brachiopoda which up to this time have been described from the shales at Waldron, Indiana, and comprises forty-two species and varieties, ascribed to twenty-four genera. It also shows whether material has been obtained which furnishes data for tracing the developmental changes.

The majority of the species which have afforded no young specimens are rare forms even in their adult state. Among the actually abundant species of which there are no means accessible of tracing the life-history, *Uncinulus Stricklandi* is a noticeable example, and it is really the only common species which has afforded no young shells. *Meristina Maria*, another abundant form, furnishes a series which is notably incomplete, as the youngest individual observed, which can with certainty be referred to it, has a length of 6 mm. Likewise the inarticulate species have yielded almost no immature specimens.

List of the Brachiopoda occurring in the Niagara Shales at Waldron, Indiana.

Crania siluriana Hall	One embryo.
Crania setifera Hall	No young shells obtained.
Crania spinigera Hall	" " "

Species	
Lingula gibbosa Hall	No young shells obtained.
Pholidops ovalis Hall	" " "
Rhipidomella hybrida Sowerby . .	Numerous inchoate specimens.
Dalmanella elegantula Dalman . .	" " "
Orthis subnodosa Hall	No young shells obtained.
Bilobites bilobus Linnæus	" " "
Orthothetes tenuis Hall	" " "
Orthothetes subplanus Conrad · . .	Full series showing development.
Leptæna rhomboidalis Wilckens . .	" " "
Stropheodonta profunda Hall . . .	No young shells obtained.
Strophonella striata Hall	Full series, showing development.
Strophonella semifasciata Hall . . .	No young shells obtained.
Plectambonites transversalis Dalman .	" " "
Mimulus waldronensis Miller and Dyer	One embryo.
Chonetes nova-scoticus Hall	No young shells obtained.
Chonetes undulatus Hall	" " "
Dictyonella reticulata Hall	Young shells not rare.
Clorinda fornicata, var., Hall . . .	No young shells obtained.
Anastrophia internascens Hall . .	Young shells not rare.
Camarotæchia neglecta Hall . . .	Young shells very abundant.
Camarotæchia acinus Hall	" " "
Camarotæchia indianensis Hall . .	" " "
Camarotæchia Whitii Hall	" " "
Uncinulus Stricklandi Sowerby . .	Mature form abundant; no young shells obtained.
Rhynchotreta cuneata Dalman . . .	Young shells common.
Atrypa reticularis Linnæus . . .	Young shells abundant.
Zygospira minima Hall	No young shells obtained.
Atrypina disparilis Hall	Young shells common.
Homæospira evax Hall	Young shells very abundant.
Homæospira sobrina sp. n.	Young shells not rare.
Nucleospira pisiformis Hall * . . .	" " "
Meristina rectirostris Hall	Full series, showing development.
Whitfieldella nitida Hall	Numerous inchoate specimens.
Meristina Maria Hall	Incomplete series showing development.
Spirifer eudora Hall	No young shells obtained.
Spirifer crispus Hisinger	Full series, showing development.
Spirifer crispus, var. *simplex*, Hall .	" " "
Spirifer radiatus Sowerby	" " "
Reticularia bicostata, var. *petila*, Hall .	" " "

* The mature characters of this species are assumed so early that the youngest forms observed show no important differences from the adult. On this account no discussion of its characters is given in the ensuing pages.

The method of illustration which has been adopted is one which seems most readily to furnish a means for comparison of characters. The embryonic shells are represented as enlarged, usually to the size of an adult, and accompanying the enlargements are natural-size representations of the final result of normal growth. Where the mature forms have been too minute to show satisfactorily the details of structure, both the developmental stages and full-grown shell have been enlarged to a convenient size. Thus the incipient stages and mature specific form are presented together. In the delineation of special features, such as the hinge, the writers have sometimes enlarged the earlier phases to a size corresponding with the same structure in the mature form, or have increased all on a uniform scale, so that both the particular characters and their comparative size are presented.

The enlarged drawings have been made by the writers, principally from the microscope; the camera-lucida was employed to insure accuracy in outline. The illustrations of the mature specimens are largely taken from the Twenty-eighth Annual Report of the New York State Museum and from the Eleventh Report of the State Geologist of Indiana, which may be consulted for a more ample representation of the adult characters of the species occurring at Waldron.

The drawings on Plates XV to XXI have been reproduced on stone, in a most satisfactory manner, by Mr. Philip Ast, and the writers wish to express their appreciation of the skill and labor he has bestowed upon the work. The illustrations given on Plate XXII were made from photographic reproductions of the actual series of specimens, and, although not serviceable for purposes of detailed study, show distinctly the nature of the material used and the almost insensible gradations obtained, representing the life-history of these species. The same completeness of material is furnished by the majority of forms described in the following pages.

The arrangement of the subject-matter in the discussions of the species may not seem to be in accordance with the usual method employed in tracing the life-history of organisms. In this case fossil organisms are to be dealt with, and, in order to insure accuracy of results, it is necessary to begin with the known and established facts and gradually descend to minute and strange forms, thereby connecting the extremes of growth. Under the caption "Developmental Changes," however, an endeavor has been made to trace the history of each feature of the shell, from its inception to maturity.

DISCUSSIONS OF THE SPECIES.

Crania siluriana Hall, 1863.

(PLATE XV, figures 1, 2.)

—— —— Hall. Twenty-eighth Ann. Rept. N. Y. State Mus. Nat. Hist., p. 148, pl. 21, figs. 3–7, 1879.
—— —— Hall. Eleventh Ann. Rept. State Geol. Indiana, p. 282, pl. 21, figs. 3–7, 1882.

But a single embryo of this species has been found in all the material examined. It is of an incipient stage of growth, measuring but 1 mm. in height, and 1.5 mm. across the aperture. Compared with the mature form, the average size of which is about 9×20 mm., it shows a relatively greater elevation and a more regularly conical form. Otherwise all the few essential characters of the adult shell are present at this early age.

Dalmanella elegantula Dalman, 1827.

(PLATE XV, figures 3–11.)

Orthis elegantula Hall. Twenty-eighth Ann. Rept. N. Y. State Mus. Nat. Hist., p. 150, pl. 21, figs. 11–17, 1879.
—— —— Hall. Eleventh Ann. Rept. State Geol. Indiana, p. 285, pl. 21, figs. 11–17, 1882.

Both species of *Orthidæ* occurring at Waldron (*Dalmanella elegantula* and *Rhipidomella hybrida*) are very abundant. In the later stages of growth the former species is readily distinguished from the latter by its flatter and shallower dorsal valve and deeper ventral valve, features which usually hold good for purposes of discrimination; but in extremely early stages of growth the nearly equivalve form of the shell makes the separation of the species very difficult, perhaps even impossible. Between the dimensions of .5 × .75 mm. and 18.5 × 18 mm. (which is a little in excess of average mature size) has been found every gradation in size and development. The minute shell which serves as a starting-point for the series may quite as well be taken as the incipient shell of *R. hybrida*, as both its valves have the same depth, while the cardinal areas and beaks show the same character of development. As there can be no doubt of this fact, it becomes impossible to determine whether a given embryo, could it have grown to maturity, would have developed into *R. hybrida* or *D. elegantula*. Until the embryos reach a size of 2 or 2.5 mm. in length, their specific value is undeterminable, and the specific individuality of *D. elegantula* can be established only with the increasing depth of the ventral valve from this point upward toward adolescence.

Unless these observations are at fault (and they have been made with great care), evidence here is very positive that the diagnostic characters of species of this group may not be assumed until the earlier stages of the existence of the shell have passed. Indications of similar character are found among the species of *Camarotœchia* and *Spirifer*. The importance of the fact is apparent and its significance will be appreciated.

Specific Characters.

Mature Form (Plate XV, figures 10–12 *a*). — Outline subcircular; hinge-line short, about one-half the width of the shell, straight.

Ventral valve elevated along the dorsum, which is arched and slopes more rapidly toward the lateral than toward the anterior margin; greatest width below the hinge-line, about half-way down the valve. Beak full, arched, incurved, and projecting over the cardinal area sufficiently to conceal the foramen. Cardinal area broadly triangular, low, incurved; foramen triangular; deltidial plates absent.

Dorsal valve shallow, nearly flat, slightly rounded over the umbo, but depressed toward the margins. A sharply defined sinus starts near the apex, but by widening with the growth of the shell, it becomes nearly obsolete before reaching the margins. Cardinal line straight; cardinal area narrow, elongate triangular; beak inconspicuous. Foramen triangular and filled by a tripartite cardinal process which passes into, without filling, the foramen of the opposite valve.

Surface of the shell closely covered by fine thread-like striæ which increase by intercalation; concentric growth-lines rare, except near the margin, where they appear as wrinkles.

Incipient Form (Plate XV, figures 3, 3 *a*). — The initial shell of the present series, measuring .5 mm. in length and .75 mm. in width, has valves of equal depth and convexity. The length of the hinge-line nearly equals the greatest width of the shell. The cardinal area is high, and equally elevated on each valve. Beaks erect; foramina large, triangular, open, and marginate. On the *ventral valve* is a single median stria, representing the dorsum of the mature shell, accompanied by one and indications of a second on each of the lateral areas, making five striæ on the valve. On the *dorsal valve* a low and wide median depression is apparent, bounded by two central striæ, these being accompanied by two accessory pairs upon the latera, making six striæ in all. It is very probable that this form represents nearly the actual initial stage in the development of the shell; and if this is the case, the inception of the plications on the surface, which become so numerous at maturity (from one hundred to one hundred and thirty on each valve), is synchronous with the formation of the rudimentary shell, while in the pauciplicate species here discussed they appear to be of secondary growth.

Developmental Changes.

General Form and Outline. — In the growth of the shell a change becomes manifest in its outline and relative proportions. The young stages have the width greater than the length, but the more rapid axial growth of the shell reverses these proportions at maturity. Moreover, in the incipient stages, the valves, as already noticed, are of nearly equal depth and convexity. In the next stage the depth of the ventral valve has noticeably increased over that of the dorsal, and, as in the latter valve the median sinus has become distinctly developed, the difference in this respect becomes emphasized. The divergence of the valves in convexity becomes increased until maturity, and this growth is accompanied in the ventral valve by a correspondingly increasing incurvature of the beak.

Beaks. — In the incipient shell the beaks are erect and distant, but not prominent. By the development of the broad sinus on the dorsal valve, the beak of this valve becomes relatively less prominent and apparently more closely appressed to the cardinal line. On the opposite valve every increase in convexity is accompanied by a corresponding increase in the incurvature of the beak; and as the shell approaches maturity, the incurvature becomes so great that it has been necessary, in the drawings which are here given showing the features of the cardinal area, to represent the beak as broken away.

Foramen. — The earliest stages of growth show a remarkable feature in the triangular, marginate, sub-equal fissures on the valves. This character may prove of a high taxonomic value. The foramen upon the ventral valve is, in every stage of development, open and free for the protrusion of the pedicle. Deltidial plates are absent in every stage of growth. In a secondary stage a cardinal process begins to form in the apex of the dorsal foramen, soon widening and becoming tripartite. As age increases, this process is projected into the ventral foramen, never quite filling it, always leaving room for the protrusion of the pedicle. In immature conditions the cardinal process is attached to the shell only at the apex of the foramen, but with maturity it comes in contact with the sides of the foramen, and at this stage entirely fills the dorsal aperture. With the increasing incurvature of the ventral beak and cardinal area, the apertures of the two valves change their mutual angle, constantly lessening it as growth advances.

Plications. — As noticed above, the earliest stages of growth observed show the striæ to be already developed on the shell, five on the ventral and six on the dorsal valve. These plications are rapidly multiplied by interstitial addition, and at maturity number from one hundred to one hundred and thirty on each valve.

Rhipidomella hybrida Sowerby, 1839.

(PLATE XV, figures 13–18.)

Orthis hybrida Hall. Twenty-eighth Ann. Rept. N. Y. State Mus. Nat. Hist., p. 149, pl. 21, figs. 18–25, 1879.
—— —— Hall. Eleventh Ann. Rept. State Geol. Indiana, p. 285, pl. 21, figs. 18–25, 1882.

Rhipidomella hybrida passes through primary developmental stages which are essentially identical with those already described for *Dalmanella elegantula*. Sufficient has been said in that connection in regard to the similarity and probable identity of the earlier embryonic stages of the shell of both species, the origin of the entire specific difference which is so apparent in the later and mature periods of development lying in the unequal growth of the valves in convexity. This increase is relatively greater in the dorsal valve of *R. hybrida* than in that of *D. elegantula*, and less in the ventral valve of the former than in that of the latter species. Thus *R. hybrida* is a more discoid, lenticular shell, showing but slight evidence of a median fold and sinus and carrying on its surface at maturity just about as many plications or striæ as its associate.

There is an obese variation from the normal form of *R. hybrida*, which was noticed by Professor Hall (*loc. cit.*), and this appears early in the development of the species, with a size of 3.5 mm. in length and 4.5 mm. in width, and reaches a maximum growth with dimensions of 14 × 13 mm. This variation is due to internal thickening and increase in convexity, and is accompanied by abundant concentric growth-lines which are as rare in the normal form as in *D. elegantula*. A representative series of this species affords variations between the following limits of size: .5 mm. in length × .75 mm. in width (minimum), and 17 mm. in length × 20 mm. in width (maximum).

Leptæna rhomboidalis Wilckens, 1769.

(PLATE XVI, figures 1–13.)

Strophomena rhomboidalis Hall.   Twenty-eighth Ann. Rept. N. Y. State
    Mus. Nat. Hist., p. 151, pl. 22, figs. 4–10, 1879.
    —— —— Hall.   Eleventh Ann. Rept. State Geol. Indiana, p. 288. pl. 22,
    figs. 4–10, 1882.

This well-known species, although extremely abundant in the mature state, is correspondingly rare in its undeveloped condition.   The young specimens which have been found are nearly all more or less broken, and it is evident that while young the shell was thin and delicate, consequently few of their remains have been preserved.   The series which has been selected is, however, very complete in its representation of the distinct phases of growth through which the individuals pass in their development from youth to maturity. The initial form, without radiating striæ;   the second phase, a shell radiatingly striate, without undulations;   the third state, striated and concentrically undulated, but without the angular geniculation of the valves in front;   and the last phase, with the full form and characters of maturity, offer a series of changes, not often traceable in Silurian brachiopods.

The development of the characters of the hinge-area is also very satisfactorily demonstrated, and affords some interesting points of comparison with certain forms of *Orthothetes* and *Strophonella.* These features are noticed at the end of the description of the species *Strophonella striata.*

Specific Characters.

Mature Form (Plate XVI, figures 4, 4 *a*, 10, 13). — Shell semi-elliptical or semi-circular in outline.

Dorsal valve flat or slightly concave in the upper part, with the marginal portions abruptly curved upward in front; beak small, carrying on its inner side a large, prominent, triangular callosity, grooved along its summit and nearly filling the area of the opposite valve.

Ventral valve usually convex in the upper part, becoming flat or concave below, and with the marginal portion produced and abruptly bent downward, geniculating with the dorsal valve; beak small, usually perforated with a small circular foramen; hinge-line often 50 mm. in length, equaling or greater than the width of the shell below; cardinal extremities twisted and often much extended; cardinal area narrow, edges parallel, formed by both valves; deltidial area broadly triangular, occupied by the grooved callosity under the dorsal beak.

Surface marked by regular, rounded, radiating striæ. From the beaks to the curtain, or geniculated portion, the shell is ornamented with regular, strong, concentric undulations or corrugations.

This species varies greatly in size and form, in the different horizons and localities where it is found. In many places the mature shells are about half the size of the specimens from Waldron.

Incipient Form (Plate XVI, figures 1, 1 *a*, 11). — The smallest entire specimen yet detected has a length of 1.25 mm. The outline is semi-oval, with the greatest width near the middle, and about one-fourth greater than the length. *Dorsal valve* convex in the upper part, becoming concave toward the front. The hinge-area of this valve is very narrow and linear, and carries beneath the beak a small grooved callosity. *Ventral valve* convex, sloping in all directions from near the foramen, around which the surface is slightly depressed. There is also a depression extending along the middle of the valve to the anterior margin. The place of the beak is occupied by an exsert, conical pedicle-tube, which partly protrudes beyond the cardinal margin of the valve and extends down to, and embraces the dorsal callosity. Cardinal area of the ventral valve comparatively broad, narrowing rapidly from the pedicle-tube to the extremities. *Surface* smooth, except along a narrow zone around the margin, which shows incipient radiating striæ.

Developmental Changes.

The form of this species being somewhat complex, the development of the shell may be conveniently sub-divided into four stages, briefly characterized as follows: —

1st Stage. Length of shell .4–1 mm.; surface smooth.

2d Stage. Length 1–2 mm.; shell radiatingly striated, without undulations.

3d Stage. Length 2–20 mm.; shell radiatingly striated, and concentrically undulated.

4th Stage. Length 20–30 mm.; entire shell radiatingly striated, concentrically undulated in the upper part, abruptly produced and geniculated in front.

The changes taking place in the form and character of the shell from one stage to another can be best shown and used for comparison in the following tabulation, where the conditions incident to each stage of growth in the various parts of the shell are briefly described: —

Development of Leptæna rhomboidalis.

	1st Stage. Initial.	2d Stage. Infantile.	3d Stage. Adolescent.	4th Stage. Mature.
Size	.4 mm.—1 mm. in length, .4 mm.—1.5 mm. in width.	1mm.— 2 mm. in length, 1.5 mm.—2.5 mm. in width.	2 mm.—20mm. in length, 2.5 mm.—40 mm. in width.	20 mm. — 30 mm. in length, 40 mm. — 50 mm. in width.
Form	Transversely semi-oval; cardinal extremities obtusely angular.	Transversely semi-elliptical; cardinal extremities angular.	Longitudinally semi-elliptical; cardinal extremities augular, becoming produced, not twisted.	Longitudinally semi-elliptical; cardinal extremities acutely angular, extended and twisted.
Contour	Convex.	Depressed convex.	Very slightly convex.	Geniculate, making the shell highly arched longitudinally.
Dorsal valve . .	Convex, concave on the margin; umbo prominent.	Concave, except on the umbo.	Concave, except on the umbo.	Flat or concave on the body of the shell; abruptly produced and curved upward around the margins.
Ventral valve .	Convex.	Convex, semiconical with the beak at the apex.	Convex in the upper part, flat or concave on the margin.	Convex in the upper part, flat or concave in the middle, and abruptly bent downward below.
Surface	Smooth.	Radiatingly striate.	Radiatingly striate and concentrically undulate.	Entire surface radiatingly striate, concentrically undulate in the upper part only.

Development of Leptæna rhomboidalis. — Continued.

	1st Stage. Initial.	2d Stage. Infantile.	3d Stage. Adolescent.	4th Stage. Mature.
Cardinal area .	Ventral high; dorsal very slender.	Ventral high; dorsal very slender.	Ventral narrow; dorsal narrow.	Both narrow, sub-equal.
Pedicle-tube ..	Exsert, full height of the area.	Not exsert, full height of the area.	Nearly full height of the area.	Obsolescent or obsolete.
Foramen	Present, circular, elevated.	Present, circular.	Present, circular.	Usually present.
Dorsal callosity.	Small, grooved.	Small, grooved.	Larger, grooved.	Very large and deeply grooved.

Among the mature shells the greatest variation is to be found in the development of the anterior *curtain*, or geniculate and sloping marginal area of the valves. In some specimens this is so excessively developed that the posterior or concentrically undulated portion of the ventral valve is at right angles to the plane of the margin. Also, in many specimens the curtain is obscurely plicate, and the radiating striæ are often irregular and sometimes fasciculate, while on the upper part of the valves these striæ are very uniform in their arrangement. No specimens have been noticed which are so strongly quadriplicate as those illustrated by Mr. Davidson, on Plate XXXIX of the "British Silurian Brachiopoda."

Senile specimens usually have the valves very much thickened from internal growth, and the margins show strong varices. It is noticeable that nearly all the old shells are covered with a growth of Cranias, Bryozoa, Favosites, etc., and it is very difficult to free the shell from this overgrowth. In consequence of this, many of the shells are scarcely recognizable, and resemble agglomerations of Bryozoa and corals.

The only other species of Brachiopoda at this locality commonly thus overgrown and involved is *Atrypa reticularis*.

Leptæna rhomboidalis is cosmopolitan and has been discussed by many authors, who have shown its great variation and wide distribution. So far as known, the youngest specimen heretofore figured is one represented by Mr. Davidson.[*] This is an individual belonging to the third stage of development, having a length of nearly 6 mm. and a distinct circular perforation of the beak.

Orthothetes subplanus Conrad, 1842.

(PLATE XVI, figures 14–20.)

Streptorhynchus subplanum Hall.   Twenty-eighth Ann. Rept. N. Y. State
   Mus. Nat. Hist., p. 151, pl. 21, figs. 26–33, 1879.
—— —— Hall.   Eleventh Ann. Rept. State Geol. Indiana. p. 288, pl. 21,
   figs. 26–33, 1882.

The series selected to represent the development of this species comprises fourteen specimens ranging from 1.5 mm. to 26.5 mm. in length. The external features of form and surface ornament are remarkably constant from the young to the mature shells. There is, however, a slight progressive modification in the relative convexity of the valves. The dorsal valve of young and half-grown individuals is nearly flat, while the ventral is moderately convex. In old specimens both valves are convex, with the dorsal somewhat more so than the ventral. The most marked changes due to advancing growth are those which take place in the hinge. Some mention of these is made under the description of *Strophonella striata*, where it is stated that the pedicle-tube retains its embryonic form and size nearly up to maturity, after which it is obscured by the internal thickening of the shell; also, that the callosity under the beak of the dorsal

---

[*] British Fossil Brachiopoda, III, Devonian and Silurian, 283, 284, pl. xxx, fig. 6. The same. — General Summary to the British Fossil Brachiopoda, 289.

valve uniformly increases in size from the youngest forms to full-grown specimens.

Specific Characters.

Mature Form (Plate XVI, figures 15, 15 *a*, 17, 20). — Shell semi-circular or semi-elliptical, depressed convex; hinge-line longer than the width of the shell; cardinal angles flat and extended.

Dorsal valve moderately and uniformly convex except at the cardinal angles; umbo not defined; beak small.

Ventral valve convex on the umbo, less convex below, and in many specimens the marginal portion is flat or slightly concave; beak small, somewhat arched. Hinge-area nearly equal in both valves, usually appearing as a deep angular groove along the cardinal margin. Under the beak of the dorsal valve is a large triangular callosity, grooved on the inside, and nearly filling the fissure of the opposite valve. Deltidium of the ventral valve broadly triangular, extending to just below the beak, and margined on each side by two narrow areas in the form of scalene triangles, which may represent the deltidial plates of other genera. Beak imperforate.

Surface marked by from fifty to one hundred (according to the size of the shell) regular, rounded striæ, with equal interspaces, increasing in number by interstitial additions. The entire shell is also ornamented with very fine, regular, sharp, concentric striæ. A large specimen has a length of 26 mm., and the width, measured along the hinge-line, is about 38 mm.

Incipient Form (Plate XVI, figures 14, 14 *a*). — The smallest specimen measures 1.5 mm. in length by 2.3 mm. in width along the hinge-line. The outline is semi-elliptical, with the cardinal angles slightly extended. *Dorsal valve* concave in the upper part, and slightly convex below. *Ventral valve* convex; beak prominent, projecting beyond the hinge-line.

The hinge characters are not well preserved in this individual. The first specimen in the ascending series which shows the hinge distinctly has a length of 2.25 mm., and will be described in the development of this part.

The *surface* of the incipient shell is marked by seventeen alternating, narrow, elevated radiating lines, with wider interspaces, and also shows several lines of growth near the margin.

Developmental Variations.

No marked changes occur in the general form of the shell other than the gradual increase in the convexity of the dorsal valve and in the extension of the cardinal angles. The dorsal valve is usually quite flat in specimens having a length of 10 mm. or less. The radiating lines increase in number by interstitial additions, from the youngest form to maturity, and the fine concentric striæ appear on all the specimens, including the initial individual in the series, where they are developed around the margins of the valves.

The earliest phase of the hinge yet noticed is found in a specimen having a length of 2.25 mm. The dorsal valve shows a foramen in the cardinal area under the beak, margined by a slight thickening of the shell. The ventral valve preserves a small perforate pedicle-tube at the apex, extending about two-thirds of the distance down to the hinge, below which is a triangular deltidial opening of the same width as the dorsal foramen.

A specimen 4 mm. in length (Plate XVI, figure 19) shows a more advanced development of the same parts. The dorsal callosity has nearly filled the sinus under the beak and has a narrow groove in the centre. The fissure of the ventral valve has increased considerably in size and relative height, showing narrow marginal plates or defined areas. The pedicle-tube is still perforate, but has not increased in size beyond the initial stage.

From this point to maturity the hinge increases in width, the dorsal callosity grows rapidly and nearly fills the fissure of the opposite valve. The pedicle-tube is obscured, and the perforation obsolete. The marginal plates, or lateral areas, are clearly defined, and have the form of narrow scalene triangles.

No important variations have been noticed among the mature specimens. Occasionally an individual diverges from the normal form by having mucronate cardinal angles, or a senile specimen shows strong imbricating varices of growth; but, as a whole, the form and surface ornaments in this species are very uniform.

Strophonella striata Hall, 1843.

(PLATE XVII, figures 1–8.)

Strophodonta striata Hall.   Twenty-eighth Ann. Rept. N. Y. State Mus. Nat. Hist., p. 152, pl. 23, figs. 1–6, 1879.
—— —— Hall.   Eleventh Ann. Rept. State Geol. Indiana, p. 290, pl. 23, figs. 1–6, 1882.

The present form is one of the most delicate and fragile species of Brachiopoda at Waldron. Individuals are not of rare occurrence, but the majority of them are more or less broken. The upper portion of the shell, or that along the hinge, being thicker and stronger than the remainder, is more often preserved, and the series is only complete in the representation of this portion, although there are several small specimens which are sufficiently entire to show the early form of the shell.

As in the other species which in their mature proportions depart from the type of structure in the group, the incipient shell is found to revert to the primitive form. The full-grown examples of this species are concavo-convex, the concave valve being the ventral; while in the young the ventral valve is the more convex. This change in the relative convexity of the valve does not begin until the individuals are about half grown, and is produced by the gradual deflection of the margin with the increase in the size of the shell.

The development of the features of the hinge is very characteristic, and, as in the other strophomenoid forms, is of primary interest. Both the dorsal callosity and pedicle-tube continue to increase in size with the growth of the shell, from the incipient form to maturity.

Specific Characters.

Mature Form (Plate XVII, figures 2, 2*a*, 8). — Shell semi-elliptical, wider than long, the greatest length being along the hinge. The body cavity is very shallow, and the shell has a concavo-convex form.

Dorsal valve flat in the upper part, moderately convex in front. *Ventral valve* slightly convex on the umbo, and concave over the remainder of the valve. Hinge-area formed by both valves. Ventral area the wider, carrying in the centre a small conical pedicle-sheath which is usually minutely perforate at the apex. Dorsal area linear, with a callosity in the middle, under the pedicle-tube of the opposite valve.

Test thin, surface ornamented by about fifty alternating radii, with three or four fine filiform striæ in each interspace; also crossed by fine irregular striæ of growth.

Two specimens measure, respectively, 19.5 mm. and 14 mm. in length, and 23 mm. and 16 mm. in width, at the hinge.

Incipient Shell (Plate XVII, figures 1, 1 *a*, 3). — The form is nearly plano-convex. *Dorsal valve* convex on the umbo, flat below. *Ventral valve* moderately convex, with a prominent pointed beak. Hinge narrow, with a small cylindrical perforated pedicle-tube in the centre of the ventral area, and a small callosity in the dorsal area. In the smallest specimen observed the surface is marked by eleven radii on the ventral valve, but is otherwise apparently smooth. Length 2.25 mm.; width in the centre 3 mm.

Developmental Changes.

On account of the imperfection of the material, it is impossible to trace any minor changes in the outline of the valves, and the specimens indicate that no considerable transformation took place. The modifications in the convexity of the valves is of more importance in this species, and can be readily observed. In the young individuals, up to about one-third full size, the ventral valve is slightly convex and the dorsal valve nearly flat. Further growth of the shell changes these relations, by the gradual deflection of the margin, until the general form of the ventral valve is concave and the dorsal valve is convex.

The radii appear very early in the growth of the shell, the smallest individual having eleven on the ventral valve, the majority of which extend to the umbo. They probably first appeared in pairs, and are found to increase in number afterward by simple intercalation.

The hinge-area is developed upon both valves in all stages of growth, although in the early stages the cardinal area of the dorsal valve is very narrow, but gradually increases, until at maturity it is nearly equal to the area of the ventral valve.

The pedicle-tube is at first cylindrical and short. Advancing in the series, it is found to become conical from growth and from the widening of the fissure, until in full-grown specimens it is wider than high. A careful examination reveals the perforation in all stages of the development of the shell. It is, however, very minute, and it is not probable that the extremely small peduncle could have performed its full function. Indeed, it may be surmised that in none of the three strophomenoid species here described was the fleshy arm sufficiently strong in mature individuals to serve as a secure support to the shell. In the embryonic forms it was a more important organ.

The hinge of the young shell illustrated in figure 4, Plate XVII, shows an excessively elongate, cylindrical pedicle-tube, of which more than one-half the length is projected above the beak. It must be considered as a supra-calcification about the peduncle, and apparently indicates a more complete functional extension.

The grooved dorsal callosity appears in the beginning of the series, and gradually increases in size, and detrudes so that the groove shows on the exterior, but just before maturity it is either filled or introverted into the deltidial cavity.

The features of the hinge, fissure, and callosity, in the Strophomenidæ, and their embryological development, seem to be peculiar to the group. They are of special interest, both on this account and also because the family has no living congeners. Although the separate characters have been presented in detail in each of the preceding descriptions, a brief review of the hinge characters is here given, showing more clearly their intimate relationships.

In the three species, *Leptæna rhomboidalis*, *Strophonella striata*, and *Orthothetes subplanus*, the initial form of the hinge is the same. Each shows a slender callosity under the beak of the dorsal valve, and a perforate pedicle-sheath in the ventral valve, which does not entirely close the deltidial opening. From this initial stage development proceeds in a different manner for each of the three species. *Leptæna rhomboidalis* and *Strophonella striata* develop in a parallel series until the individuals are about one-third grown in the first species, and two-thirds full size in the second. That is, the dorsal callosity and pedicle-sheath each increase uniformly in size up to these periods. Beyond this the divergence is rapid and marked. In *Leptæna rhomboidalis* the grooved callosity increases in size so as to nearly fill the broad fissure in the ventral valve, while the pedicle-sheath ceases growth, is atrophied and lost, although in many cases the perforation persists. *Strophonella striata* continues its hinge development without change, except that at full maturity the groove on the callosity becomes introverted into the pedicle-sheath.

The third mode of development is exhibited by *Orthothetes subplanus*, in which the pedicle-sheath does not increase beyond its initial size, while the dorsal callosity develops up to the maturity of the shell, and, as in *Strophonella striata*, the groove is on the inner side.

The function of this groove in the callosity of the dorsal

valve in the strophomenoids has not been satisfactorily determined, its existence having sometimes been considered as evidence of the perforation of this valve.* In all young shells it is evident that the passage of the pedicle is not through this groove in the dorsal callosity, but through the apex of the ventral valve by means of the channel which has been here termed the pedicle-tube or sheath. In growth-stages where there can be no question of the functional activity of this sheath, the dorsal callosity is already grooved or sinuate. It might be surmised that the purpose of the groove was to avoid compressing the pedicle when the valves were open, and this it may have been to some extent; but the evidence furnished by both recent and fossil species indicates that the valves of the articulate brachiopods could be opened only to a very slight degree. The groove persists in species after the true pedicle perforation in the ventral valve is closed and functionally useless. Its origin appears to be due to the organic deposition about the bases of the two interior cardinal processes, the interstitial area of slower deposition being represented by a fissure, groove, or sinus.

Mimulus waldronensis Miller and Dyer, 1878.

(Plate XVII, figures 9, 10.)

Spirifera? *waldronensis* Miller and Dyer. Contributions to Palæontology, Jour. Cin. Soc. Nat. Hist., April, 1878.
Triplesia putillus Hall. Trans. Alb. Inst., vol. x, abstract, p. 16, 1879.
—— —— Hall. Eleventh Ann. Rept. State Geol. Indiana, p. 298, pl. 27, figs. 19–22, 1882.

This species is among the rarest of the Waldron Brachiopoda, and it is impossible to present a series representing the variety and progress of development, as in some of the more common forms. There were but two specimens, both adults, discovered in the State Collection at the time of the publication of the "Descriptions of New Species of Fossils from the

* *Eleventh Rept. State Geologist Indiana,* 288, 289, 1882.

Niagara Formation at Waldron, Indiana."* Fortunately, there has been more recently detected a young individual of about one-fourth the normal adult size, which offers some interesting details in its form and characters.

The asymmetry of the shell is manifest even at this early stage of growth (although the median fold is not developed), and is evinced by the position of the beak of the ventral valve and by the contour of the margins. It is probable that in a still earlier phase of growth the two valves are symmetrical, or nearly so.

In the young individual under consideration (Plate XVII, figures 9, 9 *a*, 9 *b*), the outline is nearly circular. The beak of the ventral valve is very much elevated, projects beyond the cardinal line, and is directed toward the left side of the shell. The apex is truncated, and the opening is confluent with the area below.

The cardinal area is high, forming a large triangular fissure which is apparently not closed by deltidial plates. The beak of the dorsal valve is depressed, and limited by a slight furrow on each side. No lines of growth are visible, but the surface is somewhat granulose, as in many small shells of other species.

The principal differences to be noted in comparison with the adult individuals are the sub-circular outline of the shell, the depressed valves, the absence of a median fold, and the large deltidial area.

Dictyonella reticulata Hall, 1868.

(PLATE XVII, figures 11–13.)

Eichwaldia reticulata Hall. Twenty-eighth Ann. Rept. N. Y. State Mus. Nat. Hist., p. 169, pl. 26, figs. 50–54, 1879.
—— —— Hall. Eleventh Ann. Rept. State Geol. Indiana, p. 312, pl. 26, figs. 50–54, 1882.

Very few of the earlier growth-stages of this species have been observed, and these show but comparatively little varia-

* James Hall. Read before the Albany Institute, March 18, 1879.

tion from the features of the normal adult. On Plate XVII is given a figure of the youngest example found, which has a length and width of 3 mm., while the usual adult is about 16×16 mm., varying in relative proportions with the increase of senile obesity. The change in outline during growth is from sub-circular to sub-triangular, and in earlier stages the ventral fold and sinus are very ill defined. The peculiar triangular exfoliation of the shell on the umbo of the ventral valve is evidently a constant feature in every stage of growth after the shell becomes attached. The nature of this peculiarity was indicated by Billings in the original diagnosis of the genus *Eichwaldia* (*Ann. Rept. Canadian Geol. Survey*, 1857–58), and was demonstrated more fully in *Dictyonella* by Professor Hall, in the Twentieth Report on the Condition of the New York State Cabinet of Natural History (pp. 274–278, 1867). This area is underlaid by an internal shelf or diaphragm attached along its lateral margins, and having fully or rather more than the width of the median sinus. Through the space thus left between the shell and the internal diaphragm, communication is afforded with the outside world. Mr. John Young has called attention to the fact that in *D. Capewelli* the margins of the external reticulated layer of the shell about the umbonal bare spot are rough and ragged, the superficial hexagonal cells being without finish along these edges, suggesting therefrom that the animal was attached to marine objects by the substance of the shell, and afterward broken away from its attachment. (See Davidson, General Summary, pp. 355, 356.) It is true that the anterior edge of this area may be rough and uneven, but the lateral edges appear invariably straight and diverge at an essentially constant angle. The latter represent the lines of attachment of the internal plate to the interior of the valve, and if the shell has been broken in detachment from foreign bodies, the fracture in these directions has been guided by these lines, but on the unsupported anterior margin it has been rough and irregular. Upon the hinge-line of the ventral valve there exists no aperture for the protrusion of the pedicle; by the peculiar development of the articulating processes of both valves, the entire cardinal margin is closed, and therefore the passage between the internal plate and the surface of the valve may have been for the use of this organ; or, it may be suggested that as this space is rather too narrow and explanate for such a purpose, *Dictyonella* may have been attached by the substance of the shell, the internal shelf acting as a support to the strain upon the umbo and as a protection to the animal in case the shell were broken from its attachment.

Anastrophia internascens Hall, 1879.

(PLATE XVII, figures 14–16.)

———— —— Hall. Twenty-eighth Ann. Rept. N. Y. State Mus. Nat. Hist., p. 168, pl. 26, figs. 41–49, 1879.
———— —— Hall. Eleventh Ann. Rept. State Geol. Indiana, p. 311, pl. 26, figs. 41–49, 1882.

In tracing the development of this species the principal feature to be noticed is that the elemental shell conforms with the type of an ordinary brachiopod, such as *Camarotœchia;* that is, the dorsal valve, although somewhat the more convex, is smaller than the opposite valve, while in the mature state the dorsal valve is considerably larger and projects beyond the beak of the ventral valve. It is the development of this character which constitutes the most conspicuous change in the shell in its growth from the young to the fully mature condition.

Specific Characters.

Mature Form (Plate XVII, figures 15, 16, 16 *a*). — Shell ventricose. Outline transversely sub-elliptical, sometimes nearly as long as wide.

Ventral valve convex, depressed in front, forming a more or less defined sinus which carries four or five of the plications; beak short, acute; area short, broadly triangular, usually not exposed.

Dorsal valve gibbous, with the central portion elevated, frequently presenting a broad undefined median fold; beak incurved under the beak of the opposite valve; umbo prominent.

Surface marked by about fifteen strong, simple, elevated, rounded or angular plications on the body of the shell, and smaller bifurcating plications on the latera. Occasionally intercalated plications are present on the middle of the valves. The plications are crossed by fine arching striæ of growth, which are sometimes aggregated, forming conspicuous concentric lines or varices of growth.

Mature shells measure from 11 to 17 mm. in length, and from 12 to 19 mm. in width. The depth of the conjoined valves varies from 9 to 12 mm.

Incipient Form (Plate XVII, figures 14, 14 *a*). — The smallest shell observed has a length of 2 mm. and a width of 2.25 mm. The *dorsal valve* is slightly more convex than the ventral, and is a little shorter. Eight rounded plications are shown, five of which extend to the umbo of the valve. A short plication is intercalated in the middle, and there is also a short one on each side of the valve. Ventral beak small and elevated, with a broad, triangular, open area below.

Developmental Changes.

The series of specimens selected to represent the development of this species contains fifteen normal individuals, varying from a length of 2 mm. to a length of 17 mm. The proportions of length and width remain nearly constant throughout, the width being somewhat the greater.

In the smallest specimen the depth of both valves is less than one-half the length of the shell. This relation gradually changes as the shell becomes larger and more convex, until, in mature individuals, the depth is equal to three-fourths or four-fifths the length of the shell, and in extremely obese specimens this ratio is often exceeded.

The dorsal valve is more convex than the opposite valve, in all the stages of growth which have been observed, although in the elemental shell the difference is scarcely perceptible, while in the mature form it is a conspicuous feature. This valve is also shorter than the ventral in specimens up to a length of about 7 mm. From 7 to 12 mm., both valves are of nearly equal length. Further growth causes the umbo of the dorsal valve to protrude beyond the beak of the opposite valve, and the beak is incurved and penetrates the area. It seems evident that if the true initial shell were studied the dorsal valve would be found not only smaller but less convex than the opposite valve.

The fold begins to be apparent in individuals having a length of about 10 mm., and is expressed by the arching of the anterior margin. It does not sufficiently develop to become a characteristic feature, and is more or less undefined, even in many full-grown specimens.

The plications increase both by bifurcation and interstitial addition. The smallest number observed is eight, and this is gradually increased with the growth of the shell, until there are about fifteen principal plications on the body of the shell, and several smaller ones just below the cardinal extremities. The concentric striæ are not often preserved, and the plications therefore form the only conspicuous character of the surface ornamentation.

The delthyrium or triangular opening becomes completely filled by the incurved beak of the dorsal valve.

Camarotœchia acinus Hall, 1863.

(PLATE XVIII, figures 9–11.)

Rhynchonella acinus Hall. Twenty-eighth Ann. Rept. N. Y. State Mus. Nat. Hist., p. 306, pl. 26, figs. 7–11, 1879.
———— Hall. Eleventh Ann. Rept. State Geol. Indiana, p. 306, pl. 26, figs. 7–11, 1882.

Were *Camarotœchia acinus* a rare species, it might readily be confounded with the variety of *C. indianensis* which bears but a single plication in the ventral sinus. It appears, however, to have been very prolific, and its abundance serves to emphasize its specific independence. The liability to confuse it with any of the associated species arises only among forms of immature growth. Beginning with a shell which is apparently in the actual initial stage, measuring 1.2 × .8 mm., the present series is very evenly consecutive up to maturity, when the average dimensions are 8 × 6 mm.

Specific Characters.

Mature Form (Plate XVIII, figures 11–11 *b*). — Shell small, longitudinally ovate, sub-attenuate toward the beak, and truncate in front. Cardinal margins long and rapidly sloping, extending more than half-way across the shell; sides flattened, slightly excavate. Valves sub-equally convex.

Ventral valve full and rotund on the umbonal region, flattened at about the middle, thenceforward sinuate; beak incurved, but not procumbent; foramen generally concealed, or when slightly exposed, elongate or sub-triangular.

Dorsal valve more flattened in the umbonal region and in the middle, whence a low fold proceeds to the margin.

Surface marked by low rounded plications. The ventral sinus bears a single plication which is generally faint, often nearly obsolete. On each side of the sinus are four plications, those abutting on the cardinal margins being indistinct. On the dorsal valve the low, flattened fold bears two plications which are the strongest upon the shell; these are accompanied by three plications on each latus, making the whole number on this valve eight. No concentric growth-lines are apparent. Average dimensions 8 × 6 mm.

Variations from the Normal Adult. — Two plications sometimes occur in the sinus, and in such cases they are each stronger than the single sinal plication in the normal adult. The addition of the plication to the sinus increases the number in the fold to three, and the total number of plications on the shell by two.

Initial Shell (compare Plate XVIII, figures 9, 9 *a*, 9 *b*). — Two individuals, one measuring 1.2 × 8 mm., the other 1.4 × 9 mm., apparently indicate the initial stages in the growth of this shell. Neither of these examples has served well for illustration, on account of the lack of well-defined details, but they may be described as follows: Attenuate, sub-spatulate. *Ventral valve* with erect, straight beak, cardinal area high, convex, with a prominent dorsum. *Dorsal valve* flattened or slightly sinuate. Features of the cardinal area not discernible; from analogy, the foramen would be triangular and unobstructed. In figures 9–9 *b*, which show a secondary stage of growth in the shell, the portion included within the first growth-line will represent very well the characters of the primitive shell.

General Developmental Characters.

The gradual incurvature of the beak and consequent concealment of the ventral foramen may be assumed from the foregoing. It harmonizes with the associated species of the same genus in the slight variation in the form and proportions of the foramen in consecutive stages of growth, as well as in the reversal of the embryonic fold and sinus to the mature sinus and fold. The plications of the latera seem to appear simultaneously after the first varix, as shown in the figures referred to, and their number does not change materially until maturity. The embryonic sulcus on the dorsal valve, correlate with the ventral dorsum in the primitive stage, is continued at maturity into the median sulcus separating the two plications of the dorsal fold.

Camarotœchia neglecta Hall, 1852.

(PLATE XVIII, figures 3, 6–8.)

Rhynchonella neglecta Hall. Twenty-eighth Ann. Rept. N. Y. State Mus. Nat. Hist., p. 162, pl. 26, figs. 1–6, 1879.
—— —— Hall. Eleventh Ann. Rept. State Geol. Indiana, p. 305, pl. 26, figs. 1–6; pl. 27, fig. 3, 1882.

For a species so abundant as this in the Waldron fauna, the diagnostic features are retained with unusual persistence within very narrow limitations. Unlike its associate, *C. indianensis*, which it almost equals in numerical representation, there are no well-established and perduring variations from the normal adult form, and these observations are therefore limited to an essentially unvarying phase.

Specific Characters.

Mature Form (Plate XVIII, figures 8, 8 *a*). — Shell small, transversely sub-ovate; umbo scarcely prominent. Cardinal slopes long and flattened, rounding to the anterior margin which is nearly straight.

Ventral valve with the umbo elevated and slightly incurved at the tip, overhanging an elongate sub-triangular foramen. Umbonal region slightly convex, the convexity extending for one-third the length of the shell; thence forward the shell is rapidly depressed medially to form a deep sinus, which makes a high quadrangular extension on the margin; lateral portions depressed.

Dorsal valve with the umbo low and inconspicuous; apex concealed within the foramen of the opposite valve. The shell becomes rapidly elevated medially to form the fold, the latera being full and convex.

Surface covered with regular, sharp, and prominent plications, which do not vary in number at normal maturity, and which, in the growth of the shell, are increased only from the cardinal margins. Of these plications, the fold bears four, the sinus, therefore, three, and each of the latera five, those nearest the cardinal margins being obscure. This makes in all for the ventral valve, thirteen, and for the dorsal, fourteen plications. Dimensions of an average example, length, width, and depth, 9, 8, and 5 mm.

A single individual presents the only important abnormality noticed; namely, a failure to produce the requisite plications upon the latera, the dorsal valve bearing but five, and the ventral, six. Of these, three are on the fold, two in the sinus. It is interesting to notice that in the umbonal region the normal number of plications had been formed in their regular arrangement; their disappearance on the latera and irregular disposition in fold and sinus took place abruptly upon the completion of a growth-line 2 mm. from the apex.

This is a marked instance of reversion after the assumption of certain adult features.

Incipient Form (Plate XVIII, figures 6, 6 *a*). — The example with which the present series opens measures .75 × .5 mm. It is elongate sub-triangular, with the ventral beak elevated and erect, the cardinal margins sloping for two-thirds the length of the shell; foramen triangular, slightly, if at all, encroaching upon the apex, without deltidial plates, margins thin; dorsal beak rounded, inconspicuous. At one-third the distance from the apex to the anterior margin, fine thread-like plications appear, four upon the dorsal, and three (five ?) upon the ventral valve. The median sulcus on the dorsal valve is broader and deeper than any other, forming the embryonal sinus, and is accompanied by a correlatively strong plication on the opposite shell.

Developmental Variations.

General Form and Outline. — The form of the shell varies from dimensions in which the length is one-quarter greater than the width, to those of maturity when the width is slightly greater than the length. The depressed, sub-spatulate embryo eventually becomes convex and deep. The embryonal sinus and fold on the dorsal and ventral valves, respectively, are never so prominent as in *C. indianensis*, and soon become lost, the former in a sulcus, and the latter in one or more sulci upon the reversed fold and sinus of maturity.

Beak and Foramen. — The erect and acute beak of the elementary stages of growth becomes, at maturity, but slightly incurved, and never procumbent on the dorsal umbo. The foramen, at the outset triangular, subsequently has its margins thickened, and develops small and obscure deltidial plates at its base, which at maturity leave the foramen elongate and not circular. In respect to these features, the development of the species is identical with that of *C. indianensis*.

Plications. — In the first observed stadium only the umbonal area is smooth, and from the analogy of *C. indianensis* it would appear that the initial growth-stage is still wanting. At a size of 2.5 × 2 mm., the number of plications has increased from four to ten on the dorsal, and from three (five ?) to eleven on the ventral valve; and this, added to a pair of extremely obscure plications near the cardinal margins, is the normal number for maturity.

Camarotœchia Whitii Hall, 1863.

(PLATE XVIII, figures 1, 2, 4, 5.)

Rhynchonella Whitii Hall.   Twenty-eighth Ann. Rept. N. Y. State Mus. Nat. Hist., p. 164, pl. 26, figs. 23–33, 1879.
—— —— Hall.   Eleventh Ann. Rept. State Geol. Indiana, p. 307, pl. 26, figs. 23–33, 1882.

Like *Camarotœchia neglecta*, this species is subject to very slight variations at maturity, and its specific expression is well marked, but a certain embarrassment attends the first endeavor to separate the immature individuals from those of allied species.   This, however, disappears with a careful eye properly estimating the essential characters of the species. The earliest stage of growth found measures 2.75 mm. in length by 2 mm. in breadth, and from this size upward to that of 13 × 13 mm. all variations are present.

Specific Characters.

Mature Form (Plate XVIII, figures 2, 2 *a*, 2 *b*). — Shell transversely sub-elliptical; length and width about equal.

Ventral valve shallow; beak high, acute, somewhat attenuate, with the apex slightly incurved, but not concealing the triangular unclosed foramen which reaches entirely across the cardinal area.   At its apex the foramen encroaches slightly upon the umbo, and is narrowed somewhat toward the base by the imperfectly developed deltidial plates.   A median depression makes its appearance at about one-third the distance from the umbo to the anterior margin, and soon develops into a deep sinus with sharply sloping sides.

Dorsal valve deeper and more gibbous; beak inconspicuous,

and incurved beneath the ventral foramen. A strong median fold corresponds in development with the median sinus of the opposite valve.

Surface marked by strong, simple, sub-angular plications, invariably two upon the fold and one in the sinus, with six on each of the latera, making thirteen on the ventral and fourteen on the dorsal valve. Of these the plications near the cardinal margin are low and incipient, but the full number becomes permanent early in the history of the individual. Faint concentric growth-lines are sometimes visible. Dimensions of average adult 11×11 mm.

Abnormalities at Maturity. — The variations from the normal mature form are, as far as observed, wholly due to continued internal growth after individual maturity has been attained, and this is to be regarded as the concomitant evidence of senescence. There may be either a *marginal thickening*, which gives the shell a truncate appearance, or a general *internal thickening*, making the shell unusually gibbous, and forcing the ventral beak over upon the dorsal umbo.

Incipient Form (Plate XVIII, figures 1, 1 *a*). — The youngest individual observed measures 2.75×2 mm.; outline subovate, valves regularly rounded, the ventral being the more convex. *Ventral valve* with an erect, straight beak; apex acute, cardinal margins sloping rapidly forward, and slightly excavate. Foramen simple, triangular, free from deltidial plates, encroaching at its apex slightly upon the umbo; foraminal margins somewhat thickened. Dorsal beak erect but inconspicuous, full and rounded. *Dorsal valve* depressed anteriorly along the median line, this depression corresponding with the broad and low dorsum of the opposite valve. *Surface* of each valve marked by eight single, rounded plications, which extend two-thirds the distance from the anterior margin of the beak, leaving the circumbonal area smooth.

Developmental Variations.

General Form and Outline. — As growth advances, the development is more rapid transversely than longitudinally, and, consequently, the sub-ovate incipient shell becomes, at maturity, broadly transverse. The prominent dorsum of the ventral valve in the embryo is manifest at maturity only in the rounded and prominent beak, and the embryonal sinus in the dorsal valve becomes so thoroughly obsolete at maturity as to be unnoticeable. In stages of development between the dimensions 3.5 × 3 mm. and 6 × 5.5 mm., the ventral valve still retains a slightly greater convexity, but the anterior margin is entire.

Beak and Foramen. — The erect and straight beak of the incipient shell becomes slightly incurved toward maturity, but the cardinal area remains high, exposing the triangular foramen at all stages of growth. Deltidial plates make their appearance early, but never develop sufficiently to meet and enclose the pedicle-aperture, a feature indicative of arrested development, and equally true of the other members of the genus here discussed.

Plications. — The fact that the eight plications on each valve of the incipient shell do not reach the umbones indicates that the initial shell may have been smooth, as it has been shown to be in *R. indianensis*. The subsequent addition of plications takes place slowly and from the cardinal margins.

Camarotœchia indianensis Hall, 1863.

(PLATE XVII, figures 17–28.)

Rhynchonella indianensis Hall. Twenty-eighth Ann. Rept. N. Y. State
    Mus. Nat. Hist., p. 163, pl. 26, figs. 12–22, 1879.
—— —— Hall. Eleventh Ann. Rept. State Geol. Indiana, p. 306, pl. 26,
    figs. 12–22; pl. 27, figs. 4–6, 1882.

Camarotœchia indianensis is, beyond a doubt, the most prolific species in the rich fauna of the Waldron beds, and by virtue of this fact it has been possible to ascertain the developmental phases through illustrative series of exceptional completeness. It is noteworthy that the mature shell of this *Camarotœchia* presents variations from the adult type, which are so great that in a certain sense they might be regarded as passing the limitations of specific identity; however, the general form and expression of the shell are characteristic, so that, in spite of these variations, no confusion with allied species of the same fauna can arise, nor need there be any hesitation to assign to the different forms a varietal significance only. Probably ten thousand individuals of this species have passed under the observation of the writers, and of this large number fully one-half have been immature forms.

Specific Characters.

Normal Mature Form; containing two plications in the sinus of the ventral valve (Plate XVII, figure 21). — Shell sub-triangular or broadly ovate; length nearly equal to, sometimes slightly exceeding the width. Umbo prominent, sub-acute; cardinal slopes extending one-half the length of the shell, and flattened.

Ventral valve depressed convex, rounded at the beak; apex pointed and slightly incurved, exposing beneath it the elongate, narrow foramen and the inconspicuous deltidial plates. Dorsum for the first one-third the length of the shell rounded, thence, anteriorly, gradually becoming depressed. The sinus thus formed bears two strong, rounded plications which are of later origin than the pair which forms its lateral boundaries. The latera each bear three plications with traces of a fourth, making eight (ten) on the entire surface of the valve.

Dorsal valve somewhat deeper than the ventral, flattened above, depressed near the beak along the median line (embryonal sinus), thenceforward becoming gradually elevated into a fold which bears three strong rounded plications. Four similar plications are discernible on each of the latera, making in all eleven plications on the entire valve. Umbo inconspicuous, apex concealed within the foramen of the opposite valve. Concentric growth-lines obscure, or absent. Average dimensions 12 × 12 mm.

These are assumed as the normal characters of adult growth on account of the great predominance of specimens bearing *two* plications in the ventral sinus.

Variations from the Normal. — (A) *Forms with one plication in the ventral sinus*. This variation does not attain quite the size of the average normal adult, but retains the same proportion of length and breadth (size 10×10 mm.). The surface bears ten plications on the dorsal and nine on the ventral valve. In this form the embryonal sinus, visible on the earlier portion of the dorsal valve, is distinctly continuous with the strong sulcus separating the two plications on the fold in the later and marginal portions of the valve. This variation is not of uncommon occurrence, and immature individuals in various stages of development prove that it is a well-established genetic difference, and not merely an occasional monstrosity.

(B) *Forms with three plications in the ventral sinus*. The size and proportions of the normal are retained in this variety, but the shell bears usually three, sometimes four plications on each of the latera, making ten (twelve) plications for the dorsal, and nine (eleven) for the ventral valve. This form is of comparatively rare occurrence, and is not often noticed in an immature stage of growth.

(C) *Forms with four plications in the ventral sinus*. This variation is met with very infrequently, but two individuals having been obtained. While agreeing in size with the normal adult, the crowding of the sinus with plications tends to obliterate both it and the fold upon the opposite valve. Both individuals show the interesting fact that upon the dorsal valve where the fold bears five plications, that is, four sulci, the embryonal sinus is continuous with the *third* of these sulci, in one instance numbering from the right, in the other numbering from the left. Of the five plications which are thus separated into groups of three and two, it is noticeable that the outer member of the group of three is both less elevated and shorter than any other upon the fold.

Monstrous Forms. — The sole evidence of monstrous growth observed is an asymmetrical development of the plications upon the dorsal fold. Examples bearing three plications upon the fold, in rare instances have one of the plications very

large and two quite small, making one broad and one
narrow sulcus upon the fold. The phenomenon may be due
to the strongly developed tendency of the embryonal sinus to
maintain its continuity with a median sulcus even at the expense of the symmetry of the shell.

Initial Shell (Plate XVII, figures 17, 17 *a*). — The initial
shell in this series of *Camarotœchia indianensis* measures .65
mm. in length by .54 mm. in width. It is broadly ovate or
sub-pyriform in aspect, convex posteriorly, and depressed
toward the anterior margin. *Ventral valve* with the umbo
prominent, the beak elevated and erect, with the apex rounded;
cardinal margins rapidly sloping. Foramen sub-triangular,
apical portion broader than usual in the incipient stages of
plicate shells; margins not thickened; deltidial plates absent.
Dorsal valve with a rounded, inconspicuous beak. *Surface* of
both valves quite smooth. A median depression is noticeable
on the dorsal valve near the anterior margin, making this
margin sinuate. This embryo is the smallest that has been
found for any of the series of Rhynchonellidæ; and not only
on account of its minuteness, but also from the entire absence
of plications on its surface and from the elementary character
of the cardinal area, it may be regarded as the actual elemental
or initial shell.

Developmental Variations.

General Form and Outline. — The adult variations from the
normal noticed above seem to be in most instances, and probably would prove to be in all, preceded by well-defined
embryonic series leading up to them. This must be the
case, as the character of these variations, *i. e.* variation in
the number of plications on the median portions of the shell,
is such that they cannot be assumed after the attainment of
the adult condition, as is possible in certain other forms of
variation. But it is not to be assumed that the conformation
of the embryo which eventually produces any of these results
manifests them in the earliest stages of the growth of the
shell; rather, that the shells, under whatsoever variations at

maturity, all have the same unspecialized starting-point. Hence the fact that some of these variations have not shown a complete series of immature stages must be due to the insufficiency of the material, rich as it has been.

Limiting these considerations now to the normal form with *two plications* in the ventral sinus, it may be seen that the initial shell is smooth, and obcordate in outline, with beak erect, while the mature shell is strongly plicate, strongly ovate, and with the beak sharply incurved. The transition from one extreme to the other is through stages of growth between the limits .65 ×.54 mm. (initial) and 12 × 12 mm. (average adult) In growth-stages below 7 × 5 mm. dimensions, the shell is very depressed-convex, the dorsal valve up to about this point retaining a low, broad, median depression, accompanied by a similarly low and broad median elevation on the opposite valve. It is not always possible to determine with accuracy how many plications are carried by this embryonic fold and sinus, on account of not being well limited; but their eventual reversion, in the adult shell, into sinus and fold respectively, marks the feature as an interesting one, to which attention is called more at length in the description of the species *Rhynchotreta cuneata* and *Atrypa reticularis*. Rare instances occur of individuals assuming all the characters of maturity before attaining a length of 6 mm., and from this point up to the normal size for adult growth, mature dwarfs are frequently found.

Beak. — In the initial shell the beak of the dorsal valve is rounded and inconspicuous, and so remains in all stages of growth. In the opposite valve the beak is at first high, erect but not acute, the cardinal margins sloping abruptly; and with increasing age the beak becomes fuller, more and more incurved at the apex, but is never closely procumbent upon the dorsal umbo, as is the case at maturity with most of the plicate species here described.

Foramen. — At the outset the pedicle-aperture is narrowly sub-triangular, reaching to and encroaching upon the apex, free of deltidial plates and with the lateral margins unthickened, *i. e.* elemental in every respect. In the second stage of growth (after the appearance of plications on the surface, dimensions 1.5×1.1 mm.), the apertural margins have become thickened, and directly thereafter the deltidial plates begin to develop, gradually narrowing the aperture at the base. The symphysis of these plates with the valve is marked by distinctly elevated lines. In maturity the deltidial plates have developed sufficiently to close completely the lower part of the aperture, coming together behind the beak of the dorsal valve, and giving to the foramen an elliptical outline constricted toward the apex, where it encroaches upon the umbo. The fact that the development of the foramen is thus interrupted before it reaches the circular outline normal to the adult of most Paleozoic species indicates an embryonic character in the adult, and therefore a subordinate taxonomic position for the species.

Plications. — These appear only after the first stage of growth is passed and after the first growth-line has been formed. As in *Homœospira evax,* they appear over the entire surface of the shell below the growth-line all at once, and from this stage onward to maturity no increase is made in the number, except by intercalation along the margin of the fold and sinus.

Rhynchotreta cuneata Dalman, 1827,
var. *americana* Hall, 1879.

(PLATE XVIII, figures 12–22.)

—— —— —— Hall. Twenty-eighth Ann. Rept. N. Y. State Mus. Nat. Hist., p. 167, pl. 25, figs. 29–38, 1879.
—— —— —— Hall. Eleventh Ann. Rept. State Geol. Indiana, p. 310, pl. 25, figs. 29–38, 1882.

The individuals of this species do not so readily separate into three groups of long, normal, and broad forms, as do those of *Camarotœchia neglecta, Homœospira evax, Whitfieldella nitida,* and others. This seems to be due to the uniformity in the number of plications, and also in the number carried on the fold and sinus. The long and broad varieties do exist, however, but are of such infrequency as to suggest that they are not genetic variations from the typical form occurring in this locality.

The specimens from the Wenlock shales of Dudley show a considerable variation from their American congeners in having more numerous plications, of which a greater number is raised on the dorsal fold and depressed in the ventral sinus. In other respects it is believed that the description here given of the development of the shell will apply to the British form.

Rhynchotreta cuneata, although considered as abundant in the mature state, does not approach, in the number of young specimens, *Camarotæchia Whitii*, *C. neglecta*, *C. indianensis*, *Homœospira evax*, *Whitfieldella nitida*, *Spirifer crispus*, *Sp. crispus*, var. *simplex*, *Atrypa reticularis*, *Rhipidomella hybrida*, and *Dalmanella elegantula*. The entire number of young individuals examined is about one hundred and fifty, ranging in size from .8 to 1.5 mm. in length. The mature forms average about 17 mm. in length.

In the several series selected from the material at hand, it is evident that the shell assumed the characters and form of maturity when reaching a length of about 10 mm. At this period of growth the fold of the dorsal valve becomes elevated, and the sinus of the ventral valve depressed (figure 15, Plate XVIII). Previous to this stage, the dorsal valve is depressed and transversely concave, and the plications of the opposite valve are raised along the median line of the shell.

Specific Characters.

Mature Form (Plate XVIII, figures 14, 14 *a*, 14 *b*, 22, 22 *a*). — Shell triangular, cuneiform, widest across the pallial region. Length equal to about twice the depth of the valves. Beaks compressed laterally, attenuate and pointed.

Ventral valve moderately convex, sub-angular along the latera, marked by a deep sinus, which commences near the middle of the length, and becomes very marked in front, depressing three plications, of which the middle one is detruded more than the others.

Dorsal valve convex, gibbous in the posterior part, with the latera elevated and sub-angular; marked in front by a prominent fold which begins near the beak as a depression carrying four plications, of which the two central ones are usually much more elevated than the other pair.

Area high, closed by two triangular deltidial plates. Perforation of the ventral beak ovate, truncating the apex, and limited below by the deltidial plates.

Surface marked by from eight to ten strong, angular plications, which are crossed by very fine, regular, sharp, concentric striæ. Mature specimens usually measure from 10 to 17 mm. in length.

Incipient Form (Plate XVIII, figures 12, 12 *a*). — The youngest shell detected has a length of 1.5 mm., is flattened, and nearly circular in outline. The dorsal valve is depressed in the middle and carries four plications. The beak of the ventral valve is broadly triangular, exsert, and elevated, with a triangular, open area without deltidial plates.

Developmental Changes.

Contour. — In the earliest stages yet noticed the shell is nearly circular. At a length of 2 mm. it is broadly oval, and at 2.5 mm. it is ovate. The beak in the next advanced stage is more elongate, and when the length of 4 mm. is reached, the shell has a decidedly triangular or cuneate form, which becomes more pronounced up to maturity. All the young and adolescent shells are depressed, the characteristic fulness of the valves not being developed until after the assumption of the features of maturity, and when the shell approaches its normal size.

Fold and Sinus. — The smallest individual shows a slight depression in the dorsal valve, co-existing with the plications, beginning about one-fifth the length of the shell in front of the beak, widening rapidly, and becoming more defined upon approaching the margin. The latera are nearly flat. The depression, or sinus, becomes more pronounced with the advance in growth, until a length of 4.5 mm. is attained. After this period the four bottom plications gradually elevate, the sinus grows shallower, and the front margin of the conjoined valves becomes nearly straight. Upon reaching a length of 9 mm., the two central plications are sufficiently elevated to define the fold, which is hereafter the principal feature of the dorsal valve. The development from this point to full-grown individuals is principally directed to reaching a maximum prominence in the fold, and increasing the shell by increment on the lateral margins of the valves.

The development of a sinus in the dorsal valve, its subsequent obliteration, and the final elevation of the plications into a strong median fold are shown in figure 15, *1–10* of Plate XVIII, in which the undulating lines represent the anterior junction of the valves.

Beak. — The apex of the dorsal valve is strong and pointed, and is visible in all specimens up to a length of about 14 mm. After this stage the shell becomes obese, and the consequent greater inclination of the beak forces it into the foraminal cavity, where it becomes hidden by the deltidial plates.

The ventral valve is uniformly convex in all incipient specimens. The sinus develops at the same period, and in conformity with the fold of the opposite valve. The beak of the initial shell is broadly triangular, perforate at the apex, and directed outward. It gradually becomes narrower and less oblique with advancing growth, and lies in the axis of the shell in full-grown specimens. The initial perforation is a small truncation of the beak, confluent with the open area below. (See figure 16, Plate XVIII.)

Surface Ornaments. — The prevailing number of plications is eight, although it varies from seven to ten in some specimens. The entire number appears at an early period of growth, and in this respect the species offers a marked difference from some of the forms of *Camarotœchia* already considered, in which the plications increase by pairs. In a specimen 1.5 mm. long, they first appear at about one-fifth the length of the shell from the beak. Four plications are included in the depression of the dorsal valve in the incipient stages, and the two central ones finally become elevated, forming the fold in the full-grown shell. Upon approaching maturity, three of the plications in the ventral valve are depressed, the middle one ultimately much more than the others, forming the single strong plication at the bottom of the sinus.

No concentric striæ are shown on the initial shell of this series. In a specimen 3 mm. in length, these begin to develop over the outer third of the surface, as shown in figure 13, Plate XVIII.

Cardinal Area. — The foramen is at first a broad triangular opening, wider than high, with sharp margins, and truncating the beak of the ventral valve. The lateral margins are thickened in a specimen 3 mm. in length (Plate XVIII, figure 13), and the height and width of the area are equal. These proportions of height and width are preserved to maturity, although in some specimens the area is higher than wide. No deltidial plates have as yet appeared, but in the next stage, including individuals having a length of 4.5 mm., there are two narrow deltidial plates developed from the sides of the foramen (Plate XVIII, figure 18). A specimen 5 mm. in length shows the still further increase in the size of these plates, although they do not come in contact, but leave an oval opening extending from the ventral beak down to the beak of the dorsal valve (Plate XVIII, figure 19). The increase in the growth of the deltidial plates along their inner margins brings them in contact under the dorsal beak, in specimens having a length of about 7 mm. (Plate XVIII, figure 20). Further growth truncates their inner angles, thus shortening the deltidial opening. In individuals about 12 mm. long (figure 21), the opening extends but little more than half the length of the area, and the lower margin of the opening is thickened and slightly deflected. Fully matured forms, having a length of from 15 to 17 mm., have a perfora-

tion less than one-half the height of the area, which truncates
the beak more strongly than in younger shells, and the deltidial plates show a defined thickened area below the perforation, often extending to the dorsal beak (Plate XVIII,
figure 22).

Variations. — As already stated, the elongate and broadly
flabellate shells appear to be, in this species, neither common
nor genetic variations. Among the extraordinary develop-
ments are specimens with duplicate plications in the sinus,
and one showing but seven plications on the shell. Another
individual has the initial shell strongly defined by a varix of
growth, and shows on this portion ten plications, but in the
subsequent growth only eight plications are continued, these
alternating at the varix with those of the embryonic shell.

Atrypa reticularis Linnæus, 1767.

(PLATE XX, figures 12–20.)

—— —— Hall.   Twenty-eighth Ann. Rept. N. Y. State Mus. Nat. Hist.,
   p. 162, 1879.
—— —— Hall.   Eleventh Ann. Rept. State Geol. Indiana, p. 304, 1882.

The abundance of this well-known species at Waldron has
afforded the means of studying its developmental stages with
very satisfactory results. All the individuals, from the earliest observed stage upward, agree in contour, there being no
such variation in this respect as has been noticed in some
other species (*e. g.*, *Homœospira evax*, *Whitfieldella nitida*) in
which appear deviations from the normal, producing a long
type and a broad type. The youngest individual detected
has a length of 2.25 mm. and a width of 2 mm., though this
may not be regarded as the initial shell on account of the
presence of partially developed deltidial plates. From this
stage of growth to maturity, the material has afforded every
variation in size and structure.

Specific Characters.

Mature Form (Plate XX, figures 13, 13 *a*, 20, 20 *a*). —
Atrypa reticularis is so widely distributed, historically and

geographically, in Paleozoic faunas, and is so familiar to paleontologists, that a detailed description of its mature form is here unnecessary. It is sufficient to remark that the prevailing expression at this locality does not precisely conform to the type of *A. reticularis*, but is more nearly that variety described by Professor Hall (*Pal. N. Y.*, vol. ii, p. 271, 1852) under the name *Atrypa rugosa*. This is evident from the development of the varical lamellæ, which over the plications are infolded into nearly tubular processes, sometimes produced at a strong angle from the shell to a length of a millimetre or more. On the varices the plications are covered by fine concentric wrinkles. The average size of mature individuals, 25 × 25 mm., is less than that usual to the species when occurring in later, especially Devonian, faunas.

Incipient Form (Plate XX, figures 12, 12 *a*, 15, 15 *a*). — The initial shell, or the actual inchoate period in its formation, is not at present known. The incipient shell of the series here studied is very small, and can be but a few removes from the initial stage. As just observed, it measures 2.25 mm. in length by 3 mm. in width, and shows but two concentric striæ or growth-varices, with a correspondingly slight development of the deltidial plates, so that this shell may be regarded as but two stages advanced from the actual inception of the shell. The test is flat, both valves being shallow and depressed toward the anterior margin; the ventral beak high and erect, the dorsal beak inconspicuous and rounded. The foramen, which is undoubtedly triangular in the initial shell, has, at this stage, its basal angles slightly rounded by the faintly developed deltidial plates. The plications are six in number on the ventral, and five on the dorsal valve, the middle one of the latter not reaching as far toward the beak as those adjoining it, and toward the anterior margin being depressed below the lateral portions of the shell. General outline sub-circular or sub-pentagonal, as in the full-grown shell.

Developmental Variations.

General Form and Outline. — Embryos of less than 3 mm. in length are more nearly circular in outline than at any subsequent period of the existence of the individual. Directly thereafter the hinge-line represents the greatest diameter of the shell, and the outline becomes sub-pentagonal, a feature which is more apparent in young individuals having between 3 and 10 mm. length, as the increasing rotundity of the shell with the approach of maturity has a tendency to obscure, in a measure, this outline. At the earliest stage studied, the dorsal valve is distinctly depressed along the median line, forming a sinus containing a single plication which does not reach to the beak (Plate XX, figure 14, *a*). This sinus gradually becomes shallower, and the plications are increased by intercalation until they are three in number (figure 14, *b*). In the next stage all evidence of a sinus upon the anterior margin disappears, leaving it even and straight as shown in figure 14, *c;* then the anterior edge becomes reflexed, showing, in subsequent stages of growth, a fold where there had previously been a sinus, this fold bearing at first three, then five, and eventually, in the mature individual, seven plications (figure 14, *d, e, f*). This very remarkable reversion of the fold and sinus relatively to the valves which bear them is also seen in the species *Rhynchotreta cuneata*, and in all adult specimens may be clearly traced upon the earlier or embryonal portions of the valves.

Beak. — In the first stage the ventral beak is high and slightly resupinate, exposing the foramen in an inclined plane. It gradually shortens and becomes erect, and when the shell attains a length of 8 mm. it is bent forward, the cardinal area being slightly incurved. Thereafter the inflection of the area increases, concealing first the deltidial plates, and finally the foramen, until at maturity the beak lies appressed upon the embryonal sinus of the dorsal valve.

Foramen. — In the initial shell this is undoubtedly triangular and free from deltidial plates. With the starting-point of the present series, however, plates have begun to develop, thus narrowing the pedicle-aperture, and rounding its basal angles. With the growth of the plates more rapidly along the lower portion of their inner edges, the foramen shortens quickly, while narrowing but slowly, assuming in the second stage (figure 16) a lanceolate, in the third stage (figure 17) an oval, and in the fourth stage (figure 18) a broadly circular outline. In the last two of these stages the deltidial plates have come in contact with each other above the apex of the dorsal valve, and the pedicle-aperture itself has, from the second, if not from the first stage in the series, encroached upon the apex of the valve, so that, as it attains a circular outline, one-half its periphery is formed by the substance of the valve itself, and the other half by the deltidial plates. From this stage upward there is no apparent change in the actual dimensions of the foramen, and therefore with the growth of the shell it becomes relatively much smaller. It appears, however, that with the incurving of the cardinal area and the concealment of the deltidial plates the foramen becomes more and more enclosed by the apical portion of the valve, and it may be that actual contact with the deltidial plates in the last stage of development is lost. In this final stadium, with the procumbent position of the ventral beak upon the dorsal valve, the plane of the foramen is parallel to the surface of the dorsal valve, and the aperture is therefore lost to sight, or visible only at its upper edge.

Plications. — Of the five and six plications visible upon the youngest member of the series, three or four appear to exist on that portion of the shell included within the earliest growth-line, that is, presumptively, the initial shell, and they increase by intercalation until, in the adult, the average number is about sixty for each valve. Concentric lines of growth follow each other with unusual rapidity, particularly in early life.

Summary. — *Atrypa reticularis*, in the development of its beak, foramen, and deltidial plates, is in essential harmony with the other uniforaminate shells here discussed. The reversal of the fold and sinus is an interesting but not unique feature, and by the time it has been completely effected many of the characters of maturity have been assumed. From the degree of exposure of the foramen, it may be judged that the animal remained attached by its pedicle up to adult growth, but with full maturity and the approach of senility the pedicle must have become atrophied and the animal set free.

Homœospira evax Hall, 1863.

(PLATE XIX, figures 1–9.)

Retzia evax Hall. Twenty-eighth Ann. Rept. N. Y. State Mus. Nat. Hist., p. 160, 1879.
———— —— Hall. Eleventh Ann. Rept. State Geol. Indiana, p. 302, 1882.

In this species the superficial features have been found of much more permanent character than is usual in the plicate Brachiopoda from this horizon. Not far from three thousand individuals have been examined, and these show a variation in size from a length of 1 mm. and a width of .8 mm. to a length and width of 25 mm. Throughout the younger stages in this series of variations the feature of primary importance in distinguishing the embryo of this from those of other species, notably *Homœospira sobrina, Camarotœchia indianensis,* and *Camarotœchia Whitii,* is the sinus which exists on both ventral and dorsal valves; and of much accessory value, the comparatively slight variation in the number of the plications on the lateral portions of the valves. These features will be presently adverted to more at length.

Specific Characters.

Mature Form (Plate XIX, figures 2, 2 *a*). — Shell ovate, generally longer than wide, both valves almost evenly convex, and of about the same depth. Anterior margin generally slightly emarginate, on account of the median sinus which exists on both valves. In rare instances a low median fold is developed near the margin of the dorsal valve.

Ventral valve with the beak much elevated above the dorsal, and incurved, so that the plane of the foramen is parallel to the axial plane of the shell. Foramen circular, or slightly sub-triangular; deltidial plates generally obscure on account of the infolding of the beak.

Dorsal valve regularly arcuate, except at the posterior extremity, where the beak is closely incurved beneath the ventral umbo. The median sinus usually carries from three to five plications, but in advanced growth sometimes becomes filled up by the crowding of these plications. Ventral valve with a well-marked sinus, generally bearing three plications. The sinal plications on both valves take their origin in front of the beak, and are of interstitial growth, a fact which does not hold true for any of the other plications. The surface is marked by from eight to twelve rounded, continuous plications on each side the sinus of either valve, all of these extending to the beak, with the possible exception of the more obscure ones on the cardinal slopes. Only in rare instances and abnormally do these plications increase by interstitial addition. Imbricating lines of growth are often present, and fine concentric striæ are sometimes discernible.

The mature individuals of *Homœospira evax* divide themselves into three groups, based on their relative proportions:

(*a*) Normal form, in which the length and width are equal.

(*b*) Long form, in which the length is greater than the width.

(*c*) Broad form, in which the length is less than the width.

In frequency of occurrence the form (*b*) almost equals the normal, while the form (*c*) is more rarely met with. The form (*b*) is also of remarkable persistence, and starts so early in the life of the individual as to suggest a distinct genetic impulse.

Variations from the Normal Development. — These are to an unusual degree very slight, and may be classed as follows:

Obesity, which apparently occurs only when normal full growth has been attained.

A tendency to asymmetry in the development of the sinal

plications. A marked illustration of this is afforded by an individual which, in repairing an injury to its shell, has abruptly developed six plications on one side of the sinus, in continuation of three and to correspond with three on the opposite side.

The absence of plications in the sinus. This is a feature of rare occurrence, and is undoubtedly an infantile character retained in later stages of growth. A single individual of immature growth affords an illustration of a peculiar abnormality, indicating a reversal in the growth to an embryonic condition. This shell (Plate XIX, figures 3, 3 *a,*) has grown to a certain size and normally developed its plications, but an abrupt period has been placed to their development, and over the entire anterior portion of the individual, in front of a stout varix, the surface of the shell is almost smooth. This is the exact counterpart of that mode of growth to which attention is called under other species, where the smooth embryonic condition of the shell seems to be prolonged for more than the usual period of immaturity, and the mature features are thereupon abruptly developed after the formation of a sharp growth-line.

Developmental Variations.

The series of individuals illustrating the embryological changes in this species is so complete as to show by almost imperceptible gradations the entire chain of development from very near the starting-point up to maturity. This series begins with an individual measuring 1 mm. in length and .8 mm. in width, and at this stage of growth the incipient shell has manifestly not received much increment. That this, however, is not the actual primitive shell seems proved by indications of two very indistinct concentric growth-lines, and by the presence of faint radiating plications near the anterior margin, between the second growth-line and the margin itself. It is very probable that the incipient shell consisted of that portion of the individual (Plate XIX, figure 1) lying within the first growth-line, and as this would make its size about .5 × .4 mm., this fact in itself is sufficient apology for the writers' not having detected the earliest stage of its development, even if other causes had permitted its preservation.

Beaks. — In all normally developed individuals less than 5 mm. in length the beak of the ventral valve is erect and exsert. At about this stage of growth a tendency to apical incurvature is manifested, which increases up to maturity, when, under normal development, the entire umbo is evenly incurved, concealing the deltidial plates and often much of the foramen. On the dorsal valve the beak is quite obscure in the youngest forms, and in later stages of growth is concealed beneath the deltidium or incurved beak of the opposite valve.

Foramen. — This appears first as a simple triangular opening, its apex reaching to, but not truncating the apex of the umbo, and it is retained in this condition until the shell attains a length of at least 3 mm. of normal growth. At this age the deltidial plates begin to form, making their first appearance as two minute triangular laminæ, taking their origin in the basal angles of the foraminal triangle, and giving the foramen a lanceolate outline.

By increments to their internal edges, these plates presently come in contact with each other, truncating the interior basal angle of each, the plates being from this period onward in progressive symphysis. The increments to these plates are made more rapidly at and about their interior angles, and, as a result, the foramen assumes successively an elliptical, an oval, and a circular outline. The circular curve of its upper extremity is caused by a slight encroachment upon the beak, and this in mature age is so considerable that the primary or incipient shell is undoubtedly wholly absorbed. The plane of the foramen remains, except in rare instances, always vertical, although the deltidial plates become slightly bent by the incurving of the beak. A striking exception to this rule is represented on Plate **XIX**, figure 9, where an individual which has reached early maturity shows the senile feature of a beak incurved to such a degree as almost to obscure the foramen. On approaching maturity the deltidial plates appear to become anchylosed along their exterior edges, with the shell itself, the line of union being marked with a low ridge, and seem never to be displaced by any distortion of the shell, as so often occurs in *Atrypina disparilis.*

Sinus. — As already noticed, in the elementary shell the sinus begins as a low, smooth depression, equally strong on both valves, and extending almost to the beak. It gradually becomes filled by the radiating plications, which appear first at the sides, and increase toward the middle, never becoming, normally, more than six. None of these plications reach the apex of the shell.

Plications. — On the latera of the shell these seem to appear simultaneously, as shown in figure 1, where three on each side make their appearance at the same stage of growth. This number is subsequently increased to six or eight on each side in mature forms, sometimes becoming greater in old age.

Internal Apparatus. — The brachial supports in this species consist of spirals, coiled in the transverse axis of the shell, with their bases facing each other. In the mature individual the number of coils is from eight to ten. The spirals are connected by an angular loop, the branches of which take their origin on the dorsal limb of the basal coils, and are directed ventrally and backward beyond the axis of the interior cavity, forming at their junction not a simple angle, but a miniature *saddle,* from the posterior extremity of which extends straight backward a little spiniform process. The number of coils in these spires varies with the age of the shell. In preparing a series to show the development of these structures, it appears that the shelly ribbons composing the spirals not only make fewer coils in early life, but that these are of exceeding tenuity in the primary stages of development. The accompanying figures show the extremes of development noticed in these respects, figures 126 and 127 representing the character of the supports in the mature condition, and figure 128 the spirals as developed in an individual having a length of 2.5 mm., where the ribbon makes but two revolutions. As far as can be ascertained, the *loop* undergoes no essential modification in these early stages, though its precise character in the example from which this drawing has been made was not determined, but has been drawn in. The same arrangement, however, has been seen in an individual of but slightly larger growth.

The growth of these spirals consists, primarily, in the addition to the number of coils, and, secondarily, in the thicken-

126   127   128

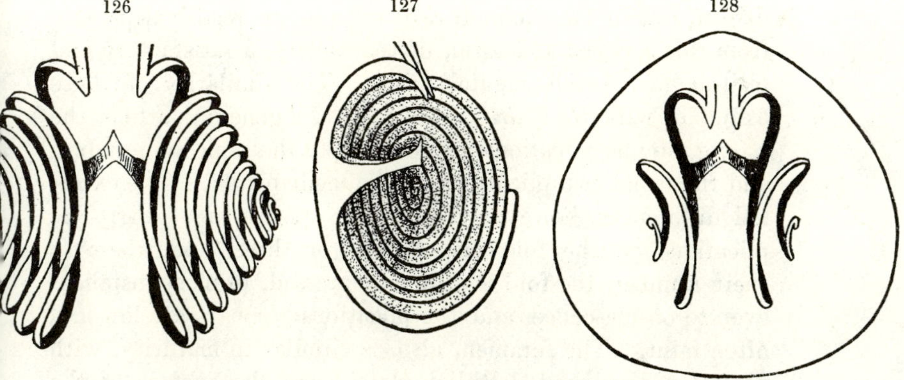

FIGURES 126–128. — Development of internal apparatus in *Homœospira evax* Hall.

ing of the ribbon. In the first case the increase in number must take place by addition to the apices of the coils, and therefore the embryonic or primary coils of the ribbon must be wholly concealed by later depositions upon them, both in length and width. The apparent looseness of the coils in their primary condition, however, must be regarded as largely due to magnification, the distance between successive coils being actually not so great as the distance between the apical turns of the ribbon in the mature spiral.

Homœospira sobrina sp. nov.[*]

(Plate XIX, figures 10–16.)

Rhynchonella Whitii Hall, *in part.*

In the examination of a large number of the specimens which have usually passed under the name of *Camarotœchia Whitii,* the writers have become convinced that, aside from the individuals which agree with the types and the description of the species, there is a series of shells which, in the mature state, may be readily confounded with immature stages of *C. Whitii,* but in their immature condition are readily separable from this species, and form, of themselves, a satisfactory and well-defined developmental series. The similarity of these examples with *C. Whitii* is found in the general outline, the strong, simple plications approximately the same in number, and the usual two plications on the median fold. The external differences, however, in the new species are these: The plications on the fold may be one or three, and whatever their number, the fold is always depressed, in most instances even to obsolescence, and the plications upon it are low and often faint. The foramen, also, is circular in maturity, with perfectly developed deltidial plates, and the surface of the valves usually conspicuously marked by fine, crowded, concentric growth-lines. Internally the difference is more emphatic, as carefully prepared specimens show well-defined spirals having their apices near the lateral margins, as shown on Plate XIX, figure 12. While disavowing the intention of describing new species as remote from the purposes of this paper, the writers have, for convenience in utilizing this form for their work, to which it makes no unimportant contribution, designated it as above, as no doubt exists of its specific value.

Camarotœchia sobrina, one of the more abundant of the Waldron Brachiopoda, is itself subject to some variation, more considerable, indeed, than that noticed in either of the species *Camarotœchia Whitii* and *C. neglecta.* The material

[* Originally referred to the genus *Retzia.*]

represents all developmental stages between the limits of these dimensions: 2 × 1.6 mm. (incipient shell) and 7 × 6.5 mm. (maturity). In its youngest stages it shows a certain degree of similarity with *Homœospira evax*, especially in the sinus on both valves and in the sinal plications. The greater number of the latter in *H. evax*, as well as the more numerous lateral plications, will serve to obviate confusion here.

Specific Characters.

Mature Form (Plate XIX, figures 11, 11 *a*, 11 *b*). — Shell small, rotund, in outline broadly ovate to sub-pentagonal. Valves of equal convexity.

Ventral valve with umbo prominent, attenuate, erect, and slightly incurved at the apex; cardinal margins not excavate, sloping with a faint curve to the sides, whence they round to the anterior edge, which, in the sinal region, is nearly straight; cardinal area distinct; foramen circular; deltidial plates prominent.

Dorsal valve sub-circular in outline, arched in the umbonal region; beak well-defined, apex concealed. In the umbonal region the median portions of both valves are slightly more convex than elsewhere, but this prominence disappears toward the margins, the valves becoming slightly flattened and depressed on the median region near the anterior margin, making a low sinus on the ventral, and a low, depressed fold on the dorsal valve. Both fold and sinus may bear one, two, or three small, often faint and unsymmetrically developed plications, the strongest of which may have its origin in the umbonal region, while the others rarely extend more than half-way across the shell. On each of the latera are four or five strong, angular, simple plications, making thus from nine to thirteen plications on each valve. The increase in these takes place altogether on the fold and sinus, the full quota of lateral plications appearing early in the history of the individual. The plications are covered by numerous fine, concentric growth-lines more noticeably developed near the margin, and at intervals becoming varicose.

The spiral brachial supports each make about five revolutions, which come to an apex near the lateral margins.

Variations from the Normal Mature Form. — An *elongate* form with an unusually high and straight beak is not of rare occurrence, and is the resultant of a very completely represented series of embryonic stages. The species is also subject to *obese* growth, resulting from two sources: (*a*) general internal thickening of the shell, (*b*) marginal thickening. Both are the result of post-adolescent or senile growth, the former producing a round, full, plethoric shell, the latter giving the shell a truncate appearance.

Incipient Form (Plate XIX, figures 10, 10 *a*). — The first stage of growth represented in this series measures 2 mm. in length by 1.6 mm. in width. The valves are sub-equally convex, somewhat depressed anteriorly. Outline broadly ovate.

Ventral valve with beak high, erect, and sub-acute; cardinal slopes broad, steep, and slightly excavate; pedicle-aperture sub-triangular, rounded at the apical angle, and also slightly at the base, by the already developed deltidial plates.

Dorsal valve sub-circular, beak full, rounded, sides slightly appressed, apex concealed. Surface of the dorsal valve marked by two thread-like plications which take their origin medially, just below the umbonal region; thenceforward they rapidly diverge, forming the embryonal sinus, which is, however, soon filled by two small plications. The latera each bear one plication, of earlier age than the sinal, and later than the primary plications. On the ventral valve the plications number the same, but the embryonal fold or dorsum is more strongly marked than the dorsal sinus. On both valves indications of the mature fold and sinus begin with the appearance of the sinal plications.

Developmental Variations.

General Form and Outline. — There is a gradual increase in convexity and diameter with each successive growth-stage, until maturity is reached. The elongate variation from the normal seems to have retained through all stages of growth the proportions of the normal embryo.

Beak and Foramen. — The erect, straight, sub-acute beak of the incipient shell in later growth becomes rounded and slightly arched or incurved. The cardinal area in all stages of development is, however, high, exceptions being made for the more extreme cases of obesity, where the deltidial plates may be concealed, but the foramen is always exposed, and the beak is never procumbent on the opposite umbo. The plates arising from the base of the thickened foraminal margins meet in such a manner as to leave the foramen sharply acute below, and sub-triangular. By their subsequent upward growth and more complete union the foramen becomes circular, the lines of symphysis with the valves still remaining thickened.

In occasional instances a probable slight displacement of the plates outwardly along the median suture gives them the appearance of sloping away from the median to the lateral sutures.

Atrypina disparilis Hall, 1852.

(PLATE XIX, figures 17–23.)

Cœlospira disparilis Hall. Twenty-eighth Ann. Rept. N. Y. State Mus. Nat. Hist., p. 162, pl. 25, figs. 39–43, 1879.
—— —— Hall. Eleventh Ann. Rept. State Geol. Indiana, p. 303, pl. 25, figs. 39–43, 1882.

Although this species is one of the less abundant members of the Waldron fauna, several hundred immature individuals have been found, the youngest of which has dimensions of 2.5 mm. in length by 2 mm. in width, the greatest size at maturity being 6.5 mm. in length by 6 mm. in width. The species being in its mature size quite small and in its surface features quite simple, it does not afford such scope for variations through the later embryonic stages as many of its associated species, and hence it will be noticed that in surface sculpture a permanency of character is retained through all stages of growth.

Specific Characters.

Mature Form (Plate XIX, figures 19, 19 *a*, 19 *b*). — Shell small, sub-oval or cordate, often sub-pentagonal; plano-convex, greatest width along the hinge.

Ventral valve convex, depressed at the sides; beak exsert, and in old forms arcuate.

Dorsal valve flat and sometimes depressed near the beak. The ventral valve is marked by two very prominent plications which pass along the deep median sinus, and are accompanied by two less distinct plications on each of the latera, making in all six plications, of which the two nearest the hinge-line are sometimes obsolescent. The dorsal valve bears a prominent median fold, and two well-marked plications on the lateral portions of the shell. Toward the margin are concentric laminæ of growth. An average adult individual measures 5.5 mm. in length by 5 mm. in width.

Variations in Outline. — The individuals divide themselves into two groups according to their outline:

(*a*) Normal form.

(*b*) Long form.

The first of these groups includes the great majority of all individuals found, which are characterized by a relatively broad figure and sub-circular outline. Members of the second group are comparatively few in number, and are elongate or sub-ovate individuals. The long form (*b*) is well defined in immature growth-stages, and appears to be a permanent varietal difference.

Abnormalities. — A variation in adult shells, noticed only in rare instances, is a tendency to an asymmetrical development in the plications, as shown on Plate XIX, figure 18, where, by unequal growth upon the lateral portions of the shell, the median plication on the dorsal valve is deflected to one side, and the corresponding median sinus on the ventral valve displaced, the axial line of the shell being occupied by one of the strong plications bounding the sinus. Another form of this asymmetry is manifested in the intercalary addition of a single plication on one side of the median plication of the dorsal valve.

A tendency to obesity is often manifested by the shell, at or before reaching the average dimensions of maturity, when it may be supposed that the full growth of the individual has been attained. This obesity is produced by a rapid thickening of the shell at the margins, making the anterior face truncate and forcing the ventral beak over the dorsal in the same manner as if the valves were forced to open along the hinge. It is, therefore, only in individuals which have reached this obese condition that the ventral beak is incurved.

Developmental Changes.

The character of the primal or elemental shell may be seen from a single example (Plate XIX, figure 18), in which the plications are abruptly developed at a distance of 1.5 mm. from the apex, and, presumably, that portion of the shell within this limit represents approximately the size of the original embryonic shell. This portion of the individual is quite smooth, and shows but a trace of the median fold and sinus. As already observed, there is a marked permanency in the surface features of the species from early youth to maturity. The smallest individual obtainable bears, as in the mature condition, six plications on the ventral and five on the dorsal valve, though those near the hinge-line are quite faint.

The *beak* is prominent and exsert, except in obese shells, where it is incurved. In the earliest stage where the character of the *foramen* is well preserved the individual has a length of 4 mm. and a width of 3.5 mm. Here it is seen to be elongate-oval, the deltidial plates having formed to such a degree as to be in contact with each other and to have anchylosed, so that the median suture is detected with difficulty. The lateral sutures always remain distinct even to maturity, and it is evident that the union of the plates with the shell along these joints has not been as firm as in many species, as it is not infrequently found that the plates have been displaced and lost.   Careful search among the smallest individuals has shown no trace of the inceptive triangular outline of the pedicle-groove existing in other species before the formation of the deltidial plates.   It is an important fact that the foramen begins to assume its mature condition so early in the history of the shell, although its development was evidently in conformity with the general type.

The subsequent development of the deltidial plates changes the form of the foramen to that of a circle, as shown in figures 22 and 23.   In the early life of the shell the plane of the foramen is in, or parallel to, the axial plane; at maturity, before any obesity or senile thickening takes place, the foramen, in becoming less elongate, truncates the apex of the valve, and makes a large angle (sometimes almost 90°) with the axial plane; subsequently, with increase in obesity, it becomes again more nearly parallel to this plane.   In the last condition the deltidial plates are curved inward, and often to a large degree concealed.

Meristina rectirostris Hall, 1882.

(PLATE XXI, figures 4, 5, 11–13.)

Meristella rectirostra Hall.   Eleventh Ann. Rept. State Geol. Indiana, p. 301, pl. 27, figs. 10–14, 1882.

This small species is one of the less abundant of the brachiopods of this fauna, and probably has often been confounded with undeveloped individuals of *Whitfieldella nitida*. It presents, however, adult features which will not allow it to be confounded with that species, and although some difficulty arises in separating the diminutive forms of the two species, *M. rectirostris* is characterized by the absence of deltidial plates in every stage of its existence.

The series representing this species does not include stages of growth as early as in some of the others, but is sufficiently complete to permit the statement that, were younger forms accessible, they would probably add little to a knowledge of the developmental changes.   The series begins with individuals having a length of 2.5 mm. and a width of 1.75 mm., the adult form measuring 9 mm. in length by 7.5 in width.

In all stages of growth earlier than that approximately indicated by a size of 6 × 4 mm., it is very difficult, and from the present observations impossible, to draw the line of separation between this species and *W. nitida*, and the fact which has been demonstrated for *Dalmanella elegantula* and *Rhipidomella hybrida;* namely, that in the earliest growth-stages no specific differences are manifest, will be probably found to hold good for these two species also. And in the latter case a considerably larger size is attained by the embryonic forms than is reached by the former species, before the differential characters are assumed. This is due to the fact that these two species, when mature, have essentially no surface sculpture, and differ less in general form and outline than do the mature individuals of *Dalmanella* and *Rhipidomella*.

Specific Characters.

Mature Form (Plate XXI, figures 12, 12 *a*, 12 *b*). — Shell sub-pentagonal or ovate; beak erect, acute, and prominent, rapidly widening toward the base. Lateral margins nearly straight for about one-third the length of the shell, thence rounding to the anterior margin. Valves about equally convex, giving the shell a sub-lenticular contour.

Ventral valve with attenuate, straight, or slightly arcuate beak. Foramen triangular and without deltidial plates.

Dorsal valve more nearly sub-pentagonal in outline; beak incurved into the foramen of the ventral valve.

Surface smooth, or in rare instances showing a faint pseudo-punctate appearance which is entirely superficial. Dimensions of average adult 9 × 7.5 mm.

Incipient Form (Plate XXI, figures 4, 4 *a*). — Shell measuring 2.5 × 1.75 mm. Oval, proportionally longer and narrower than in the adult state. Beak elevated, acute, straight. Foramen of the ventral valve very broad, triangular, extending to the apex. Dorsal beak full, rounded, and inconspicuous.

Shell convex just below the beak, becoming depressed toward the anterior margin.

Developmental Variations.

General Form and Outline. — In the incipient stadia of growth the shell is extremely elongate and quite perfectly oval; the beak of the ventral valve is relatively broad, its lateral margins having a slight outward curve. With growth the shell broadens, and the ventral beak becomes more attenuate, while the greatest width of the shell, instead of being at or below the middle, comes nearer the hinge-line.

Beak. — From being erect, straight, and relatively broad in the ventral valve, at the outset, it becomes, at maturity, narrow, attenuate, and slightly incurved toward the apex.

Foramen. — In the earliest observed stage the foramen is a broad, triangular opening, covering nearly the entire cardinal area, reaching, but not encroaching upon, the apex of the valve. In subsequent stages of development this opening narrows with the narrowing of the beak but, as at no stage deltidial plates are developed, the contraction is due to the encroachment of the cardinal portions of the valve along the foraminal margins. The interesting fact of the persistent absence of deltidial plates throughout the entire existence of the individual may be interpreted as a retention to maturity of a character embryonic in allied species; the small size of the mature shell and the very slight incurvature of the ventral beak also contribute to the embryonic expression of the species.

Whitfieldella nitida Hall, 1843.

(Plate XXI, figures 6–10.)

Meristina nitida Hall.   Twenty-eighth Ann. Rept. N. Y. State Mus. Nat. Hist., p. 160, pl. 25, figs. 1–7, 1879.
—— —— Hall.   Eleventh Ann. Rept. State Geol. Indiana, p. 300, pl. 25, figs. 1–7, 1882.

Whitfieldella nitida is a very abundant and characteristic fossil in the Niagara fauna of central Indiana, reaching a much greater development both in size and numbers than in

the New York outcrops of the formation. The individuals
vary in size from a length of 2.5 mm. and a width of 1.75
mm., to a length of 25 mm. and a width of 22 mm., and the
series representing these variations is, on account of the
abundance of specimens, very complete within these limits.
It is, however, a noticeable feature of the species that in
most respects, except size, the characters of maturity are
assumed early in the life of the individual, and as the form
is essentially devoid of surface sculpture, the interest in
its development rests to a larger degree than usual upon
abnormalities in individuals either mature or approaching
maturity.

Specific Characters.

Mature Form (Plate XXI, figures 9, 9 *a*, 9 *b*). — Shell broadly
sub-pentagonal to ovoid; beaks extended and more or less
prominent.

Ventral valve with the greatest convexity at about one-
third the length of the shell in front of the beak: Beak
arched, incurved over the dorsal valve; apex evenly truncated, the circular foramen lying in a vertical plane. Cardinal slopes extending for more than one-half the length of the
shell. A low median depression is noticeable on the younger
portions of the valve.

Dorsal valve with a similar convexity; beak incurved and
concealed. A very low and inconspicuous median elevation
corresponds with the depression on the opposite valve.

Surface smooth or with fine concentric striæ and a few
conspicuous lines of growth toward the anterior margin.
Average individuals measure 20×16 mm., large examples
not infrequently 25×21 mm.

Variations in Outline. — Two very distinct groups of forms
are evident in this species, in one of which, (*a*) *normal*, the
outline of the mature shell is obcordate or sub-pentagonal.
When immature, the anterior margin is evenly circular, but
in all cases the proportion of length to width is essentially
the same. Probably five-eighths of the specimens found
belong to this group.

In a second group, (*b*) *long form*, the shell is elongate-spatulate, and proportionately deeper than the normal, but, with a single exception, individuals have not been observed to exceed a length of 10 mm. and a width of 7 mm. This variation is so persistent that it appears to be well founded genetically, and not merely an occasional occurrence. Tracing backward from the mature shell to the earlier stages of development, both this and the normal form are found merging into each other; hence both have had a similar starting-point. The long form, however, reaches maturity of development at a very early age, and never approaches the size or proportions of the normal adult. A tendency to *obesity* is especially noticeable in the group (*b*), the majority of such individuals being below the normal full-growth.

A single adult example shows traces of broad, rounded plications on each side of the fold and sinus, a singular condition in a species uniformly non-plicate.

Incipient Form (Plate XXI, figures 4, 4 *a*, 10, 10 *a*). — Shell 2.5 mm. in length by 1.75 mm. in width; elongate-oval; beak elevated, straight, acute. Pedicle-aperture of the ventral valve very broad, triangular, extending to, but not encroaching upon the apex. Dorsal beak full, rounded, but inconspicuous. Valves convex just below the beaks, becoming depressed toward the anterior margin. The shell is proportionally much narrower than the adult form.

The starting-point of this series is precisely the same form of shell as that taken for the incipient stage in the species *Meristina rectirostris*. Under the discussion of that species reference has been made to the impossibility of separating these two species, in their earlier stages, and the impression of the specific characters may be regarded as of subsequent development.

Developmental Variations.

The surface characters being unvariable, the important changes in development are confined, as far as observable, to the pedicle-aperture and deltidial plates. As already observed, the beak is incurved so early in the history of the individual that these embryological changes can be observed only in very young specimens. This incurvature of the ventral beak appears to become fixed earlier in the normal than in the elongate form. For example, figure 10, Plate XXI, represents an elongate individual with a length of 2.5 mm. and a width of 1.5 mm., with an open triangular foramen, and no apparent development of the deltidial plates; but the normal form of the same size has the plates developed, the foramen nearly circular, and the beak incurved. In the condition represented in this figure, the embryos of this species are readily confounded with *Meristina rectirostris*, in which the triangular aperture is retained until maturity. The latter species is, however, distinguishable in all the later stages of its existence by the body of the shell being broader and the ventral beak narrower and more attenuate.

Individuals which show the deltaria in their different phases are difficult to obtain on account of the tendency of the beak to incurvature as soon as the plates begin to form. An individual is represented in figure 7, Plate XXI, of somewhat abnormal height of beak, showing an intermediate stage of growth in the plates and the formation of the foramen; and in figure 8 an individual of the same size, with the foramen circular and the deltaria completed and concealed by the infolding of the beak.

Meristina Maria Hall, 1863.

(PLATE XXI, figures 1–3.)

Meristella Maria Hall. Trans. Albany Inst., vol. iv, p. 212, 1863.
Meristina Maria Hall. Twenty-eighth Ann. Rept. N. Y. State Mus. Nat. Hist., p. 159, pl. 25, figs. 8–12, 1879.
—— —— Hall. Eleventh Ann. Rept. State Geol. Indiana, p. 299, pl. 25, figs. 8–12, 1882.
Whitfieldia tumida * (Dalman sp.) Davidson. Supp. Brit. Sil. Brach., p. 107, 1882.

This species may be regarded as presenting a general external form and effect diametrically opposed to that in

* The late Mr. Davidson identified the Waldron species with the *Atrypa tumida* of Dalman, the type of his genus *Whitfieldia*.

Meristina rectirostris. To the erect, attenuate, acute beak, open pedicle-aperture, shallow valves, and asinuate anterior margin of the latter, the full, rounded, incurved beak, concealed cardinal area, ventricose valves, and strongly sinuate anterior margin of *M. Maria* are strongly contrasted. Between the mature characters of these two species, *Whitfieldella nitida* is conspicuously mediate.

Immature specimens of *M. Maria* are far from abundant. Indeed, the present series shows only about thirteen different grades of development, and the smallest individual which can be referred with certainty to the species measures 6 × 6 mm. (adult 29 mm. in length by 32 mm. in width).

The writers have, however, assigned to the species a minute embryo measuring .75 × .75 mm., and if this is correctly done, the embryos of this species in the earliest stages of growth differ from those of the other non-plicate species here discussed, in a much stronger tendency toward a circular outline.

The beak of the ventral valve becomes incurved, and the cardinal area obscured very early, so that the discussion of the development of these parts is necessarily much curtailed.

Specific Characters.

Mature Form (Plate XXI, figures 3, 3 *a*). — Shell comparatively large, ventricose, broadly ovate or sub-pentagonal.

Ventral valve gibbous in the umbonal region, with a low, broad dorsum extending from the umbo to near the middle of the valve, where it becomes flattened, sinuate, and at the anterior margin is reflected dorsally into a linguiform extension. Beak closely incurved over the dorsal valve, fully concealing the foramen. Cardinal slopes angulate and slightly excavate.

Dorsal valve evenly convex, somewhat gibbous, strongly arcuate transversely along the dorsum, which becomes elevated into a low fold, deeply emarginate in front for the reception of the extension from the opposite valve. Beak obtuse, incurved, and concealed.

Surface smooth, marked by concentric growth-lines near the margin.

Occasionally individuals of large growth show a greater length than breadth, presenting an elongate form, but this variation seems to be due to more rapid axial growth after the attainment of adult size, and does not manifest itself in the incompletely developed shells.

Incipient Form (Plate XXI, figures 1, 1 *a*, 2, 2 *a*). — The minute shell which appears to have been the initial form for the species has a circular outline and depressed convex valves. The ventral valve is evenly convex, with the beak erect, short, and broad. The cardinal area is low, the foraminal aperture triangular, reaching to, but not encroaching upon the beak. The deltidial plates are absent. Dorsal valve with the beak not incurved, but inconspicuous. Neither valve bears any trace of a median elevation or depression.

Developmental Variations.

General Form and Outline. — An inclination toward a lenticular form and circular outline is noticeable in all immature individuals. Until a size of about 18 × 18 mm. is attained, there is rarely, if ever, any trace of the strong marginal fold of maturity.

Beak. — The low but erect ventral beak of the initial shell has, in the next stage of growth, become inflected and obtuse, not, however, so as to conceal the foramen, which remains apparent above the apex of the dorsal valve, until the rapid increase in convexity, which immediately precedes maturity, sets in. Thereafter the ventral beak becomes more closely incurved, and thrust over upon the dorsal valve, to the loss of all external trace of the cardinal area.

Foramen. — The elemental hiatus is shown in the initial shell, and the subsequently developed deltidial plates appear in the next growth-stage. In the latter case the foramen has become nearly if not quite enclosed, and has also encroached upon the apical portion of the valve, which forms about one-half its periphery. In all subsequent stages of growth the deltidial plates are concealed, and whatever portion of the foramen appears thereafter above the dorsal valve is enclosed by the circumbonal tract. With the approach of maturity this gradually disappears, and at full growth every trace of it has become obliterated.

Spirifer crispus Hisinger, 1826.

(PLATE XX, figures 6, 7.)

Spirifer crispus, var. *simplex* Hall, 1879.

(PLATE XX, figures 4, 5.)

Reticularia bicostata Vanuxem, 1842, var. *petila*
Hall, 1879.

(PLATE XX, figures 1–3.)

Spirifera crispa Hall. Twenty-eighth Ann. Rept. N. Y. State Mus. Nat.
　　Hist., p. 157, pl. 24, figs. 6–12, 19, 1879.
—— —— Hall. Eleventh Ann. Rept. State Geol. Indiana, p. 295, pl. 24,
　　figs. 6–12, 19, 1882.
Spirifera crispa, var. *simplex* Hall. Twenty-eighth Ann. Rept. N. Y.
　　State Mus. Nat. Hist., p. 157, pl. 24, figs. 1–5, 1879.
—— —— —— Hall. Eleventh Ann. Rept. State Geol. Indiana, p. 286,
　　pl. 24, figs. 1–5, 1882.
Spirifera bicostata? var. *petila* Hall. Trans. Alb. Inst., vol. x, abstract,
　　p. 15, 1879.
—— —— —— Hall. Eleventh Ann. Rept. State Geol. Indiana, p. 297,
　　pl. 27, figs. 8, 9, 1882.

The three forms which are here treated together are closely allied in all their general characters. It is in their initial stages, however, that the resemblance becomes more than superficial, for, in young shells of less than 2 mm. in length, it is difficult, and sometimes impossible, to refer them to any one of the three groups. A general expression of the common characters is furnished by the young of *Spirifer crispus*, var. *simplex*, illustrated by figure 4, on Plate XX.

Taking *Reticularia bicostata*, var. *petila*, as the simplest form, the young shell is found to be nearly circular in outline, with a single, broad, median fold on the dorsal valve. Passing to *S. crispus*, var. *simplex*, of the same size, the outline is seen to be broader, and there is an incipient plication on each side of the median fold. The outline is still broader in *S. crispus*, becoming decidedly sub-elliptical, and the two lateral plications on the dorsal valve are nearly equal in strength to the median fold. The surface ornamentation consists of fine spinulose, or granulose, concentric striæ, differing very little in any of the three species.

In tracing the development of *R. bicostata*, var. *petila*, the shell is found to retain its embryonic characters up to full growth, neither materially changing its form, nor adding to the primitive number of plications. Likewise, *S. crispus*, var. *simplex*, changes very little except to increase in width and add a pair of plications at maturity. Individuals of *S. crispus* develop parallel to the variety *simplex*, up to a length of 5 mm., or until about two-thirds the size of full-grown examples is attained. Subsequently, more plications are added, increasing the number from three or five to eleven, but otherwise the general features of the shell are unchanged. Even the relative convexity of the valves remains the same at all periods.

In the incipient forms the cardinal line extends for about one-fourth the width of the shell, and at maturity measures three-fourths of this width. The foramen does not develop at the same rate; at first it occupies one-half or one-third of the ventral area, but advancing growth gradually diminishes this ratio, until it is one-fourth or one-fifth the size of the hinge-area. Two narrow, triangular, deltidial plates are present in full-grown individuals, but they do not serve to close the fissure, which remains open in all stages of growth.

S. crispus, var. *simplex*, reaches a width of 8 mm., and *S. crispus* often measures 22 mm. in width. Occasionally a specimen of *S. crispus* of the usual size is found with but seven plications on the dorsal valve, suggesting a very large example of the variety, or that the characters of the smaller and simple form are sometimes continued far beyond the period when they usually disappear. Also, the features both of the species and variety may be combined in a single specimen, as one abnormal example has three plications on one side of the median fold and four on the other.

Spirifer radiatus Sowerby, 1825.

(PLATE XX, figures 9–11.)

Spirifera radiata Hall. Twenty-eighth Ann. Rept. N. Y. State Mus. Nat. Hist., p. 157, pl. 24, figs. 20–30, 1879.

———— ———— Hall. Eleventh Ann. Rept. State Geol. Indiana, p. 296, pl. 24, figs. 20–30, 1882.

The series of specimens representing the gradation in size from very young to mature forms is quite complete, but, unfortunately, the characters of the most interesting feature, the deltidium, are not well shown. The foramen is usually but partially closed when the shell reaches nearly its full dimensions, and at this period the beak of the ventral valve is so incurved and thickened that the detailed development of the deltidial plates is obscured, and rendered difficult of interpretation.

This species has been so fully discussed in all its aspects, on account of its wide geographical distribution and varied physical conditions, that a diagnosis of the adult form is unnecessary in this place (*vide* Plate XX, figures 11, 11 *a*).

Incipient Form (Plate XX, figures 10, 10 *a*). — The smallest example yet detected has a length of 1.5 mm. The specimen is not well preserved, and the one used for illustration and description is somewhat larger, measuring 2 mm. in length. The differences appear to be so slight that the characters of the larger may well be applied to the smaller individual.

The shell is nearly circular and flattened, with the beaks not incurved but directed outward. The area of the ventral valve is broad, triangular, open, and extends nearly the entire length of the cardinal line. The incipient dorsal fold and ventral sinus extend nearly to the beaks, and on each side there are about ten radiating striæ. Radii are also present on the fold and in the sinus.

Developmental Changes.

The changes in the shell from advancing growth are principally the gradual widening of the valves, on account of the extension of the cardinal line and extremities, and the incurving of the beaks, from the progressive increase in the depth of the valves. From being circular in outline, the shell slowly widens until it is one-seventh wider than long. The ventral beak in old specimens is so arched over the area as nearly to conceal it, and prevent the opening of the valves to any extent. In the early stages the depth of the conjoined valves is about half the length of the shell, while in obese mature forms the depth is equal to the length.

The deltidial plates first appear as narrow elevated laminæ along the sides of the fissure under the ventral beak. A specimen about half-grown shows them as represented in Plate XX, figure 9, consisting of triangular plates approximately as in figure 3′ of the following diagram. They are subsequently united along their inner margins, and rarely, in the material at hand, can any appearance of a foramen be discovered. In old shells the growth and thickening of the pseudo-deltidium makes it rugose, and it nearly closes the area.

From an examination of a number of species of *Spirifer* showing considerable variety in the mode of development of the pseudo-deltidium, it is believed that there is no essential difference, and that all intermediate conditions between the features represented in *Spiriferina* by Deslongchamps (see Summary) and the characteristic mode of development in *Terebratula* and *Rhynchonella* occur in this group. The genus *Spirifer* presents all these stages. In some species the area is apparently closed by growth from the apex, and in others by the meeting of the deltidial plates at the base of the area and inclosing a foramen as in *Rhynchonella. Spirifer niagarensis, S. perlamellosus,* and *S. cumberlandiæ* are examples of the former mode, and *S. sulcatus* and approximately *S. radiatus* represent the latter. Both conditions are reached by accretion along the inner edges of the deltidial plates. The initial state is represented by a narrow elongate lamina on each side of the triangular area. Further growth produces a triangular plate, and to the form of the triangle is due the apparent growth of the pseudo-deltidium from the apex of the fissure downward, or from the base of the fissure toward the beak of the ventral valve.

The accompanying diagrammatic outlines (figure 129) serve to illustrate the changes and the final results.

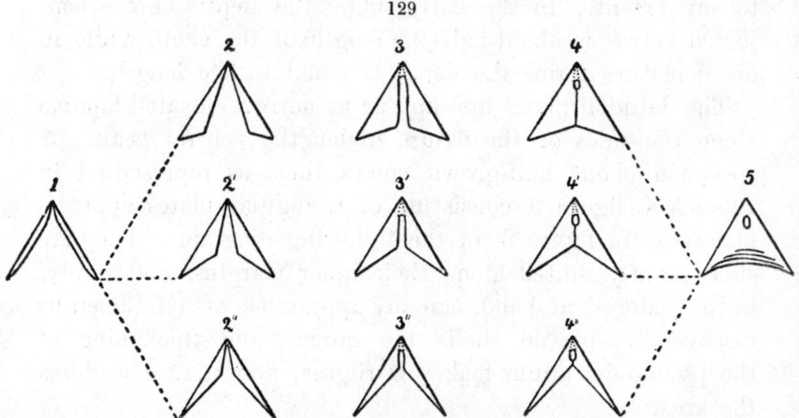

FIGURE 129. — Deltidial development in *Spirifer*.

Figure 1 represents an area in an early stage of growth, with a narrow deltidial plate on each side, alike for each series.

Figure 2 shows scalene triangular plates, with the shortest side at the base of the area.

Figure 2′ shows plates with the two free edges more nearly equal.

Figure 2″ presents narrow triangular plates as in figure 2, but with the shortest edges in the apex of the area.

In figures 3, 3′, 3″, the growth has continued in the direction initiated in the preceding stage, and the apex of the area has been partially filled from the internal thickening of the beak.

Figures 4, 4', 4" show the completed deltidial plates with the circular perforation. The plates in figure 4 nearly close the area, while in figure 4" the opening is nearly as high as wide. Further growth can now take place only along the lower free edges of the plates.

Figure 5 represents the results of subsequent growth and thickening, which have obliterated the evidences as to the mode of development, and unified all three cases. The position of the foramen below the apex of the area does not appear to be due to the approximation and union of the deltidial plates, but to the lowering of the actual cavity of the beak from the natural thickening of the shell, so that the foramen, as in other genera, is at the real termination of the ventral umbonal cavity.

It is seen that the manner of development is alike in each case, varying only from differences in the form of the plates in the earlier stages. The finished pseudo-deltidium is also the same, although the methods of attaining the result differ in each.

Figures 1, 2, 3, 4 are represented by *S. sulcatus*,* and vary in no important particulars from the mode of development in *Terebratula* and *Rhynchonella*.

Figures 1, 2', 3', 4' are partially represented by *S. radiatus*, although in this species the circular foramen is usually obliterated by subsequent thickening and growth. (See figures 9, 10, 11, Plate XX.)

Figures 1, 2", 3" are well shown in *S. niagarensis* (figure 8, Plate XX), and the subsequent stages appear in mature forms of *S. perlamellosus* and *S. cumberlandiæ*. Other forms, notably those with elevated areas (such as *S. macronotus*, *S. medialis*, together with *Cyrtina* and *Cyrtia*), present considerable differences in the completed pseudo-deltidium, due principally, it is believed, to the internal thickening of the beak and the growth of the transverse septum.

The pseudo-deltidium of *Spirifer* thus appears to be the

---

* State of New York, Report of the State Geologist for the year 1882, published 1883, pl. ix, figs. 1, 2, 3.

exact homologue of the deltidial plates in *Terebratula, Rhynchonella*, etc., and to be radically different from the deltidium of *Strophomena, Stropheodonta, Orthothetes*, and allied genera.

Summary of Developmental Changes.

Size and Contour. — Although the species described in the preceding pages present a wide variation in form and general appearance, the nature of the changes which take place in the development of the shell is remarkable in its uniformity.

In nearly every species the inceptive state is represented by a shell having a sub-circular outline, with valves of slight convexity. This phase usually disappears before the individual reaches a length of 1 mm., after which the specific characters are assumed, and are progressively emphasized with each succeeding increment.

On comparing the incipient stage in these fossil shells with that of recent brachiopods, as given by Mr. E. S. Morse for *Terebratulina* and by Mr. W. K. Brooks for *Glottidia*, it is found that, in respect to actual size, there is a slight, though perhaps unessential difference. At the earliest stage of growth figured by Morse,[*] the shell has a length of about .3 mm. and in the next stage represented, of approximately .6 mm.

The first two stages of the shell figured by Mr. Brooks [†] represent free animals, and measure .24 and .3 mm. in length, respectively. The shell became attached by the pedicle only upon attaining a length of 2.5 mm.

Most of the fossil forms have furnished evidence, either from actual elemental specimens or from the apical portions of subsequent incipient stages, that the true initial shell did not reach a size of more than .5 mm. in length. Soon after

[*] On the Early Stages of *Terebratulina septentrionalis*. *Mem. Boston Soc. Nat. Hist.*, II, pl. i. figs. 2, 3, 1869.

[†] The Development of *Lingula* and the Systematic Position of the Brachiopoda. *Johns Hopkins University, Chesapeake Zoöl. Lab.*, pls. i and ii, 1879.

this period the characters of each species become developed
and impressed upon the shell more or less gradually.

Even such distinct groups as *Camarotœchia, Spirifer, Athyris, Rhynchotreta, Anastrophia, Nucleospira*, and the meris-
toids, in their initial stages, approach one another so closely
that they can be determined only from comparatively trivial
features. They are alike in form, contour, convexity, beaks,
and cardinal area, and the only marked differences are to be
found in the faint indications of plications, striæ, folds, and
sinuses. For species of some genera, as *Dalmanella, Rhipidomella, Meristina*, and *Spirifer*, even these characters are
not determinative, and it is impossible to refer certain
embryos to their proper places.

From the foregoing statements it would be naturally
inferred that the species which at maturity present characters abnormal to the typical structure, have been diverted
from the harmony which existed in the incipient stages, with
the other members of the group. This has been shown to be
the case in all the foregoing reversed species examined, belonging to the genera *Anastrophia, Strophonella*, and *Mimulus*.

Beginning with the initial shell having a circular outline
and depressed valves, it is found that subsequent growth takes
place about the periphery, and in the majority of species the
convexity is gradually increased until maturity is reached.
This assertion does not hold true for such forms as the
Strophomenidæ, which vary in convexity, either very slowly
or not at all, up to individuals about half-grown, when the
valves become more or less deflected and often concave.
Such reversion in the shell is in conformity with the degeneracy which is traced in the development of the cardinal
area and pedicle-sheath, mentioned on a subsequent page.

The observations of Brooks and Morse, in the works cited,
show that in both the hingeless and the hinged brachiopods,
as represented by *Lingula* and *Terebratulina*, the early
stages of the shell approach a sub-circular outline, and Brooks
remarks (*op. cit.* p. 43), that "the recent and fossil shells of
the various species of *Crania, Lingula, Lingulella*, and *Obolus*,

and other hingeless brachiopods, furnish a series of adult forms representing all the changes through which the outline of *Lingula pyramidata* passes during its development."

In these respects, then, uniformity is established in the embryology of the ancient Silurian types and their modern descendants.

Valves. — The dorsal valve in young shells is smaller than the opposite, and usually more depressed. These relations, as a rule, are continued up to adult size, except that the ventral valve often increases more rapidly in convexity, producing a consequent incurving of the beak over the cardinal area; as in *Dalmanella* and *Camarotœchia indianensis*. Some species present both beaks as incurved, a condition well represented in *Meristina Maria*, *Dictyonella reticulata*, *Camarotœchia acinus*, and *C. neglecta*. In *Anastrophia* the comparative relations of the valves become reversed from their initial condition, on account of the more rapid increase in the depth of the dorsal valve, so that, at maturity, the dorsal beak is much incurved, and often the umbo extends beyond that of the other valve, although the beaks preserve their normal condition of superposition.

Several of the species show an embryonal sinus in the dorsal valve, with a corresponding fold in the ventral, beginning soon after the initial stage of the shell has been passed, and disappearing before the shell is half grown. Those forms presenting this feature to a marked degree are *Rhynchotreta cuneata*, *Camarotœchia Whitii*, *C. indianensis*, *C. neglecta*, *C. acinus*, *Atrypa reticularis*, and *Homœospira sobrina*. In *Rhynchotreta cuneata* and *Atrypa reticularis* (Plate XVIII, figures 12–15, and Plate XX, figures 12–14), the gradual inception of this sinus, its maximum development and obsolescence, and, finally, its reversion into a fold which thereafter persists and usually increases in prominence in all the succeeding stages of growth, have been shown. The embryonal sinus is not present in *Spirifer*, *Anastrophia*, *Dictyonella*, *Meristina*, *Whitfieldella*, *Dalmanella*, *Rhipidomella*, nor in the Strophomenidæ. Such of these as show a dorsal fold or ventral sinus have them developed early in the growth of the shell, and they usually increase regularly to the time when the full size of the shell is attained.

Beaks. — The beak of the ventral valve in its earliest condition is commonly erect, pointed outward, and of a broad triangular form, while that of the dorsal valve is small, not prominent, and lies in the longitudinal axis of the shell. In all cases the subsequent deepening of the valves tends to incurve the beaks toward the cardinal area. The degree of incurvature varies greatly in the different species. *Meristina rectirostris* shows a minimum, and *Meristina Maria* or *Atrypa reticularis* a maximum, and between these limits all intermediate conditions occur. The usual degree of incurvature is presented in *Spirifer radiatus, Homœospira evax*, and the Rhynchonellidæ.

The outlines on Plate XVIII, illustrating the profiles of the beaks in a series of *Rhynchotreta cuneata*, represent an uncommon condition, for in this species the ventral beak, from its divergent initial position, gradually approaches, and at maturity attains parallelism with the longitudinal axis of the shell. It never becomes sufficiently incurved to conceal to the slightest degree the deltidial area, while the initial dorsal beak becomes more and more incurved, until, finally, it lies entirely within the ventral umbonal cavity.

Those species furnished with a circular apical perforation, as *Atrypa reticularis, Homœospira evax,* and *Rhynchotreta cuneata*, lose the initial point of the ventral beak from absorption, due to the increase in the size of the perforation or to its final terminal position. In *Atrypa reticularis*, or *Meristina Maria* even, both beak and perforation are destroyed, from the forcing of the ventral beak into contact with the dorsal umbo, produced by the great increase in the depth of the valves from growth along their anterior margins.

Cardinal Area. — Omitting for the present the Strophomenidæ and Orthidæ, the initial state of the ventral cardinal area for all other forms is a broad triangular opening beneath the beak, with simple sharp margins. This condition is never passed by *Meristina rectirostris*, which shows a uniform, open, triangular area in every period of growth.

A further advanced state of progress initiates the deltidial plates, which first appear as narrow laminæ along the sides of the area. The areal development of *Spirifer crispus*, *Camarotœchia neglecta*, and *C. acinus* ceases at this point.

In the next stadium the further growth of the deltidial plates along their free edges gives them a triangular form, and they tend to narrow the limits of the opening and define the peduncular foramen. *Spirifer niagarensis* and *Camarotœchia Whitii* represent species which are arrested at this period.

The completed growth shows the deltidial plates uniting by symphysis along a median line, and enclosing near the apex of the area a more or less circular pedicle perforation. *Rhynchotreta cuneata*, *Meristina Maria*, *Homœospira evax*, etc., after passing through all the earlier conditions, reach this limit of development.

The results of senile and extravagant growth often obliterate or degenerate the normal deltidial advancement, the plates becoming thickened and their features obscured, while in some species processes are given off, as in a number of the Mesozoic Rhynchonellidæ.

The cardinal area of the Strophomenidæ in its early phase shows a small pedicle-sheath for the ventral valve and a narrow grooved process under the beak of the dorsal. The perforation for the passage of the peduncle does not materially increase in size with the growth of the shell and often is obliterated, while the dorsal callosity usually reaches a considerable development.

Additional evidence of the degeneracy of the pedicle is afforded by many species of other genera, which have a calcareous attachment to foreign objects at the apex of the ventral valve, the pedicle, therefore, becoming functionally obsolete.

Observations having some analogy with the facts here presented have been made, in a very restricted sense and usually incidentally, by various authors. The present results, though derived from the species of a single fauna, must not be given too limited an application, for they involve nearly every important family of Paleozoic articulate brachiopods, and it may be tentatively assumed that, as a rule, the essential features of variation observed in any member of a genus will hold good of the other members. In regard to the development of the characters of the pedicle-passage, *i.e.*, the deltidial plates and the foramen, there is good reason to regard the process as substantially identical in all the genera represented, making the necessary allowance for the peculiar variation seen in the Strophomenidæ, which may not, however, prove it an exception to the general statement.*

The various terms which have been sometimes applied to the condition of the deltidial plates in the rostrate genera — as *deltidium amplectens*, when the foramen is entirely surrounded by the plates, as in various Mesozoic Rhynchonellæ (but in no Paleozoic species, as far as known); *deltidium sectans*, when the plates bound the foramen only on the lower side, the upper side encroaching on the substance of the umbo, as in *Terebratula Whitfieldella*, etc.; *deltidium discretum*, when the plates do not come into contact, as in *Terebratella*, some species of *Rhynchonella*, etc., — must be regarded as having no further significance than to express the existing condition of the foramen and deltidial plates in any given *specimen;* that is, as indicating a stage of development, not necessarily a generic or even specific character.

The observations of M. Eugene Deslongchamps upon these features are of much value, and in most respects, as far as carried out upon related forms, are in harmony with those here expressed (Note sur le développement du deltidium chez les brachiopodes articulés: *Bull. Soc. Géol. France,* 2ᵉ ser. t. xix, pp. 409–413, pl. ix, 1862), but with his conclusions

---

* See footnote, p. 393.

there are some points of difference. The investigations referred to were made upon one (or more) Mesozoic species of *Terebratula*, *Rhynchonella*, and *Spiriferina*, specific designations not given. The illustrations of *Terebratula* (figures 1 *a*, *b*, *c*, and column A, *a*, *β*, *γ*, *δ*) show in effect the characters seen in *Whitfieldella nitida*, *Meristina Maria*, and others; those of *Rhynchonella*, early stages of similar character, resulting in a *deltidium amplectens*, such, as just observed, have not been found in Paleozoic Rhynchonellæ.

In *Spiriferina*, according to M. Deslongchamps, the pseudo-deltidium is produced by the gradual development of a single plate in the apex of the triangular opening, increasing downward with age, a very distinct mode of formation from all the others, and open to verification in the species described by that author, as his figures make no allowance for a pedicle-sinus or perforation, a feature, though not of frequent occurrence in the Spiriferidæ, yet one necessary to account for.

The writers have examined specimens of *Spiriferina pinguis* Deslongchamps, *S. rostrata* Schlotheim, and *S. Walcotti* Sowerby, and find that these species, at least, develop triangular deltidial plates. Those in *Spiriferina pinguis* and *S. Walcotti* are comparable with the same parts in *Spirifer perlamellosus* and *S. cumberlandiæ*, and their form and mode of growth are expressed by the outlines 2″, 3″, on page 384, and is further shown in figure 130. Additional growth causes the plates to unite along the median line, obliterating the partially formed pedicle-perforation, and subsequent increment can naturally take place only along their lower free edges.

The remarks on *Spirifer radiatus* and *S. crispus* indicate that the development of the plates in this member of the same family is quite in harmony with the process as seen in the rostrate forms generally.

The following is the summarization of Deslongchamps's conclusions, as given by himself:

(1) The deltidium is one of the most important features in the articulated brachiopods.

(2) As far as Jurassic species are concerned, the deltidium may suffice to characterize the families.

(3) In the various stages of development of this part the aspect of the shell is entirely changed.

(4) The deltidium appears under three important modifications: (A) Development below the peduncular arm, characterizing the Terebratulidæ; (B) development above the peduncle, Spiriferidæ; (C) mixed development, surrounding the peduncle, Rhynchonellidæ.

(5) The stage at which the development is arrested or the exuberance of development may suffice to characterize sections under the families.

It has just been shown that conclusions 2, 4, and 5 are not capable of the extended application which Deslongchamps has given them.

A preceding remark, that the course of development of the deltidial characters throughout the genera

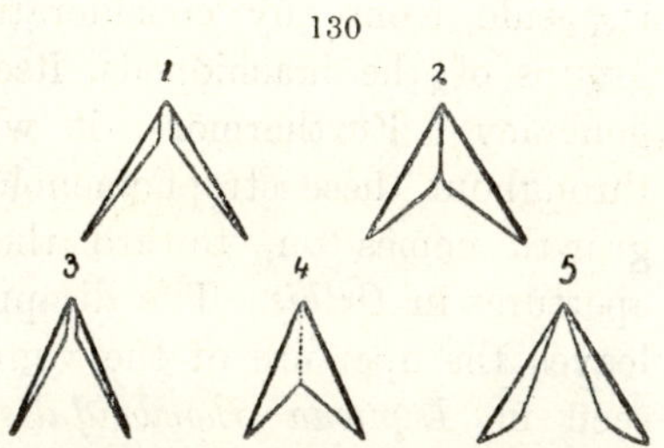

FIGURE 130. — Deltidial development in 1, 2, *Spiriferina pinguis* Deslongchamps; 3, 4, *Spiriferina Walcotti* Sowerby; 5, *Spiriferina rostrata* Schlotheim.

here discussed may be considered as fundamentally uniform, calls for explanation in its application to the Orthidæ and the Strophomenidæ.* In the latter forms it has been shown that the remarkable development of the pedicle-sheath is primary, and is invariably more or less atrophied with age, and probably functionally inactive at maturity. Hence the retention of this sheath in any species at maturity is the perdurance of what must serve as an embryonic character within the limits of this family. It cannot escape observation that the pedicle-sheath is in analogy with the entire rostrate umbo of the

[* The following correlations (*supra*) of the characters of the cardinal area were made before the true significance of the pedicle-sheath in the strophomenoid genera was understood. The subject is fully discussed (*ante*) in the second part of the Development of the Brachiopoda.]

ventral valve in the Rhynchonellæ, Terebratulæ, etc., as a specialized extension of the valve for the protrusion of the pedicle. (Compare the extreme development of the umbo in the genus *Terebrirostra.*) That these parts are also homologues, it is difficult to prove, on account of the pedicle-sheath becoming more degenerate as maturity approaches; but, assuming this homology, the sheath and its gradual disappearance may be regarded as an indication of degeneracy in the family, the presence of the sheath pointing toward a derivation from the rostrate type.

The atrophy of an organ so highly specialized as the sheath is, aside from any consideration of relationship to other groups of the brachiopods, itself confirmatory of such degeneracy. Furthermore, it will be noticed that there is, throughout these strophomenoids, an inclination, as mature growth comes on, toward the simple triangular pedicle-apertures in *Orthis.* The disappearance of the pedicle-sheath leaves the aperture of the ventral valve essentially free, as seen in *Leptæna rhomboidalis* and *Orthothetes subplanus,* while the aperture of the dorsal valve is filling *pari passu* with a callosity. In other words, the structure of these parts is actually degenerating toward maturity, to that of *Orthis,* which is the simplest, least differentiated condition among the articulated brachiopods, and serves to fortify the position of the genus at the base of the entire series. In *Orthis,* the pedicle-apertures on both valves are of the same size in early growth, and have undoubtedly acted together as a single opening, through which the fleshy arm was protruded as much on one side as the other, a fact indicative of an extreme lack of differentiation in the two valves in the articulate species, but agreeing closely with some of the inarticulate genera, as *Lingula, Leptobolus, Obolus.* The specialization which accompanies subsequent growth confines the pedicle more closely to the ventral aperture, and, as a result, the dorsal aperture is gradually filled by a callosity. Thus, also, the Strophomenidæ; but *Orthothetes subplanus* shows at maturity what has not yet appeared in *Dalmanella;*

namely, the initiation of deltidial plates, in conformity with the general course of development of the cardinal features observed in other families.

It is not well in this place to go beyond the scope of this work, and the species of Strophomenidæ here discussed for facts confirmatory of these observations. It may be remarked that the stage at which the development of the deltidial features has been arrested at maturity in this family varies with the species, not with the genus. When every trace of these features is obliterated, as is usual in *Stropheodonta*, a slight abrasion of the apical substance of the shell will often show a trace of the obsolete pedicle-tube. At times, in the same genus, this is retained at maturity as an external feature, and in such a case is usually accompanied by some indication of the sub-apical sheath. In both *Stropheodonta* and *Ortho-thetes* (especially of the later Paleozoic faunas), the cavity of the pedicle-sheath, if it is retained in any form, at maturity, has been filled by the deposition of calcareous matter about the compound cardinal processes of the opposite valve, and thus wholly diverted from its original function.

In conclusion, it is to be observed that of recent species of brachiopods a very great number show an incomplete development of the deltidial plates at maturity. Such is *Rhyncho-nella* to a large degree; also *Cistella, Kraussina, Terebratella,* and *Magasella;* and it may be assumed that the structural degeneracy which is thus indicated is the natural concomitant of the secular decline of the entire class.

It is not improbable that from an early form related to the genus *Orthis*, phylogenetic development tended in two main channels, — one leading through *Strophomena, Scenidium, Orthisina, Leptæna, Chonetes, Productus,* and *Strophalosia,* and the other in the direction of *Rhynchonella, Spirifer, Atrypa, Retzia,* and *Terebratula.*

Internal Apparatus. —The present observations upon the development of the brachial supports are limited to the species *Homœospira evax.* Here it is found that the number of revolutions of the spiral ribbon increases with age, but it is not certain what the inceptive condition of this apparatus may have been. In the early stage represented on page 365, where the ribbon has completed two revolutions, the supports must have been exceedingly tenuous and delicate, for they can be traced in the crystalline or muddy filling of the shell only by extremely faint lines, composed of minute dots of pyrite. As observed under the discussion of these features, the character or actual existence of the loop connecting the spirals was not established, but it is developed, with all normal characters, in a shell 4 mm. in length, where the ribbon makes four revolutions.

It has been shown by Morse,[*] that in *Terebratulina septentrionalis* the loop (*i. e.*, the entire brachial support) begins by the development of two acute processes from the lower moiety of the dental plate, which assume the character of crura, eventually meeting and coalescing on the dorsal side, forming the completed loop at an early stage, the ventral horns of the loop never uniting. The simple nature of the support in these shells precludes the possibility of the continued growth which obtains in the more complicated apparatus of the spiriferous species. The inception of the brachial support was observed by Morse in an individual 1 mm. in length, but the lateral processes are not conspicuously developed until a length of 3 mm. is attained, and they have not united at a length of 4 mm. It is therefore possible, from these data, that *Homœospira evax* does not have the loop completed at so early an age as that indicated by a length of 2.5 mm.

The observations by Morse are corroborated by those of Dall[†] on *Leiothyrina cubensis*.

Surface Ornaments. — Nearly all the observations upon initial shells or upon that portion situated at the apex of the beak of more advanced stages and representing the initial

---

[*] On the Early Stages of *Terebratulina septentrionalis*. *Mem. Boston Soc. Nat. Hist.*, II, pl. ii, figs. 48–55, 1869.

[†] Report on the Brachiopoda of Alaska and the adjacent Shores of Northwest America. *Proc. Acad. Nat. Sci. Phila.*, 1877, pt. ii, p. 155.

shell, seem to warrant the assertion that the surface ornaments do not appear until the second or a later period is
reached in the development of the shell.

For the plicate species nearly the full number of plications appear simultaneously, as in *Camarotœchia indianensis,
C. acinus,* and *Rhynchotreta cuneata,* or they are introduced
in pairs, as upon *Camarotœchia Whitii, ?C. neglecta,* and
Homœospira sobrina.

The striæ of *Leptœna rhomboidalis* are developed to the
full capacity of the marginal area as soon as the first growth-
line is completed. The number is increased in three distinct
manners: (*a*) by division, (*b*) by intercalation, and (*c*) by
addition at the cardinal angles. Some species present all
three of these, while others add to their striæ or plications by
any one or two of the modes.

The concentric ornamentation in such species as *Spirifer
crispus* and *Orthothetes subplanus* appears early in the growth
of the embryo, and continues to be repeated without variation, except in *Leptœna rhomboidalis* and allied forms, which
develop, during the last stage of growth, a geniculated curtain without concentric undulations.

Varieties and Abnormalities. — Varieties usually begin to
express themselves early in the development of the shell,
and the divergence from the normal form rapidly increases as
maturity approaches. Several of the species represented by
abundant material are readily separable into three distinct
groups of forms, (*a*) *long form,* (*b*) *normal form,* and (*c*) *broad
form.* The history of each may be clearly traced, and they
usually are found to unite with the line of the normal form
(*b*) several removes from the initial member of the series.
Sometimes the varieties do not reach the adult dimensions of
the normal shell and may be considered as varietal dwarfs.

The representation of varietal and of certain abnormal conditions by complete series of fossil specimens shows that in
these directions there was a distinct developmental tendency,
or genetic impulse, independent of normal growth.

Senility is always expressed by the thickening of the shell

which takes place after the individual reaches adult size. The thickening may involve the whole interior of the valves, producing a truncate appearance at the margins, or it may take place by frequent interrupted growth along the margins, giving to this portion a varicose aspect.   As a result of this senile growth, the vertical diameter of the shell is increased, and the beaks are involuted, so that they are often so closely appressed as to conceal the cardinal area and truncate the ventral beak, and in addition, the margins of the valves lose the characteristic ornamentation of the species and correspond to the gerontic stages as defined by Mr. Hyatt.*

Abnormalities frequently find an explanation in some pathological or accidental conditions which become instituted at any period in the life of the animal, and leave their impress on the shell.   The functional failure of a developing organ may cause the parts to revert to an embryonal type, and although it is difficult to apply this statement to the shelly covering of the animal, yet this condition is sometimes found. The specimen of *Camarotœchia neglecta* described on page 342 is an instance of this kind.   Another abnormal variation is noticed in certain individuals which preserve the larval features of the shell long after it has passed the early stages, and when, in many cases, it has reached the full adult dimensions.

* Values in Classification of the Stages of Growth and Decline, with propositions for a New Nomenclature.   *Proc. Boston Soc. Nat. Hist.*, XXIII, 1888.

5. DEVELOPMENT OF BILOBITES *

THE Linnæan species so well known under the name of
Orthis biloba, and so widely distributed in the Silurian rocks
of the world, represents one of the very distinct members
into which the *Orthis* group is now divided. It is much
removed from ordinary *Orthis* in general external features,
and only by means of developmental characters is it possible
to arrive at any idea of its genetic history.

After having been referred to various genera, including
Anomia, *Terebratula*, *Delthyris*, and *Spirifer*, by different
authors prior to 1848, Davidson † first showed conclusively,
from a study of the internal characters, that the true relations were with the genus *Orthis*. Its position has since
remained unchallenged, and subsequent investigation has not
brought forth any new characters, nor invalidated the results
obtained by Davidson. The additional observations here
made concerning the development of the shell, while adding
to a knowledge of the species, merely serve to bind this
form more closely to the group having the broad designation of *Orthis*. Professor King in 1850 ‡ proposed the genus
Dicœlosia for this species, on account of its characteristic
form, and authors disposed to divide *Orthis* have recognized
this name. Since then it has been shown that Linné gave
the generic term *Bilobites* to the type species of King's genus,
and this name is now generally adopted with the rank of a
sub-genus. The validity of the specific names applied to

---

* *Amer. Jour. Sci.* (3), XLII, 51–56, pl. i, 1891.
† *Bull. Soc. Géol. France*, 2d ser., V, 321, t. 3, fig. 18, 1848.
‡ Monograph Permian Fossils, 106, 1850.

variations from the typical form is not of much moment in this place, although the geologic history and interpretation of these differences are of considerable interest. Two well-defined varieties or species are recognized in Sweden, and are represented in outline by figures 2 and 28, Plate XXIII. The prevailing form in the Wenlock shales at Dudley, England, agrees with figure 28, and also represents the ordinary form from the Niagara Group of Indiana and New York. Each locality, however, presents minor differences, mainly of local interest, and seldom of varietal importance. In western New York, besides the ordinary form with both valves convex, there is found an arcuate, deeply bilobed variety, agreeing with the extreme of the Swedish *B. bilobus*, var. *Verneuilianus* Lindström, represented in Plate XXIII, figure 2. The lobes of the New York variety are commonly more divergent, as shown in the outline, Plate XXIII, figure 1. This form was recently described by Ringueberg as *Orthis acutiloba*.*

The Lower Helderberg species known as *B. varicus* Conrad, sp., presents an amount of departure from typical *B. bilobus*, as would be anticipated from the change in the chronological and physical conditions of the species, combined with its extremely prolific development at this time. The abundance and comparatively large size of individuals clearly indicate most favorable conditions for their existence and multiplication, and, also, for the assumption and transmission of any varietal forms in harmony with the environment.

Mature individuals from Dudley, England, and Gotland, Sweden, represented by figure 28, Plate XXIII, correspond in all characters with specimens of *B. varicus* which are about half or two-thirds grown. After reaching the adult *bilobus* stage, *B. varicus* continues its growth, but this subsequent increment is gerontic in its nature, although such senile features are here the conditions of simple maturity or the completed ephebic stage. Evidences of this are seen in the gradual obsolescence of the pronounced lobation of the

* *Proc. Acad. Nat. Sci. Phila.*, 134, 1888.

shell and the cessation of areal growth in the neanic period. The form known as *B. bilobus*, var. *Verneuilianus*, Lindström, from Gotland, shows a tendency to develop in the opposite direction, as the lobation becomes more and more pronounced with growth, and the shell exceeds in size the normal species. The decrease in the lobation of *B. varicus* is a degeneration towards an embryonic character, while the arrested areal development produces a condition of partial isomorphism resembling one of the higher groups of *Orthis*, such as *Rhipidomella* (*R. Michelini* L'Éveille).

From what has been stated, it seems evident that the form typified by *B. bilobus* from the Niagara was, at that time, not a very plastic type, and capable of only slight degrees of variation or departure from the normal form. Naturally, all the modifications which occur containing a differentiation of the essential idea of the genus appear in the early history of the group, and are found previous to the Lower Helderberg form. The latter species while losing, in a manner, its *bilobus* expression at maturity, degenerates into forms resembling ancestral and other groups.

The material for the basis of this paper was collected by the writer from the lower members of the Shaly Limestone of the Lower Helderberg Group, along the top of the main escarpment of the Helderberg Mountains, between Clarksville and the Indian Ladder, Albany county, New York. Half-grown and fully developed specimens of *Bilobites varicus* Conrad, sp., can still be picked up in considerable numbers in the soil formed of the decomposed limestones. The species, however, is not so abundant as formerly. Professor James Hall is authority for the statement (*Pal. N. Y.*, vol. iii, p. 493) that forty thousand individuals were collected between 1843 and 1853, and about four thousand in the four following years. The young specimens have been obtained only by carefully examining the decomposed surfaces of the limestones, and by treating with hydrochloric acid slabs of rock in which the fossils are replaced by silica.

After considerable labor and search, about a thousand individuals have been obtained. From this number it has been possible to select a series of over forty specimens, showing stages of growth ranging from shells a little less than one-half a millimetre in length to a length of nine millimetres, thus representing the development between these limits by almost insensible gradations.

Developmental Changes in Bilobites varicus.

In the youngest specimens yet detected, measuring .49 mm. in length, and semi-elliptical in outline, the dorsal valve is longer than the ventral; the hinge is equal to the greatest width of the shell; both areas are high, sub-equal, and perforated by a triangular fissure in each valve. In rare instances the pedicle covering or pseudo-deltidium is apparently retained in young shells. Figure 131 of the ventral area shows the fissure and pedicle covering, with the foramen at the apex of the beak. The covering is soon absorbed or abraded during subsequent growth, and the pedicle then emerged through the fissure below.

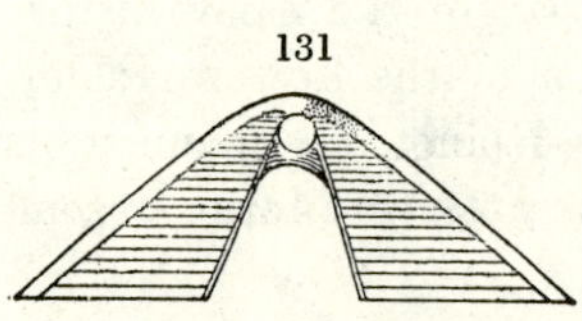

FIGURE 131. — *Bilobites varicus*, Conrad; ventral area. × 25.

None of these characters obtain in the neanic or ephebic stages, which are represented by a cordate, bilobed shell; dorsal valve shorter than the ventral; hinge-line much shorter than the width of the shell, and an inconspicuous dorsal area without a fissure.

The series of outlines (Plate XXIII, figures 11 to 26) drawn to the same scale, illustrate both the important changes which take place in the general form, and the corresponding increase in size from stage to stage. The rounded frontal margin of figures 11 and 12 becomes straight in figure 13, and in figure 14 a gentle sinus is apparent, which is pronounced in figure 15, and thereafter is the conspicuous character of the entire shell up to the ephebic stage represented by figure 23. Figures 24 and 25 show that upon reaching maturity a gerontic tendency to obliterate the marginal sinus is initiated; thus degenerating to an embryonal condition of lobation similar to figure 14.

The length of the hinge-line from an initial dimension equal to the greatest width of the shell becomes equal to but one-half the width of the shell in a specimen 3.5 mm. wide; and in a full-grown individual, as represented by figure 25, the hinge is not more than one-quarter the width of the shell. From having sub-equal areas the change is rapid, so that in a very early stage, but two or three removes from the initial one of the series, the ventral area is the larger and the fissure higher. This ratio progressively increases, and after the shell reaches a length of 1.5 mm., the dorsal area ceases to be a conspicuous feature. All areal growth and hinge extension end in the middle neanic period, and to this cause is due the great disparity between the length of the hinge and the width of the shell in ephebic individuals. The nepionic shells show some extension of the cardinal angles, but the auriculation does not become apparent until the lobation of the valves is initiated.

On account of the greater length of the incipient dorsal valve and consequent obliquity of the area, the fissure and area of that valve may be seen when the shell is viewed from the ventral side, as in figure 10, and, consequently, the ventral area is concealed from the dorsal aspect, as shown in figures 3–9 and 11–15. This is a remarkable reversion of characters, and one which appears to be of considerable significance from a phylogenetic standpoint.

The radiating striæ first appear on the lower half of the initial shell of the series, indicating that in an earlier condition the shell was smooth. The striæ appear in pairs. The first two striæ extend to the antero-lateral borders. An additional intercalated pair is next introduced, together with a single one on each side between the primary radii and the cardinal border. The number after this stage is more rapidly increased by increment in the cardinal lateral areas than in the median region.

Observations. — As shown in the ontogeny of *B. varicus*, the generic stock was derived from a radicle having, in many respects, the characters of the group represented by *Platy-*

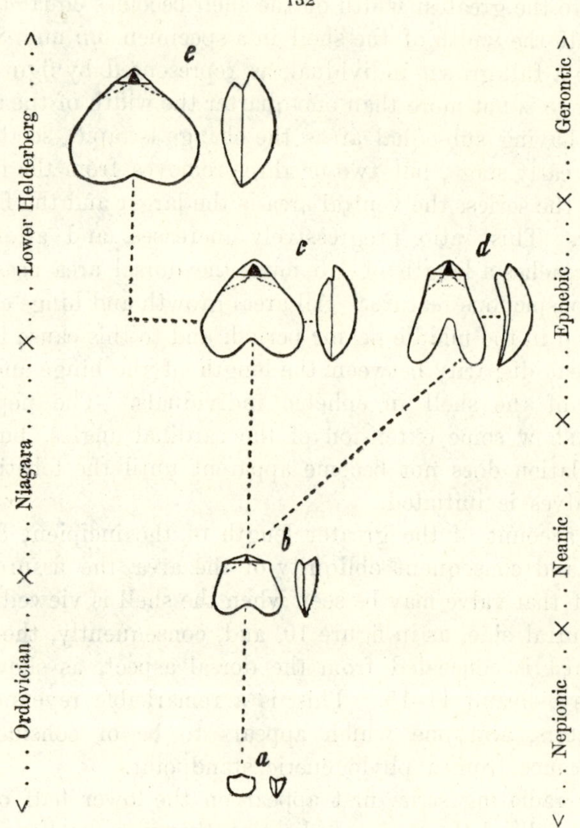

FIGURE 132. — Genesis of *Bilobites*.

a, nepionic stage. Ordovician type like *Platystrophia biforata*. × 4.
b, nealogic period at which divergence begins. × 4.
c, *Bilobites bilobus*. Niagara Horizon. × 2.
d, *Bilobites Verneuilianus*. Niagara Horizon. × 2.
e, *Bilobites varicus*. Lower Helderberg Horizon. × 2.

strophia biforata. The general proportions of the nepionic shell in *B. varicus* resemble it very closely. The length of the hinge at this period, the high hinge areas in both valves, with sub-equal triangular fissures, and the extent of the dorsal and ventral beaks, are characters very much the same as in *Platystrophia biforata.*

These features are maintained until the neanic stage represented by figure 15, after which arrested hinge extension and increasing areal growth in the ventral valve rapidly obliterate the early characters, and in addition the growing lobation of the valves emphasizes the expression of *Bilobites.*

The genesis of the species is represented in figure 132, in which it is shown that all these species are alike in their development up to an early neanic period (*b*). *B. Verneuilianus* (*d*) diverges at this point, progressively increasing its variation from the normal direct growth, as exemplified in *B. bilobus* (*c*). *B. varicus* (*e*) passes through all the *bilobus* stages, and culminates in larger individuals, with less pronounced lobation of the shell.

The direct line of development is represented by *B. bilobus,* and it is significant that this form also has the greatest geological and geographical distribution. Next, the divergent and in direct line, typified by *B. Verneuilianus* and *B. acutilobus,* is also widely distributed, but less so than the first. Finally, the gerontic form, *B. varicus,* culminated and disappeared within very narrow time and regional limits.

6. DEVELOPMENT OF TEREBRATALIA OBSOLETA Dall [*]

(Plates XXIV and XXV)

Fischer and Œhlert [†] have given a full account of the development of the brachial supports in *Terebratella dorsata* and *Magellania venosa*, from Tierra del Fuego. This work, together with that of Friele, [‡] and Deslongchamps, [§] on the northern species *Macandrevia cranium* and *Dallina septigera*, and the equatorial species *Mühlfeldtia sanguinea*, constitute nearly all that is known regarding the metamorphoses taking place in the brachial supports during the growth of an individual belonging to the higher genera of the Terebratellidæ.

It is of interest to add another species to this list, especially as it represents a northern form, the development of which has not been hitherto studied. This form offers, moreover, some additional features for comparison, and two very early stages have been discovered which are both of genetic value.

The material for this work has been kindly furnished by Dr. William H. Dall, of Washington. The specimens were dredged by the U. S. steamer *Albatross*, in 113 fathoms, at Station 2984, off Cerros Island, Lower California. In a report on some shells from this expedition, by Dr. Dall, [||] this brachiopod was described as *Terebratella occidentalis*, var. *obsoleta*. Subsequent study, however, has led him to

[*] *Trans. Conn. Acad. Sci.*, IX, 392–395, 398, 399, pls. ii, iii, 1893.
[†] *Bull. Soc. Hist. Nat. d'Autun*, V, 5 plates, 1892.
[‡] *Arch. Math. Nat.*, Bd. XXIII, 1877.
[§] Études critiques des Brachiopodes nouveaux ou peu connus, 1884.
[||] *Proc. U. S. Nat. Mus.*, XIV, 1891.

consider it as a distinct species, *T. obsoleta* (Plate XXIV, figures 6–9), and this determination is here adopted. A comparison of the two forms shows that *T. occidentalis* is much wider and less convex, the plications stronger, and the valves heavier. In the previous paper it was stated that *Terebratalia obsoleta* was morphically equivalent to *Terebratella dorsata*, and that, during growth, both went through different series of transformations to reach this final equivalent type of structure. It is now proposed to illustrate and describe in greater detail these metamorphoses of *T. obsoleta*. The generic homologies and differences are discussed in the paper already mentioned.

The earliest stage observed in this species (Plate XXIV, figure 10) has a length of .3 mm. It is comparable to an early stage of *Cistella neapolitana* described by Kovalevski.* Both agree in having an incomplete circlet of centripetal tentacles. The smaller and younger tentacles are near the front margin, while those first formed are on each side of the mouth. This condition agrees with the tentacular multiplication described by Brooks in *Glottidia*, by Kovalevski in *Cistella* and *Lacazella*, and by Morse in *Terebratulina*. Tentacles $t1$, $t1'$ were the first pair to be formed, and $t2$, $t2'$, $t3$, $t3'$, $t4$, $t4'$ followed in pairs in the order named; $t5$ has just appeared, and the corresponding one on the other side is not yet seen. Figure 10 also shows the tooth-like projections (*ds*), forming, with the cardinal angles, sockets for the reception of the teeth in the ventral valve. The adductor muscles are indicated at *ad*, and the diductors at *did;* the mouth is at *m*, with the visceral mass posterior to it. No diverticula have yet appeared to form the pallial sinuses.

When the shell has reached a length of .65 mm. (Plate XXIV, figure 11), the circlet of tentacles is complete, and the pallial sinuses appear as two slightly branched tubes extending anteriorly from the sides of the visceral mass. From the correspondence of the structure of *Gwynia* to this

---

* See C. E. Beecher, Development of the Brachiopoda, pt. ii, figs. **15, 16.** *Amer. Jour. Sci.*, XLIV, 1892.

stage of *Terebratalia* and other genera of the Terebratellidæ it is called the *gwyniform* stage.

By the time a length of 1 mm. is attained (Plate XXIV, figure 12, and Plate XXV, figure 1) a triangular septum is visible anterior to the middle of the dorsal valve. Its elevation inflects the circlet of tentacles, making the lophophore reniform or bilobed, and producing a brachial structure similar to that in *Cistella*. This stage is therefore called the *cistelliform* stage. The only advance over this condition shown in adult *Cistella* is that the band supporting the cirri is calcified and attached to the crural points. In *Terebratalia*, during the growth immediately following and still included in the *cistelliform* stage, the septum increases in height, and often two small lateral expansions appear on the posterior sloping edge (Plate XXV, figures 3, 4).

In the transformation to the next, or *platidiform*, stage (figures 5–8), the crural points appear, and there is a narrow groove along the top of the septum, which is arched over at the posterior end, forming at this point a small cylinder, nearly vertical to the floor of the valve. This is the beginning of the ascending branches, or secondary loop. The septum is quadrangular, and develops on its anterior edge several spinous processes (figure 7). The growth of the descending branches from the crura and their union with the septum bring about the *platidiform* stage, as represented in Plate XXV, figure 8; and the growth of the ascending branches produces the *ismeniform* structure (figure 9). The expansions on the sides of the secondary loop are not present in all specimens, and do not seem to be important characters. The cirrated band at this time (Plate XXIV, figures 4, 5) extends continuously from behind the mouth to the origin of the descending branches, and along them to the septum, thence obliquely backward on the septum, and over the ascending branches to the median line, where new cirri are introduced.

The ascending lamellæ from the septum already have begun to divide or separate anteriorly, and in the next stage (Plate

XXV, figure 10) they show lacunæ produced by resorption. The structure of the loop at this time agrees with that in *Mühlfeldtia*, and this period of development has been called the *mühlfeldtiform* stage.

The ends of the descending branches have continued to widen on the septum (Plate XXV, figure 11), and extend toward the ascending branches, with which they soon join (figure 12), bringing in the *terebrataliform* type of structure. The completion of this stage is accomplished by the further separation of the ends of the ascending branches, and by the resorption of the expanded ends of the descending branches to form the connecting bands with the septum (Plate XXV, figures 13, 14, 15).

It should be noted that the septum in the *cistelliform* stage is wholly anterior to the middle of the length of the dorsal valve. Septal growth takes place chiefly on the posterior end, and at the same time resorption along the anterior edge serves to move the septum backward, until by the time the *terebrataliform* stage is reached it is posterior to the middle, and in adult specimens it is in the umbonal region.

A comparison of the growth-stages of *T. obsoleta* with those in *Macandrevia cranium* and *Dallina septigera* shows considerable similarity, except, of course, in the adult condition. The general features of each are alike, and may be correlated in the same manner, stage for stage. The septum in *T. obsoleta* in the *platidiform* stage is considerably broader than in the other forms, and the descending branches join it considerably lower down. The two stages preceding the *platidiform*, which present the brachial structure first of *Gwynia* and then of *Cistella*, are of chief importance.

7. DEVELOPMENT OF THE BRACHIAL SUPPORTS IN DIELASMA AND ZYGOSPIRA*

(PLATE XXVI)

It has been shown by several authors † that the brachial supports in the Terebratellidæ pass through a series of distinct metamorphoses during the life of the animal. In the higher genera these stages may be correlated with the adult structures of lower forms, thus furnishing satisfactory data for a systematic arrangement of the genera and for their phylogenetic relations.

This kind of research naturally requires ontogenetic series of considerable completeness, and it is often difficult or impossible to obtain such material representing fossil forms. Moreover, the fossils must be exceptionally well preserved to afford a means of working out the development of a structure so delicate as the calcareous lamellæ supporting the brachia, especially in young specimens from one to five millimetres in length.

It first seemed desirable to determine the development in some genus of the Terebratulidæ from the Paleozoic, in order to ascertain whether the brachial supports, as in Neozoic and recent forms, passed through a series of transformations, and to determine the most primitive form of the loop in the Ancylobrachia. For this purpose a species of *Dielasma* (*D. turgidum*) obtained from Mr. Moritz Fischer was used. The specimens are from the St. Louis Group of the Lower Carboniferous in Kentucky. The shells are partially silici-

---

* Beecher and Schuchert. *Proc. Biol. Soc. Washington*, VIII, 71–78, pl. x, 1893.

† Davidson, Friele, Deslongchamps, Fischer and Œhlert, and Beecher.

fied, generally filled with transparent calcite, and afford very satisfactory preparations of the arm supports. It was found that the loop of *Dielasma* underwent transformations during growth, and that the earliest stage observed is like *Centronella*. This establishes the *centronelliform* loop as the simplest type of loop in the Ancylobrachia. Besides *Centronella*, other adult representatives of the same structure are *Rensselæria* and *Newberria*. They are all late Silurian, Devonian, and Carboniferous genera, but the *centronelliform* structure continues later, and is represented in the Trias by the genera *Juvavella* Bittner and *Nucleatula* (Zugmayer) Bittner.

It was at once suggested that interesting results might be obtained in studying the development of a spire-bearing brachiopod, and, as the earliest species more clearly show their phylogeny in their ontogeny, the ancient genus *Zygospira* was selected. Very complete material was accessible, collected by the writers, from the Trenton of Minnesota and Kentucky, so that series of specimens were assembled representing all stages of growth from specimens .8 mm. in length to mature size. They were prepared to show their brachial supports, and it is clearly demonstrated that the primitive arm support in *Zygospira* is a terebratuloid loop having a *Centronella*-like form, which undergoes several modifications before the growth of the spiral lamellæ, thus in so far resembling the development of *Dielasma*.

These results threw doubt on a number of Lower and Upper Silurian species described as having recurved loops, and previously referred to the higher terebratuloid genera *Macandrevia* or *Waldheimia*. The shells are impunctate, while *Rensselæria* and *Centronella* are distinctly punctate, like all other well-known Terebratulæ. Upon investigation it has been ascertained by Hall and Clarke and the writers that the species which have been referred to *Hallina* and *Macandrevia* from the Silurian are spire-bearing forms, and therefore do not belong to the Ancylobrachia.

Fischer and Œhlert have called attention to a number of recent species which have been erroneously based upon the immature stages of higher species, and in the Terebratellidæ it is evident that great uncertainty must exist in the identification of specimens not fully adult. Now, finding that Paleozoic genera of both loop and spire-bearing stocks (Ancylobrachia and Helicopegmata) in the adolescent period likewise pass through metamorphoses representing the structures of other genera and even other sub-orders, it is manifest that species cannot be referred to their proper genera, nor genera correctly defined, unless the individuals studied are adult and their characters constant for a definite period of time.

Development of the Loop in Dielasma turgidum.

The earliest stage thus far observed was found in a specimen a little over four millimetres in length (Plate XXVI, figure 1). The loop at this time is composed of two broad descending lamellæ, which begin at the ends of the crura and extend forward, curving ventrally until they unite in the median line, forming an angular ridge, acuminate in front. As previously mentioned, this structure is very similar to that of *Centronella*, and this stage is therefore called the *centronelliform* stage.

The first change in the form of the loop is brought about by a resorption of the pointed anterior portion, so that the outline is re-entrant in front (figure 2). Further resorption in the same manner results in the production of two posteriorly directed branches, as shown in figure 3. This form may be considered as an early immature *Dielasma* loop, as subsequent growth does not materially modify its general characters.

The adult loop, represented in figures 4–6, differs from the early *Dielasma* stage chiefly in the divergence of the descending branches.

In the *centronelliform* stage the lamellæ converge, and the loop extends half the length of the shell. Both of these relations gradually alter until, in the early *Dielasma* stage, the descending branches are nearly parallel, the loop extends less than half the length, and finally, when mature, the descending branches diverge and the loop is two-fifths the length of the dorsal valve.

The natural inferences to be drawn from the development of the loop in *Dielasma* are, that *Centronella* represents a larval or immature condition of the higher genera, and that the *centronelloid* loop is the primitive type in the Terebratulidæ. Therefore, as *Centronella* and the closely related genus *Rensselæria* are the only early punctate terebratuloids known, and as they have the primitive type of loop, there arises the question of the validity of the Upper and Lower Silurian species with recurved loops, referred to *Waldheimia* and *Hallina*.

Hall and Clarke (*Pal. N. Y.*, vol. viii, part ii, pp. 147–153, not yet published) describe and figure the brachial supports in *Hallina*, showing that both *H. Nicoletti* Winchell and Schuchert and *H. Saffordi* Winchell and Schuchert are provided with short spires of about one volution, connected by a transverse band, as in *Zygospira*. In removing the ventral valve and exposing the loop from that side, as is often done, the short spiral lamellæ have been overlooked. Similar observations have been made by the present writers, so that the systematic position of these forms is now established.

Specimens of *Waldheimia bicarinata* Angelin, from the Upper Silurian of Gotland, were also examined. They were found to possess well-defined spiral cones, and in other respects agreed with the diagnosis of *Dayia*. These facts indicate that the specimens described by Davidson as *Waldheimia Mawii* (Fossil Brachiopoda, Supp. vol. iv, pt. v, pl. iv, figs. 1–3) are the young of *Dayia navicula* Sowerby, sp. (*ibid.*, pl. v, figs. 1–4).

Development of the Brachial Supports in *Zygospira* recurvirostris.

The smallest specimen in which the internal structure was observed measures about 1.33 mm. in length (Plate XXVI, figures 7, 8). The brachial supports consist of two straight, ventrally concave, primary lamellæ, rapidly increasing in width from the thin crural plates to near the centre of the valve, where they unite, forming a plate with a central angular ridge. The anterior end of the plate is pointed, as in *Centronella.*

In a specimen about 2 mm. in length (figures 9, 10), the primary lamellæ are practically of the same form as in the preceding, but much of the original central portion of the loop has been resorbed, so that the lamellæ are connected by a short but comparatively wide, ventrally arched, transverse band. The lamellæ, or descending branches, are also more spreading anteriorly, and there is a slight deflection at the crural points which becomes more and more pronounced as growth progresses.

In the next stage (figure 11), which has a length of 2.33 mm., the descending branches are more diverging, and the transverse band is longer and more broadly excavated in front.

The succeeding stages here described are based upon material derived from near the top of the Trenton, where the specimens of this species are usually larger and more transverse than those from near the base of the Trenton, which is the horizon of the specimens illustrated in figures 7–11. Therefore, when the loop in figure 9 is compared with that of figure 12, it is seen that the latter is much the wider, from the greater size and breadth of the shell, which has at this stage a length of 3.33 mm., while the former is but 2 mm. long. The loop in figure 12 is somewhat more advanced than in figure 9, the transverse band being narrower and slightly elevated posteriorly, some resorption having taken place along the inner edges of the primary lamellæ. Further resorption in same direction produces the brachial support illustrated in figure 14. This form of loop in *Z. Nicoletti,* *Z. Saffordi,* and *Z. recurvirostris* from the lowest Trenton is retained to maturity. However, in specimens of *Z. recurvirostris* from the upper Trenton the posteriorly curved, transverse band is not a mature feature, since it becomes changed into the form represented in figure 15. In previous stages the transverse band is ventrally arched, but it now bends dorsally, and remains so during subsequent growth until near maturity, when the sinus of the dorsal valve causes it to assume a sigmoid curve.

The spirals next begin to develop (figures 16 and 17) as two slender converging lamellæ, curving toward the ventral valve and originating from the outer pointed ends of the loop. These lamellæ then incurve dorsally and laterally to a point just posterior to the transverse band, forming the first volution of a spiral (figure 18). In this manner further growth and elongation of the lamellæ continue until maturity is attained, when there are about three volutions in each spiral cone (figure 20). The calcareous brachial supports occupy about the same relative space in the shell cavity in all stages of growth.

Observations and Correlations.

Zygospira is the earliest spire-bearing genus known, as it is found in the Birdseye Limestone of the Trenton period. It is of considerable interest, therefore, to study the development of the spirals. From the ontogeny, it is shown that the brachial supports in *Zygospira* begin as a loop greatly resembling that of Devonian *Centronella*. Moreover, the loop passes through a series of metamorphoses before the spirals make their appearance.

The most ancient species are *Z. Nicoletti* and *Z. Saffordi*, small semi-plicate forms, in which the spirals are very rudimentary, consisting of about one volution. In the same geological horizon occurs *Z. recurvirostris*, having from two to two and one-half turns of the lamellæ in each spiral. The same species from the upper Trenton has three volutions, while in *Z. modesta* of the middle Lorraine there are from four to five whorls (figure 25). In *Z. Headi* (figure 24) a large globose finely striated species of the upper Lorraine, there are six whorls to a cone. The geological history, therefore, shows a gradual increase of from one to six turns of the lamellæ in each spiral.

The transverse band connecting the primary lamellæ also undergoes a series of changes. It has been shown that the centronelloid loop (figure 7) passes into one having the lamellæ joined by a posteriorly directed, transverse band (figure 14). This form of loop is retained as a mature feature in the brachia of *Z. Nicoletti*, *Z. Saffordi*, and in the lower Trenton varieties of *Z. recurvirostris*. Passing to the specimens of the latter species, which are geologically later, the band no longer joins the lamellæ as far anteriorly as in the older variety (figure 20). The point of connection in *Z. modesta* is variable (figures 25 and 26), but is usually more posterior than in *Z. recurvirostris*, while in *Z. Headi* it is manifestly more posterior than in any of the older species of *Zygospira*. The transverse band is now no longer arched backward, but is just the reverse (figure 24), while its position is progressively more and more posterior, and the loop is gradually shortened before the spirals make their appearance. The gradual increase in the number of the whorls in each spiral and the recession of the transverse band have gone on together.*

The family Atrypidæ includes the genera *Zygospira*, *Glassia*, *Atrypa*, and *Dayia*. It is easily distinguished from all other families comprised in the sub-order Helicopegmata, since the spirals are between the first descending branches of the lamellæ, while in the Spiriferidæ, Nucleospiridæ, and Athyridæ the primary lamellæ are between the spirals.

The gradual increase in the number of whorls in the spirals and the pushing backward of the transverse band in the Atrypidæ is carried farthest in the species of *Atrypa*. In *Cœlospira Barrandei* and *C. marginalis* the brachial supports, as worked out by Davidson and Glass, consist of about five volutions, and are similar to those of *Zygospira*, except that

---

* The extreme anterior position of the transverse band in *Z. recurvirostris* is therefore of no more than specific value, and on this account *Anazyga* Davidson cannot well be separated from *Zygospira*.

the transverse band is more posterior, since it originates near the ends of the crura. This mature condition of *Cœlospira* is seen to be a young condition of *Atrypa* (Davidson), but, as the spirals are more loosely coiled and the transverse band always continuous, this genus should be regarded as valid in the evolution from *Zygospira* to *Atrypa*. In mature *Atrypa reticularis* from the Upper Silurian, there may be as many as sixteen volutions in each spiral cone (Davidson), but more often the number is smaller. The transverse band in this species during its young stages is continuous, but in the adult condition it seems to be usually disunited in the middle. This feature becomes a distinct adult character in the Devonian specimens, which also have a greater number of whorls in the spirals, as shown in a Chemung specimen of this species in Yale University Museum, having twenty-four turns of the lamellæ in each spiral.

The ontogeny and phylogeny of the species of *Zygospira* indicate strongly that the *Atrypidæ* had its origin in a form with a centronelloid loop. A further natural conclusion from the same evidence is that the Ancylobrachia are older and more primitive than the Helicopegmata.

IV

MISCELLANEOUS STUDIES IN DEVELOPMENT

IV

MISCELLANEOUS STUDIES IN DEVELOPMENT

1. DEVELOPMENT OF A PALEOZOIC PORIFEROUS CORAL*

(PLATES XXVII–XXXI)

THE origin and affinities of many groups of Paleozoic corals are still obscure. The main elements of the recognized system of classification seem to be stable, yet so little is known of the growth and structure of a number of important groups that they occupy a different place in almost every arrangement of the genera. Each fact of development affords data which eliminate, to a degree, the want of knowledge concerning their origin and relations. Unless the growth of the organism is obscured by pronounced accelerated or degradational features, its interpretation is simple and throws much light on its ancestral history. Paleozoic types in general are least modified in their development by acceleration. They usually show some marked expression of their prototype, and also the succession of changes through which they have passed during their evolution.

The species here discussed was originally described as *Michelinia lenticularis* Hall,† from the Lower Helderberg Group of New York. If *Michelinia* is entitled to recognition, it will exclude this form, as it is without tabulæ. *Pleurodictyum*, as now defined, is more in harmony with these

---

* *Trans. Conn. Acad. Sci.*, VIII, 207–214, pls. ix–xiii, 1891.
† *Twenty-sixth Rept. N. Y. State Mus. Nat. Hist.*, 113, 1874.

features, and therefore the species *M. lenticularis* is here referred to this genus. The large calices and their constant origin at the basal epitheca are not, however, essential characters of *Pleurodictyum*. The structure and growth of this species indicate that it represents one of the simpler types of poriferous corals. For this reason its development is without the numerous modifications necessary in more complex forms, and its laws of growth are not complicated.

Development of Pleurodictyum lenticulare.

The nepionic stage is well marked. It comprises the growth of the corallum to the completion of a simple initial cell. This primitive cell, or nepionic stage (Plate XXVII, figures 1, 7, 8), has the form of an oblique inverted cone, flattened on one side. The flattened area represents the lower or attached side, and the oblique base of the cone is occupied by the aperture of the corallite. The apical portion is smooth for about one-fourth the length of the cell. Then the concentric lines of growth become apparent, and over the distal half radiating ribs are also developed. The interior of the apex is granulose. At about the middle of the cell the granules are arranged in rows, forming the beginnings of the septal lines.

The simple growth of the initial cell continues until the entire procumbent portion is completed. A thickening of the margin then takes place, and an upward growth of the corallite is initiated. At the commencement of this upward growth the first bud starts out from the lateral edge of the initial calyx, either to the right or left of the axis. This condition represents the first neanic stage. The bud resembles the parent cell in all particulars, and reaches considerable size before the second appears, as shown in Plate XXVII, figures 9, 10. The visceral cavities are confluent, as the initial apex of the bud opens into the calyx of the first cell.

The succeeding neanic stages, to the completion of the

first circle of peripheral calices, have been observed mainly from the epithecas of mature or nearly full-grown corallums, represented on Plate XXVIII, figures 1, 2. In these examples the lines of growth are so perfectly shown that all the stages are distinctly marked, and may be satisfactorily studied.

What is here considered as the second neanic stage is represented on Plate XXVII, figure 3, showing the initial corallite, with the first and second buds on opposite sides. This process of alternate gemmation from the parent cell continues until the circlet of calices is completed, as shown in figures 4, 5, and 6. In this species the normal number of peripheral calices is seven, making eight corallites in the completed neanic corallum. The last cells to be formed are (1) the sixth and seventh, budding from the anterior side of the first calyx, and (2) the eighth, or posterior cell. Plate XXVII, figure 12, represents the completed neanic corallum, with the initial cell and six well-developed peripheral calices. The eighth has just begun to fill up the space between the second and third. It will be noticed that there is a direct correspondence in the size of the calices to their relative age. The first calyx is much the largest. Then, decreasing serially, come the second, third, fourth, fifth, sixth, and seventh, while the eighth is undeveloped. An inspection of the upper surface of a mature corallum will thus usually determine the order of successive calical additions. After the appearance of the posterior, or eighth calyx, the corallum commonly grows to double the diameter of the completed neanic stage, resulting in the normal ephebic or mature condition, as represented on Plate XXIX, figures 1, 2. Nearly all the full-grown specimens found agree in this respect.

A corallum rarely presents any departure from the normal number of calices. Plate XXX, figure 1, is an example of a variation in the number of peripheral corallites, for in this specimen there are eight in the circle instead of the usual seven. A variation in the opposite direction is shown in another specimen having five well-developed corallites about the parent cell. Old-age characters are expressed in two ways: First, the cell walls become thickened around the margin of the epitheca without destroying the symmetry of the corallum, as shown in Plate XXX, figure 2; second, by the indefinite and unequal development of the peripheral cells, together with the addition of calices budding from the cells forming the primary circle. One specimen, appearing at first sight as an example of cell division or fission, is shown in Plate XXXI, figure 1. It may be explained as resulting from the abnormal growth of the second and adjacent calices, four and eight. This lateral impulse further resulted in sending off the small, peripheral, tertiary corallites numbered in the figures 9, 10, 11, 12, and 13.

It should be understood that this arbitrary expression of normal and abnormal growths applies only to the species *Pleurodictyum lenticulare.* The same numerical arrangement will not hold good for genera like *Favosites, Michelinia, Striatopora*, etc. Otherwise, it is believed, the general laws of growth here brought out will hold good for these and other related genera.

Some doubt may exist as to the propriety of referring the specimens illustrated on Plate XXVII, figures 9–11, to *P. lenticulare.* Unfortunately, material of this kind is rare and difficult to obtain. With the exception of the position and direction of the first bud (figure 10), all the characters agree, so far as can be observed, with ordinary specimens of *P. lenticulare.* The second cell of the corallum represented in Plate XXVIII, figure 1, curves rapidly backward, although at first the axis has an anterior direction. Taking this view of the specimen (Plate XXVII, figure 11), it is not difficult to see how the succeeding enlargement and curvature of the bud could extend backward, thus properly limiting the size of the eighth or last of the primary circlet of calices.

The method of determining the relative age and succession of the corallites can be seen in Plate XXVIII, figures 1, 2, and Plate XXIX, figure 2. The initial cell occupies the central position, and forms the boss or apex of the basal epitheca. The first bud is nearly on a plane with the base of the initial cell, and is the one nearest the apex. The second and successive buds are respectively more distant and at a higher level. Specimens having broad surfaces of attachment to foreign objects have these distinctive features of the epitheca obliterated, and the only guide to the order of the corallites then lies in their comparative size and position on the upper surface of the corallum.

General Conclusions.

The first feature to be noted in the development of a poriferous coral, as here described, is the simple cyathiform character of the initial corallite. This nepionic stage is without mural pores, and has an epitheca over the entire exterior of the cup. The septal lines become developed toward the end of this stage. These features are in harmony with the young of many Paleozoic corals, such as *Cladochonus*, *Aulopora*, or *Syringopora*, and clearly indicate a primitive, simple, and imperforate ancestry for the Perforata. A similar origin and development obtains in *Favosites*, as may be seen from the figure of a young colony of *F. Forbesi*, var. *occidentalis*, given by Professor Hall.*

The first neanic stage, represented by the primitive corallite with one bud, is the first transition toward both a compound and a perforate coral (Plate XXVII, figure 9). This stage has two calices making it a compound coral, and has an opening through the cell walls or connecting channel between the corallites, forming the first mural pore. The manner of growth and the structure of the corallum at this stage are suggestive of *Aulopora*, and should be given considerable significance. The visceral cavities in *Aulopora* are confluent, and rudimentary septa or lines of spinules are often present. *Romingeria* has a growth resembling *Aulo-*

---

* *Indiana Geol. and Nat. Hist.*, 11*th Rept. of the State Geologist*, pl. i, fig. 12, 1881.

pora and *Syringopora*. It is without pores on the portion where the corallites and buds are free, but when these are in juxtaposition at their bases mural pores are developed. The upward growth of the initial cell of *P. lenticulare* proceeds but a short distance before the circlet of peripheral corallites is completed. Thus at this stage there are at least seven mural pores opening into the primary calyx. If this tendency to the formation of numerous buds persists throughout the upward growth of the corallites, the non-development of the buds consequent upon the adjacent living corallites would naturally result in the production of mural pores. The basal epitheca limits the fleshy portion of the organisms, and represents an area unfavorable to the acquisition of food or for the natural development of calices. Therefore it would prevent both the maintenance of mural pores and the growth of basal buds.*

A *Favosites* in which one or more cells became inactive or dead shows in its subsequent growth the closing over of this area by the budding of the surrounding cells. Each cell is connected with the parent by an apical pore (Plate XXXI, figures 3, 4). Without this opportunity to bud afforded by the death of one or more corallites, or by their divergence, the adjacent cells would have developed only mural pores. In the figure of *Pleurodictyum problematicum* given on Plate XXXI, figure 2, three of the initial pores are indicated by dotted lines from *p*. No distinction can be made between these and the ordinary pores, except that the latter are usually not as large. This difference in size would be expected, as the primary pore represents the bud which suc-

---

* The presence of basal mural pores or openings through the epitheca has been asserted by Meek and Worthen (*Pal. Illinois*, III. 409, 1868). The specimens from which this observation was made, are from a friable sandstone, which does not usually preserve minute details with much distinctness. The depressions between the spinules on the septal lines could easily be mistaken in a cast for the fillings of mural pores, and it is believed by the writer that this interpretation should be given. *P. lenticulare* occurs as calcareous or silicified, and in the condition of casts. No basal mural pores are present. Also, none can be observed in the casts of *P. problematicum*, from Pelm, Germany.

ceeded in producing a corallite; whereas the other attempts at budding resulted no further than the production of mural pores. The conclusion to be drawn is that the mural pores in such genera as *Favosites, Striatopora, Pleurodictyum, Michelinia*, etc., are ineffectual attempts at budding, resulting only in the perforation of the cell walls. This explanation agrees with the pronounced and persistent tendency to gemmation characteristic of the genera mentioned. They also represent compound forms having individualized epithecas, and this feature naturally arises from the same system of budding obtaining in the simple corals.

Professor Verrill has shown that the presence or absence of tabulæ is of little or no importance in a natural classification.* Therefore the non-tabulate feature of *P. lenticulare* is without special consequence in a discussion of the relations of this species with *Favosites*, or other tabulate poriferous genera.

If the preceding interpretations of structure and affinity are correct, a simple, conical imperforate, non-tabulate prototype, or protocorallum, may be assumed for the Madreporaria Perforata. The next derived form, represented by the early neanic stages of *P. lenticulare*, has the structure and growth of *Aulopora*, and consists of the parent cell with one or more buds. At this stage, which may be called the *Aulopora* stage, the initial corallite has the same number of mural pores as developed buds, for each bud leads into the parent cell by a basal opening or pore. *Aulopora* may thus be considered as representing a primitive type of a poriferous coral, in which the number of pores in each corallite corresponds to the number of buds given off plus one connecting it with the parent cell. Some species of this genus are free throughout most of their growth (*A. subtenuis* Hall), agreeing closely with the erect growth of *Romingeria* and *Syringopora*. This fact removes one of the important arguments against the relations of *Aulopora* with these genera. The corallites of *Aulopora* usually send off buds before turning out of the

* *Amer. Jour. Sci.* (3), III, 287, March, 1872.

common axis of the branch or colony, after which no gemmation commonly takes place. By the explanation here advanced this lack of a tendency to gemmation in the distal portions of the corallites in this genus accounts for the absence of mural pores when such portions are in contiguity. The periods of gemmation in *Romingeria* are periodic. Several buds, often forming a verticil, are given off from the parent corallite. Considerable elongation of the tubes takes place before other series of buds are produced. The budding is prolific at these points, and here also occur the mural pores. The latter are therefore developed when the period of gemmation is in force. If pores are formed elsewhere when the corallites happen to come into juxtaposition, it may possibly be explained as the result of a stimulus produced by the contiguity of the animals. Further observations are necessary to show that pores exist at other places than the bases of the verticils or points where numerous buds are given off and where from crowding the corallites are in juxtaposition.

It therefore seems that, primarily, the development of mural pores is identical or homologous with the process of gemmation. Whether this cause is operative in such forms as *Columnopora* or *Alevopora* yet remains for investigation. The porous condition of the walls in these genera may be an inherited character without an active exciting cause, or it may be teleologically different.

2. SYMMETRICAL CELL DEVELOPMENT IN THE FAVOSITIDÆ*

(Plates XXXII and XXXIII)

The majority of compound corals included in the Favositidæ are composed of polygonal prismatic cells or corallites in juxtaposition. When, however, these cells become free, their form is cylindrical. The polygonal form of closely arranged cells is therefore explained as the natural result of crowding.

The species *Pleurodictyum lenticulare* Hall, sp., is an example of simple cell growth and multiplication. In the development of this species, as shown by the writer in the previous paper, the initial corallite is first conical. The growth of a peripheral series of buds results in changing the sub-circular section of the parent corallite into a polygon. The buds are angular on the sides in juxtaposition to the parent cell and adjacent buds, but on the free portion of their periphery they are cylindrical. The subsequent growth of peripheral buds brings the first series wholly within the corallum, and they are then polygonal in section like the parent corallite.

In compact corals with long cell tubes, as *Michelinia* and *Favosites*, there is a maximum limit to the size of the corallites. Thus the form of the cells which have reached this limit of diametral extension is that of equal hexagonal prisms. This is of course due to the well-known fact of six equal tangent circles about a central circle of the same size. Then from crowding, or from the elimination of the interstitial spaces, they assume a regular hexagonal form. The speci-

* *Trans. Conn. Acad. Sci.*, VIII, 215–220, pls. xiv, xv, 1891.

men of *Cleistopora geometrica*, illustrated by Edwards and Haime,[*] represents the maximum size of the cells and their equal development in this species. Although the tubes are not long, the calices are nearly of the same size, and regularly hexagonal.

After the completion of a circle of calices about the parent cell of the corallum enlargement takes place, (1) by buds from the periphery, and (2) by intermural gemmation. The first is not attended by any phenomena differing from the production of the primary circlet of calices about the initial cell. The second takes place under other conditions, and is the chief method of increase in the growth of large corallums having numerous corallites.

The radial arrangement of the tubes in a large hemispherical or cylindrical mass tends to make the axes of the corallites diverge. This divergence can be taken up only by an increase in the diameters of the tubes, or by the addition of new calices between the others. The latter mode is called intermural gemmation. In *Favosites* and allied genera the maximum size of the corallites is soon reached, and the expansion of the coral is mainly derived from intermural growth. The study of this method of increase properly begins after one or more rows of calices have been developed about the parent cell, and the calices have reached their full dimensions.

The following description of a symmetrical system of intermural cell multiplication was observed in a hemispherical specimen of *Michelinia convexa* D'Orbigny, from the Corniferous Limestone of the Falls of the Ohio. It shows very clearly the stages of development of the interstitial buds, and their modifications. Other corals were examined to the same end, and were found to agree in all essential particulars, whenever their growth was not irregular from their condition of fixation, or from the excessive development or death of a number of the corallites. An exact number of peripheral buds is not

---

[*] Monographie des Polypiers Fossiles des Terraines Palæozoïques, 252, pl. 17, fig. 3, 1851.

necessary to illustrate the general laws of intermural growth. The buds produced from any given cell cannot always agree with the symmetrical method here described, on account of the crowding of similar series from adjacent or neighboring corallites. After eliminating these variations, it was found that the process of intermural gemmation in general is quite uniform, and closely conforms to that in *Michelinia convexa*.

Plate XXXII, figure 1, represents diagrammatically the top of a corallum composed of a central parent cell and six equal peripheral buds, making seven nearly equal calices in the corallum. The upward growth of these corallites and the divergence due to the direction of their axes tend to separate them from the parent cell. In consequence of this separation of the corallites, they would naturally assume a cylindrical form, and there would thus appear triangular interspaces between the tangent points of any three adjacent calices. These angles, therefore, afford the only opportunities for the introduction of a set of intermural buds, and their initial triangular form is determined by the conditions of growth. The smallest number of buds which can be symmetrically placed, and compensate for the divergence of the corallites, is three, one from each alternate angle of the hexagon (Plate XXXII, figure 2).

If these interstitial cells were to grow without the introduction of others, until the original peripheral series was completely separated from the parent or central cell, there would result a corallum containing only triangular corallites. There is, however, a manifest tendency of the organism to the production and maintenance of a cylindrical form, or of a prism with nearly equal radial axes, as in a hexagonal or polygonal prism. To accomplish this, and further to take up the divergence of the corallites, three new interstitial buds are introduced at the remaining three unmodified angles, as shown in figure 3. At this stage there are six symmetrically disposed triangular buds, or intermural cells, about the central corallite, truncating its original angles, and making it a twelve-sided prism. This stage is the third toward the formation of a series of mature interstitial calices.

During the third stage the intermural buds increase in size until they completely surround the parent cell. Then further growth truncates their adjacent angles, thus adding two more sides to each bud, making them pentagonal in section. This marks the fourth stage of intermural growth. At the same time the central corallite loses six of its sides, and returns to its early hexagonal form. The axes have revolved 30°, and the original sides have now become the angles of the corallite (Plate XXXII, figure 4).

At this period of growth it is necessary to consider a series of buds on the periphery of the corallum, marked *1″*, *2″*, etc., in Plate XXXII, figures 3 and 4. They are first triangular in form like the others, and of two sizes, owing to their different ages. The growth of this series continues until they touch and truncate the angles of the first series (*1′*, *2′*, etc.), producing the fifth condition or stage. The first series of buds has now three hexagonal and three pentagonal corallites (Plate XXXII, figure 5).

In the last or sixth stage (figure 6), the further growth of all the intermural cells results in a corallum of nineteen nearly equal hexagonal corallites. The original parent cell (A) is at the centre, the first six intermural cells (*1′*, *2′*, etc.) completely surround it, and the six new peripheral corallites (*1″*, *2″*, etc.) are interposed between the members of the original circlet (*1*, *2*, etc.). The effect of this intermural growth, then, is to dissociate all the first series of corallites from the parent cell and from each other. (See Plate XXXIII.)

The changes taking place in the number and form of the cells may be tabulated as on page 433.

Buds are developed in *Favosites* and *Michelinia* whenever there is a space or opportunity for their growth, unless the corallum is affected by some abnormal condition. If this tendency to form a solid mass of corallites were not so strong, and if the process of budding took place only at comparatively remote intervals, the corallum would have the form of *Romingeria*. In *Michelinia convexa* it is evident that, if the divergence of the corallites was considerable and not wholly filled by intermural growths, there would result a verticil of corallites about the parent cell which would soon become free. The peripheral corallites, also, would become separated. Then after further growth the parent cell would give off another verticil of buds, the other corallites, likewise, develop similar verticils, and the whole form and mode of growth would be like that of *Romingeria*. From this point of view, *Romingeria* may represent an early form of symmetrical cell development in the poriferous corals. The acceleration of the periods of gemmation and consequent approximation of the corallites carrying their verticils of buds would produce all the conditions of cell growth and intermural gemmation exhibited by *Favosites* or *Michelinia*.

Stages.	Form of primary cell.	Whole No. of cells.	Number of intermural buds.	Number of sides of buds.
Nepionic.	cone.	1	0	0
First completed neanic or first condition requisite to intermural gemmation.	6-sided prism.	7	0	0
2d stage.	9-sided prism.	11	3	3
3d stage.	12-sided prism.	16	9	3
4th stage.	6-sided prism.	19	12 { 6 .. 3 / 6 .. 5 }	
5th stage.	6-sided prism.	19	12 { 3 ....3 / 3 ....4 / 3 ....5 / 3 ....6 }	
6th stage.	6-sided prism.	19	12	6

Summary.

The growth of intermural buds compensates for the natural divergence of the corallites. New cells are introduced whenever the old corallites have reached their maximum size, and when their divergence approaches a separation of the cell tubes.

28

The form of the buds is first that of a triangular pyramid or prism, and is due to the mechanical conditions of growth. During subsequent increase they touch and truncate each other, changing from triangular to five- and six-sided prisms. Complete symmetrical normal development produces a corallum with equal hexagonal calices. The process of intermural gemmation changes the sides of the parent cells to angles, and the older corallites, originally in juxtaposition, become separated from each other by new series of interstitial calices.

3. DEVELOPMENT OF THE SHELL IN THE GENUS TORNOCERAS HYATT[*]

(PLATE XXXIV)

THE leading embryonal characters of the genus *Tornoceras* have been drawn mainly from results obtained in the study of *Tornoceras retrorsum* von Buch, and allied species from the Devonian of Germany.[†] Probably the best study of any one of the species is that given by W. Branco of *T. retrorsum*, var. *typum* Sandberger.[‡] The adult features have been determined from the type *T.* (*Goniatites*) *uniangulare* Conrad, and other closely related forms. Hitherto, knowledge of this species has not been sufficient to give a reasonably full diagnosis of the genus in its developmental relations, and the results of the following study aim to supply the deficiency. The importance of this is evident, as the characters of the type are of prime consequence, and because *T. retrorsum* offers some differences in its development, and apparently belongs to one of the more advanced phases in the evolution of the generic stock. Instead of presenting a gradual growth from its simple nautiliform protoconch through several slightly diverging stages, it exhibits, to a degree, the principle of accelerated development, as will be shown hereafter; while *T. uniangulare* has a more uniform and complete growth, and is probably one of the initial and most primitive species of the genus.

---

[*] *Amer. Jour. Sci.* (3), XL, 71-75, pl. i, 1890.

[†] *Proc. Boston Soc. Nat. Hist.*, XXII, Hyatt. **Genera of Fossil Cephalopods,** 320, 1883.

[‡] *Palæontographica*, XXVII. Beiträge zur Entwickelungsgeschichte der fossilen Cephalopoden, 1880.

Besides the sutural and tubular development, the material studied illustrates the inception and growth of the surface ornaments, and as these features are rarely found, the principles involved are of more than generic application.

The specific limits of *T. uniangulare* have not been clearly defined, and many of the forms referred to *Parodiceras* (*Goniatites*) *discoideum* Hall, are evidently of the former species. A comparison of the type specimens of both with others which have been grouped with them, as figured in the Thirteenth Report, New York State Cabinet, and in Vol. V, Part II, of the Palæontology of New York, shows that the first species is really the common one, and as far as known, the second is represented only by the original types.*

The adult differences are mainly noticeable in the depth of the air-chambers and in the sutural curves. They can readily be determined by strictly limiting the characters to those first ascribed to each species. *Parodiceras discoideum* is also apparently without the narrow cone at the bottom of the annular lobe, and the ventral saddle is much less depressed.

The material for this paper is a portion of a collection presented to the museum of Yale University by Thomas G. Lee, M.D. The particular lot containing the *Tornoceras* consisted of several hundred nodular concretions of pyrite of a radiated structure, obtained from the Devonian (Hamilton) shales of Wende Station, Erie county, New York. Most of them preserved an organic nucleus, and about twenty-five species have been identified as belonging to the Trilobita, Cephalopoda, Pteropoda, Pelecypoda, Brachiopoda, and Crinoidea.

The test of the trilobites and the shells of the brachiopods are but little altered, while those of the cephalopods and pelecypods are usually replaced by sphalerite, a difference

---

* The following list is proposed as corrected references to *T. uniangulare:*

13*th Ann. Rept. N. Y. State Cab.*, 98, figs. 6 (bis) (type specimen) and 6 ?

Pal. N. Y., V, pt. ii, pl. 71, figs. 11–14 (fig. 14 = type specimen) ; pl. 72, figs. 6, 7 ; pl. 74, figs. 2, 4 ; VII, pl. 127, figs. 10, 11, 12.

Of these, pl. 71, figs. 11, 12, 13 ; pl. 74, fig. 4 and pl. 127, figs. 11, 12, were referred to *P.* (*Goniatites*) *discoideum* Hall.

evidently connected with the more soluble nature of the pearly shells of the nuculoids and cephalopods.

By carefully breaking away the outer enveloping volutions of a number of specimens of *Tornoceras*, the early parts of the shell were uncovered, and found to be well preserved, and therefore suitable for study. The drawings on Plate XXXIV were made from the microscope, with a camera lucida.

The protoconch (figures 1, 2, Plate XXXIV) has an axial diameter of about 1.1 mm., and among several specimens measured varies but little from this dimension. The vertical diameter is a little shorter, so that the general form is that of a prolate ellipsoid. The latera are prominent, and exposed as central bosses in the umbilicus of a young shell.

At what precise growth-stage the umbilicus becomes closed cannot be ascertained from the material studied, but it is evidently open during the formation of several whorls. During the concrescence of the first few air-chambers, while the diameter of the tube is diminishing, the tendency of the umbilicus is to enlarge rapidly. Subsequent increase in the tube and the greater involution of the whorls contract it, so that in adult specimens it is closed, while in large and often senile individuals a secondary deposit is made about the umbilicus, entirely obliterating it and covering the growth lines of the shell.* Evidently this formation is similar to that deposited by the dorsal lobe of the mantle in *Nautilus pompilius.*

The axial diameter of the embryo shell is somewhat greater than that of several of the succeeding air-chambers. Thus, in its growth, the tube first contracts, and does not assume the regular rate of increase until after the formation of at least the second septum. A cross section at the first septum is transversely sub-elliptical, slightly arcuate, with a longer diameter two and one-third times greater than the shorter. When a transverse diameter of 1.2 mm. is reached by the

---

* This feature is well represented in fig. 11, pl. 127, *Pal. N. Y.*, **VII**, supplement.

larval shell, the outline of a section is lunate, but the proportions of length and height are not materially changed. A section of the adjacent whorl is still more arcuate, as shown in figure 6, and in an adolescent specimen 11.5 mm. in diameter it is seen that the diametral relations have become interchanged, and that the outer whorl is elliptical in a vertical direction, and excavated by the inner whorl to nearly half its longest diameter, making the shell in all neanic and ephebic stages decidedly compressed in outline (figure 13).

The first septum (figures 1, 2, 4) is moderately concave, and extends nearly to the axis. The suture is simple, being nearly in a single plane, without apparent lobes or saddles. Occasionally the internal mould shows a siphonal lobe due to the breaking away of the extremely thin filling between the siphon and ventrum, but perfect specimens determine this to be an accidental condition.

In the section represented in figure 12, the first two septa are much thicker than those immediately succeeding, a feature also noticeable on the exterior of the internal mould. Likewise the first and second air-chambers are deeper than the three or four following. With these exceptions the septa and air-chambers are generally uniform in their progression until the adult stage.

It has been already noted that the first septum is extremely simple, without apparent lobes or saddles. In the second septum there is a well-developed sinus over the siphuncle, forming a rounded ventral lobe, and a broad lateral saddle, with a corresponding, though less prominent, dorsal saddle. The third and following septa present more and more sharply angular ventral lobes, until finally it is further extended by a siphonal fissure in post-nepionic stages. The lateral saddle is not so strongly curved from the fourth to the seventh suture, which is quite flat, but in the eighth a slight retral bend is observable. This evidently marks the inception of the lateral lobe. The septum is now divided into the leading members characteristic of the group; namely, a ventral lobe and saddle, a lateral lobe and saddle, a dorsal saddle and an annular lobe, although the two latter are less strongly marked than the others. Further growth merely serves to emphasize these features, until the neanic stadium, when the ventral lobe is extended by the siphonal fissure, and a minute cone appears at the bottom of the annular lobe.

Several specimens of the protoconch give evidence of the presence of the siphonal cæcum, and show that it was probably closely appressed to the ventral wall. Figure 3 illustrates the ovoid marking on the interior of the shell, enclosing two diverging lines which apparently represent the appressed portion of the true cæcum, while the outer curved lines limit the shelly deposit of attachment. The relative diameter of the siphon at the first and for a number of succeeding chambers is much greater than in the mature shell (figures 1, 6, and 13). From the beginning, it is situated close to the abdominal wall, and is nearly invariable in its character.

The embryonic shell is very thin, and almost smooth in its earlier portions; then fine revolving lines of granules appear, which become progressively more pronounced and arranged in transverse rows, between which the earliest of the concentric striæ are developed. With the increase in the strength of the striæ, the granules disappear, and are obsolescent before the protoconch is completed (figure 7). The striæ are sharp, elevated, and straight, forming a conspicuous feature of the ornamentation, until in the third or fourth whorl, when they become subdued, and finally are replaced by the fine inconspicuous and often fasciculate lines of growth which are present in the adult shell. No indication of a funnel is shown up to the completion of the first whorl (figure 10), as the striæ continue straight across the ventrum, but in the second (figure 11) the pronounced sinus in the striæ shows that the funnel had developed or, at least, had become of functional importance.

A comparison of the figures of *T. retrorsum* von Buch, var. *typum* Sandberger, as illustrated by Branco (*loc. cit.*, pl. v, fig. vii), with the present species shows that the former presents a more arcuate first septum, and that the second is comparable with the third or fourth of *T. uniangulare*, clearly indicating that the development has been accelerated by the skipping of at least two phases of growth in the septa. In other characters the two forms merely indicate differences which are probably only of specific importance, such as the more angular form of the lobes and saddles in *T. retrorsum*, var. *typum*, and the absence of the minute cone at the summit of the annular lobe.

V

PLATES AND EXPLANATIONS

PLATE I

SPINES OF RADIOLARIA (Page 16)

Spiniform processes of recent Radiolaria taken from the shells of the following species : —

Figure 1. — *Heliosphæra coronata.*
" 2. — *Astrosphæra stellata.*
" 3. — *Astrophacus solaris.*
" 4. — *Stylosphæra calliope.*
" 5. — *Heliodiscus glyphodon.*
" 6. — *Pripodictya triacantha.*
" 7. — *Pleuraspis horrida.*
" 8. — *Hexacontium sceptrum.*
" 9. — *Acanthosphæra clavata.*
" 10. — *Acanthosphæra clavata.*
" 11. — *Cromyodrymus quadricuspis.*
" 12. — *Hexacontium clavigerum.*
" 13. — *Orosphæra horrida.*
" 14. — *Staurocyclia phacostaurus.*
" 15. — *Tripospyris capitata.*
" 16. — *Archipera cortiniscus.*
" 17. — *Tripospyris conifera.*
" 18. — *Orosphæra serpentina.*
" 19. — *Staurolonche petusa.*
" 20. — *Astrosphæra stellata.*
" 21. — *Staurodictya elegans.*
" 22. — *Hexastylus contortus.*
" 23. — *Stephanospyris excellens.*
" 24. — *Podocyrtis magnifica.*
" 25. — *Hexancistra mirabilis.*
" 26. — *Dictyophimus Cienkowskii.*
" 27. — *Elatomma juniperinum.*
" 28. — *Castanura Tizardi.*

Figure 29. — *Pleuraspis horrida.*
" 30. — *Staurocarynum arborescens.*
" 31. — *Rhizosphæra serrata.*
" 32. — *Phænocalpis petalospyris.*
" 33. — *Aulospathis bifurca.*
" 34. — *Aulographis bovicornis.*
" 35. — *Aulographis ancorata.*
" 36. — *Aulographis bovicornis.*
" 37. — *Sphærozoum verticillatum.*
" 38. — *Cladococcus pinetum.*
" 39. — *Hexancistra triserrata.*
" 40. — *Cladococcus stalactites.*
" 41. — *Hexancistra quadricuspis.*
" 42. — *Heliodrymus ramosus.*
" 43. — *Heliodrymus dendrocyclus.*
" 44. — *Aulographis pandora.*
" 45. — *Cladoscenium ancoratum.*
" 46. — *Cladococcus scoparius.*
" 47. — *Auloscena penicillus.*
" 48. — *Circostephanus coronarius.*
" 49. — *Lychnosphæra regina.*
" 50. — *Auloscena spectabilis.*
" 51. — *Cælospathis ancorata.*
" 52. — *Octodendron spathillatum.*

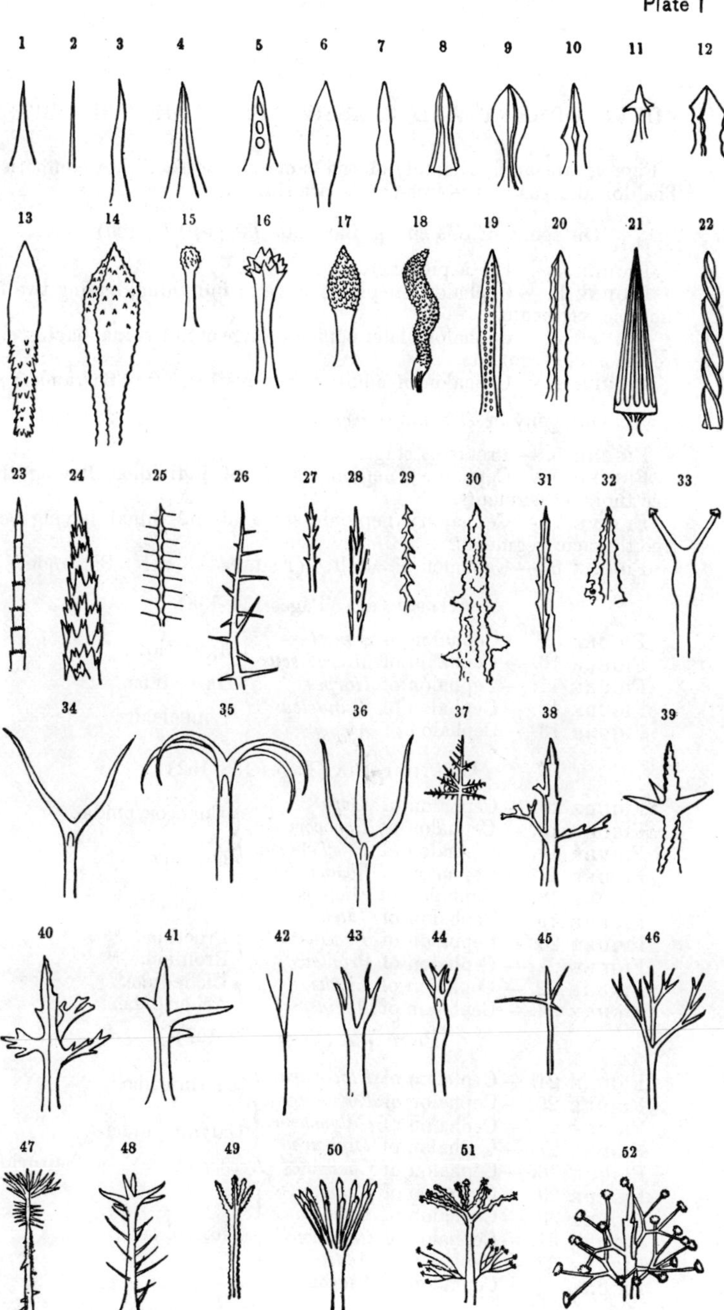

SPINES OF RADIOLARIA

PLATE II

DEVELOPMENT AND CLASSIFICATION OF TRILOBITES

Figures are approximately of the same size to facilitate comparison. The dorsal areas of the free-cheeks are shaded.

Ontogeny of *Sao hirsuta* Barrande (Pages 128, 129)

FIGURE 1. — Protaspis stage.
FIGURE 2. — Cephalon; nepionic stage of individual having two free thoracic segments.
FIGURE 3. — Cephalon; later nepionic stage of individual having eight free thoracic segments.
FIGURE 4. — Cephalon of adult.   (Figures 1–4, after Barrande.)

Ontogeny of *Dalmanites socialis* Barrande (Pages 128, 129)

FIGURE 5. — Protaspis stage.
FIGURE 6. — Cephalon; nepionic stage of individual having three free thoracic segments.
FIGURE 7. — Cephalon; nepionic stage of individual having seven free thoracic segments.
FIGURE 8. — Cephalon of adult.   (Figures 5–8, after Barrande.)

HYPOPARIA (Pages 134–138)

FIGURE 9. — Cephalon of *Agnostus*	} Agnostidæ.
FIGURE 10. — Cephalon of *Microdiscus*	
FIGURE 11. — Cephalon of *Harpes*	} Harpedidæ.
FIGURE 12. — Cephalon of *Trinucleus*	} Trinucleidæ.
FIGURE 13. — Cephalon of *Ampyx*	

OPISTHOPARIA (Pages 138–152)

FIGURE 14. — Cephalon of *Atops*	} Conocoryphidæ.
FIGURE 15. — Cephalon of *Conocoryphe*	
FIGURE 16. — Cephalon of *Ptychoparia*	} Olenidæ.
FIGURE 17. — Cephalon of *Olenus*	
FIGURE 18. — Cephalon of *Asaphus*	} Asaphidæ.
FIGURE 19. — Cephalon of *Illænus*	
FIGURE 20. — Cephalon of *Proëtus*	} Proëtidæ.
FIGURE 21. — Cephalon of *Bronteus*	} Bronteidæ.
FIGURE 22. — Cephalon of *Lichas*	} Lichadidæ.
FIGURE 23. — Cephalon of *Acidaspis*	} Acidaspidæ.

PROPARIA (Pages 152–157)

FIGURE 24. — Cephalon of *Placoparia*	} Encrinuridæ.
FIGURE 25. — Cephalon of *Encrinurus*	
FIGURE 26. — Cephalon of *Calymmene*	} Calymmenidæ.
FIGURE 27. — Cephalon of *Dipleura*	
FIGURE 28. — Cephalon of *Cheirurus* (*Eccoptocheile*)	} Cheiruridæ.
FIGURE 29. — Cephalon of *Dalmanites*	
FIGURE 30. — Cephalon of *Dalmanites*	
FIGURE 31. — Cephalon of *Chasmops*	} Phacopidæ.
FIGURE 32. — Cephalon of *Acaste*	
FIGURE 33. — Cephalon of *Phacops*	

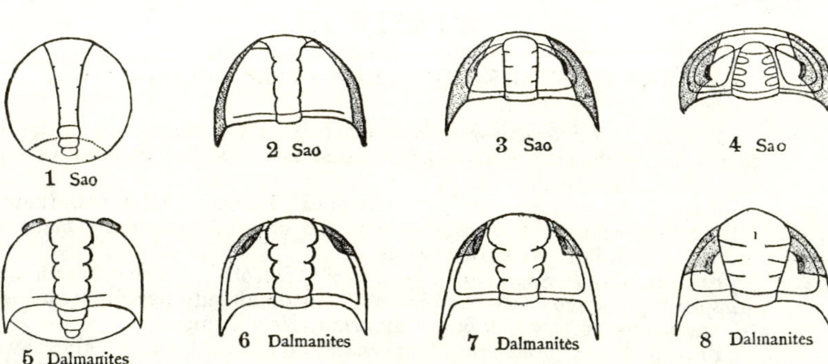

HYPOPARIA

9 Agnostus
10 Microdiscus
11 Harpes
12 Trinucleus
13 Ampyx

OPISTHOPARIA

14 Atops
15 Conocoryphe
16 Ptychoparia
17 Olenus
18 Asaphus
19 Illænus
20 Proëtus
21 Bronteus
22 Lichas
23 Acidaspis

PROPARIA

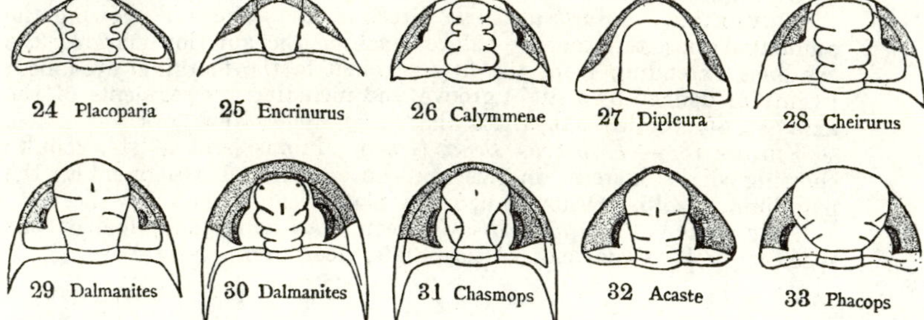

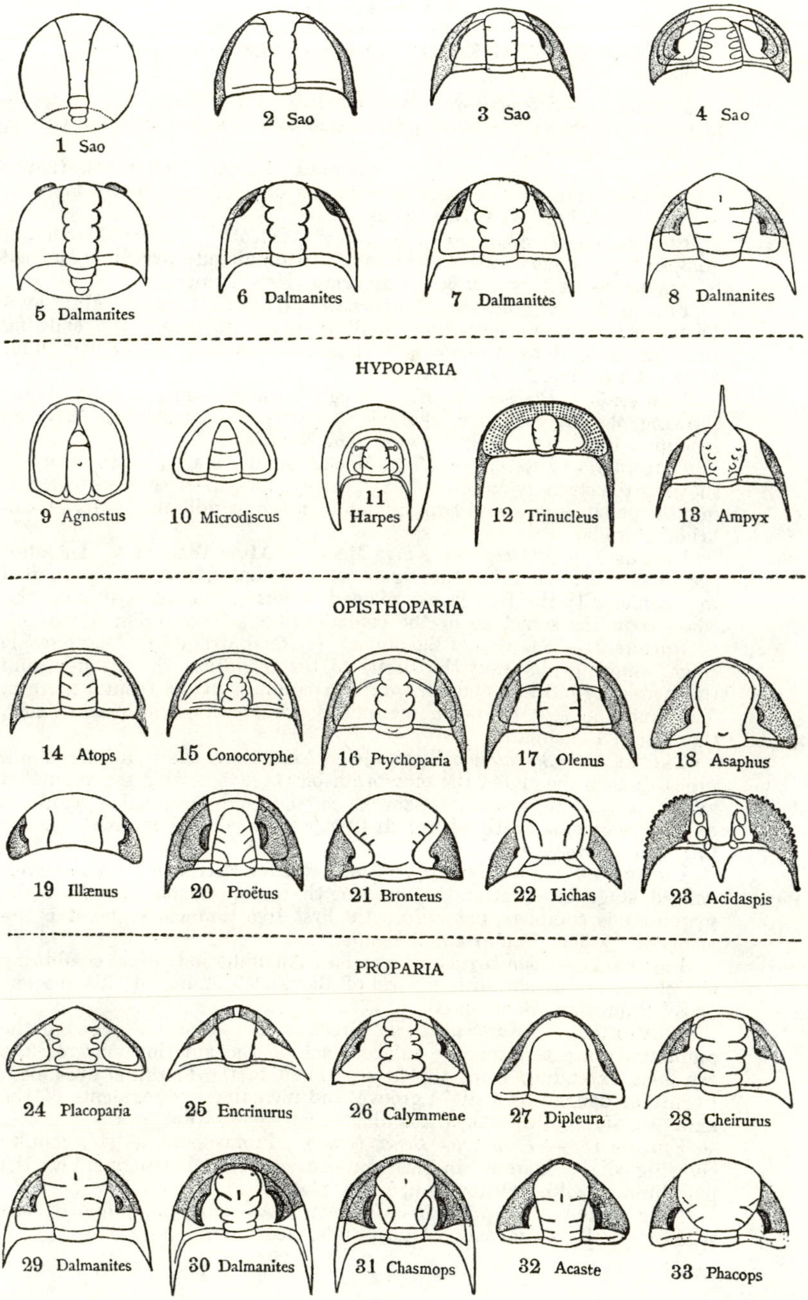

CLASSIFICATION OF TRILOBITES

PLATE III

LARVAL STAGES OF TRILOBITES (Pages 171–173)

Figure 1. — *Solenopleura Robbi* Hartt. (After Matthew.) Anaprotaspis stage; showing obscurely annulated axis. × 30. St. John Group, Cambrian, New Brunswick.

Figure 2. — *Liostracus onangondianus* Hartt. (After Matthew.) Anaprotaspis stage; the neck lobe is the only one distinctly marked. × 23. Cambrian, New Brunswick.

Figure 3. — *Ptychoparia Linnarssoni* Walcott. (After Matthew.) Anaprotaspis stage; axis slender, slightly annulated; pygidium defined by transverse furrow. × 30. Cambrian, New Brunswick.

Figure 4. — *Ptychoparia Linnarssoni*, Walcott. (After Matthew.) Protaspis; representing a later moult than the preceding, and showing stronger annulations on the axis, with an additional one on the pygidium. × 25. Cambrian, New Brunswick.

Figure 5. — *Ptychoparia Kingi* Meek. Anaprotaspis or early stage; showing obscurely defined characters, partly due to the fact that the specimen is a cast. × 45. Cambrian, Nevada.

Figure 6. — *Ptychoparia Kingi* Meek. A later stage (metaprotaspis); showing the strongly annulated axis, the eye-line, the free-cheeks including the genal angles, and two segments on the pygidium. × 45. Cambrian, Nevada.

Figure 7. — *Ptychoparia Kingi* Meek. (After Walcott.) An adult specimen. This and the other figures of adult individuals are represented in outline, with the free-cheeks shaded, to bring out more strongly the changes in the structure of the cephalon. × ½. Cambrian, Utah.

Figure 8. — *Sao hirsuta* Barrande. (After Barrande.) Anaprotaspis stage; showing obscurely the limits of the pygidium, the eye-line, and the nearly cylindrical glabellar axis expanding on the frontal margin. This and the two following specimens are preserved as casts. × 30. Cambrian, Bohemia.

Figure 9. — *Sao hirsuta* Barrande. (After Barrande.) A later moult, probably near the end of the metaprotaspis stage; showing the annulated axis expanded in front; free-cheeks narrow and marginal; pygidium of four segments, with pleura distinctly marked and grooved. × 30. Cambrian, Bohemia.

Figure 10. — *Sao hirsuta* Barrande. (After Barrande.) A more advanced stage at or after the close of the paraprotaspis, in which the pygidium is complete, but before the first free thoracic segment is developed. × 30. Cambrian, Bohemia.

Figure 11. — *Sao hirsuta* Barrande. An adult individual combining the characters as shown in several of Barrande's figures of this species. × ½. Cambrian, Bohemia.

Figure 12. — *Triarthrus Becki* Green. Anaprotaspis; showing the annulated axis, terminating before reaching the anterior margin; the eye-lines extending from the first segment to the marginal eye-lobes; pygidium defined by a slight groove, and including two segments of the axis. × 45. Ordovician, Utica Slate, near Rome, New York.

Figure 13. — *Triarthrus Becki* Green. Protaspis at a later moult; showing slight increase in size and the addition of a segment to the pygidium. × 45. Ordovician, Utica Slate, near Rome, New York.

Figure 14. — *Triarthrus Becki* Green. An adult individual of this species. × ½. Ordovician, Utica Slate, near Rome, New York.

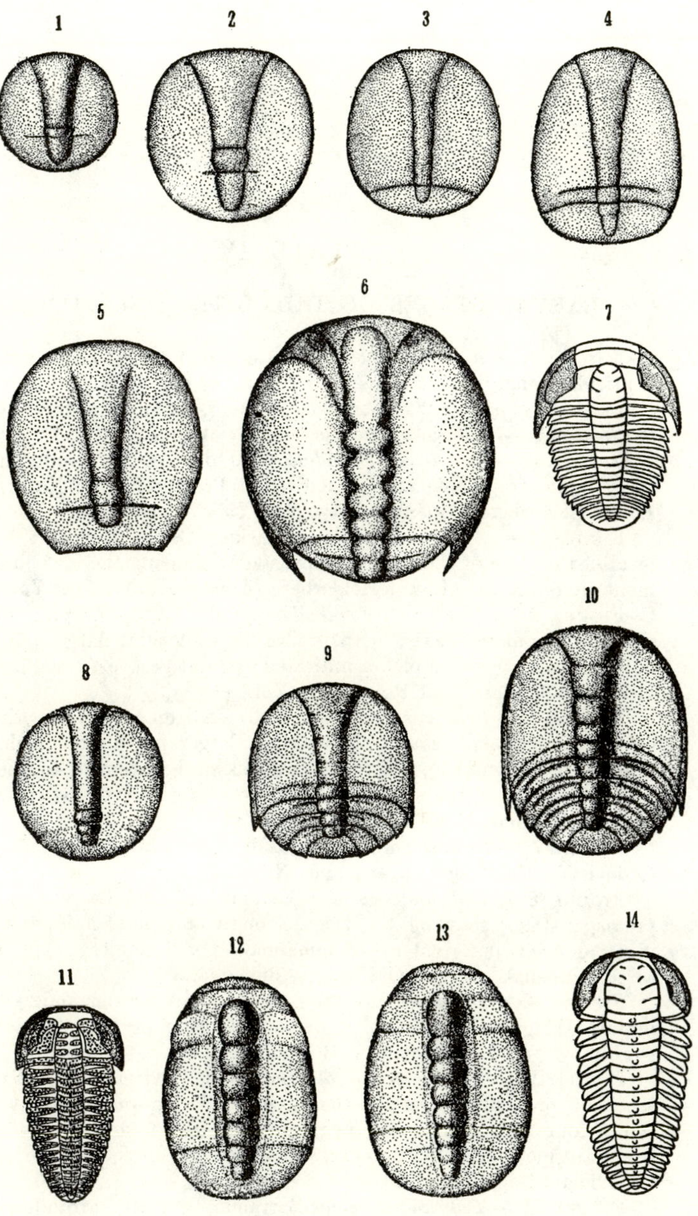

LARVAL STAGES OF TRILOBITES

PLATE IV

LARVAL STAGES OF TRILOBITES (Pages 173–176)

Figure 1. — *Acidaspis tuberculata* Conrad. Anaprotaspis; showing denticulate margin and spines on cephalon; axis strongly annulated; eyes sub-marginal. × 20. Lower Helderberg, Albany county, New York.

Figure 2. — The same; profile, slightly oblique. × 20.

Figure 3. — *Acidaspis tuberculata* Conrad. An adult individual, restored from fragments and an entire enrolled specimen. Natural size. Lower Helderberg, Albany county, New York.

Figure 4. — *Arges consanguineus* Clarke. Dorsal view of a larva at or after the close of the paraprotaspis stage; showing the form and ornamentation. × 20. Lower Helderberg, Albany county, New York.

Figure 5. — *Proëtus parviusculus* Hall. Anaprotaspis; showing strongly annulated axis, with groove at each side; large prominent anterior eyes; pygidial pleura indicated by faint grooves. × 45. Ordovician, Utica Slate, near Rome, New York.

Figure 6. — *Proëtus parviusculus* Hall. A later moult, near the close of the paraprotaspis stage; showing the larger pygidium which, however, is still incomplete, and the slight backward movement of the eyes. The right side of the specimen is restored. × 45. Ordovician, Utica Slate, near Rome, New York.

Figure 7. — *Proëtus parviusculus* Hall. An adult individual. × 2. Ordovician, Utica Slate, near Rome, New York.

Figure 8. — *Dalmanites socialis* Barrande. (After Barrande.) Anaprotaspis stage; showing the large strongly annulated axis; the prominent anterior marginal eyes; mucronate genal angles; pygidium of three segments. × 30. Ordovician, Bohemia.

Figure 9. — *Dalmanites socialis* Barrande. (After Barrande.) Metaprotaspis stage; showing the stronger definition of the pleura of the pygidium. × 30. Ordovician, Bohemia.

Figure 10. — *Dalmanites socialis* Barrande. (After Barrande.) The specimen probably represents the close of the paraprotaspis stage, and shows four segments in the pygidium and the first evidence of the backward movement of the eyes, which now indent the margin. × 30. Ordovician, Bohemia.

Figure 11. — *Dalmanites socialis* Barrande. (After Barrande.) Outline of an adult individual. × ½. Ordovician, Bohemia.

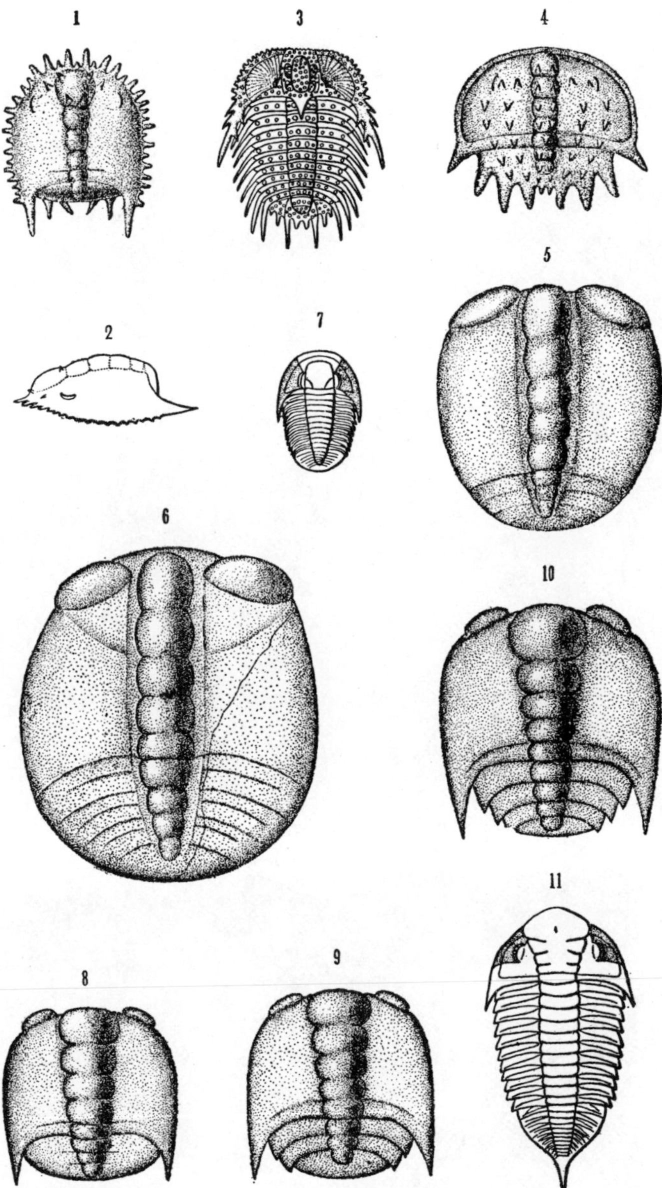

LARVAL STAGES OF TRILOBITES

PLATE V

CRUSTACEAN LARVÆ (Pages 191, 192)

The Roman numerals indicate the appendages in their consecutive order.

I, 1st pair of appendages, or antennules.

II, 2d pair of appendages, or antennæ.

III, 3d pair of appendages, or mandibles.

IV, *V*, etc., maxillæ, maxillipeds, swimming feet, etc.

ocl, unpaired eye ; *oc*, paired eyes ; *lb*, labrum.

Figure 1. — *Triarthrus Becki*. A restoration of the ventral side of the protaspis stage in accordance with the best evidence at present attainable, as explained in the text. The *VI*th and the *VII*th pairs of appendages belong to the abdomen, which is marked off by a transverse line ; *mt*, metastoma ; *g*, free-cheeks.

Figure 2. — *Apus cancriformis*. (After Claus, from Faxon.) Phyllopoda. Nauplius larva, just hatched ; ventral side. Behind the mandibles (*III*) are indications of five thoracic somites (*y*).

Figure 3. — *Apus cancriformis*. (After Claus, from Faxon.) Phyllopoda. Second larval stage (metanauplius) ; ventral side. The second maxilla (*V*) is wanting ; *f*, frontal sense organs.

Figure 4. — *Branchipus stagnalis*. (After Claus, from Packard.) Phyllopoda. Nauplius stage.

Figure 5. — *Artemia gracilis*. (After Packard.) Phyllopoda. Nauplius stage ; showing obscure segmentation.

Figure 6. — *Limnaida Hermanni*. (After Lereboullet, from Packard.) Phyllopoda. Nauplius ; dorsal side ; first pair of appendages obsolescent ; labrum (*lb*) greatly developed.

Figure 7. — *Lepidurus productus*. (After Brauer, from Bernard.) Phyllopoda. Nauplius with obscure segmentation of the trunk (*y*).

Figure 8. — *Leptodora hyalina*. (After Sars, from Balfour and Bronn.) Phyllopoda, Cladocera. Nauplius larva from winter egg ; *y*, rudimentary feet.

Figure 9. — *Daphnia longispina*. (After Dohrn, from Claus.) Phyllopoda, Cladocera. Nauplius stage of embryo, with rudimentary appendages.

Figure 10. — *Moina rectirostris*. (After Grobben, from Faxon). Phyllopoda, Cladocera. Embryo from the summer egg in the nauplius stage, developed in the brood-cavity of the parent ; appendages rudimentary.

Figure 11. — *Cyclops tenuicornis*. (After Claus, from Balfour.) Copepoda, Natantia. Nauplius, first stage. This and the next are the original forms described as *Nauplius*, by O. F. Müller, and believed at that time to be adult.

Figure 12. — *Cyclops tenuicornis*. (After Claus, from Balfour.) Copepoda, Natantia. Nauplius ; second stage ; *IV*, maxillæ.

Figure 13. — *Cetochilus septentrionalis*. (After Grobben, from Faxon.) Copepoda, Natantia. Nauplius; just hatched; ventral view.

Figure 14. — *Actheres percarum*. (After Claus, from Faxon.) Copepoda, Parasitica. Larva at the time it leaves the egg, with only two anterior unbranched pairs of appendages of the typical nauplius present. Under the skin are the rudiments of six pairs of appendages; *III*, mandibles; *IV*, maxillæ; *V, VI*, maxillæ; *VII, VIII*, swimming feet.

Figure 15. — *Balanus balanoides*. (After Hoek, from Faxon.) Cirripedia. Nauplius.

Figure 16. — *Lernædiscus porcellanæ*. (After F. Müller, from Faxon.) Cirripedia, Rhizocephala. Nauplius; ventral side; showing outline of dorsal shield.

Figure 17. — *Sacculina purpurea*. (After F. Müller, from Huxley and Balfour.) Cirripedia, Rhizocephala.

Figure 18. — *Cyprus ovum*. (After Claus, from Faxon.) Ostracoda. First larval (nauplius) stage, with bivalve shell and unbranched second and third pairs of appendages.

Figure 19. — *Nebalia Geoffroyi*. (After Metschnikoff, from Faxon.) Leptostraca. Side view of the so-called nauplius stage of the embryo within the egg. Rudiments are present of the two pairs of antennæ (*I, II*) and the mandibles (*III*).

Figure 20. — *Euphausia*. (After Metschnikoff, from Faxon.) Schizopoda. Nauplius; just hatched.

Figure 21. — *Euphausia*. (After Metschnikoff, from Faxon.) Schizopoda. Nauplius at a later stage; ventral view; *mt*, metastoma; *IV, V,* maxillæ; *VI*, maxilliped. In the next, or Protozoën, stage the appendages (*IV, V, VI*) are true phyllopodiform feet.

Figure 22. — *Mysis ferruginea*. (After Van Beneden, from Faxon.) Schizopoda. Nauplius-like embryo; side view. The appendages are unsegmented, and the third pair is quite rudimentary. A number of later metamorphoses are undergone in the nauplius skin, until the full number of appendages is developed.

Figure 23. — *Peneus*. (After F. Müller, from Faxon.) Decapoda, Macroura. Nauplius; from dorsal side.

Figure 24. — *Lucifer*. (After Brooks, from Faxon.) Decapoda, Macroura. Ventral view of embryo artificially removed from the egg; *IV, V, VI*, buds representing the two pairs of maxillæ and first pair of maxillipeds of the adult.

Figure 25. — *Palæmon*. (After Bobretzky, from Faxon.) Decapoda, Macroura. Nauplius stage of embryo within the egg.

Figure 26. — *Astacus fluviatilis*. (After Reichenbach, from Faxon.) Decapoda, Macroura. Nauplius stage of embryo.

Figure 27. — *Limulus polyphemus*. (After Kingsley.) Xiphosura. Ventral view of embryo; showing the budding of the legs.

Figure 28. — *Limulus polyphemus*. (After Packard, from Balfour.) Xiphosura. Ventral view of embryo in the egg; showing the rudiments of six pairs of legs; *m*, mouth.

Figure 29. — *Limulus polyphemus*. (After Packard, from Balfour.) Xiphosura. Oblique side view of embryo, with the mouth and rudimentary limbs on the ventral plate.

The figures of embryonic *Limulus* are introduced for comparison. They are so different from the nauplius that detailed notice seems unnecessary.

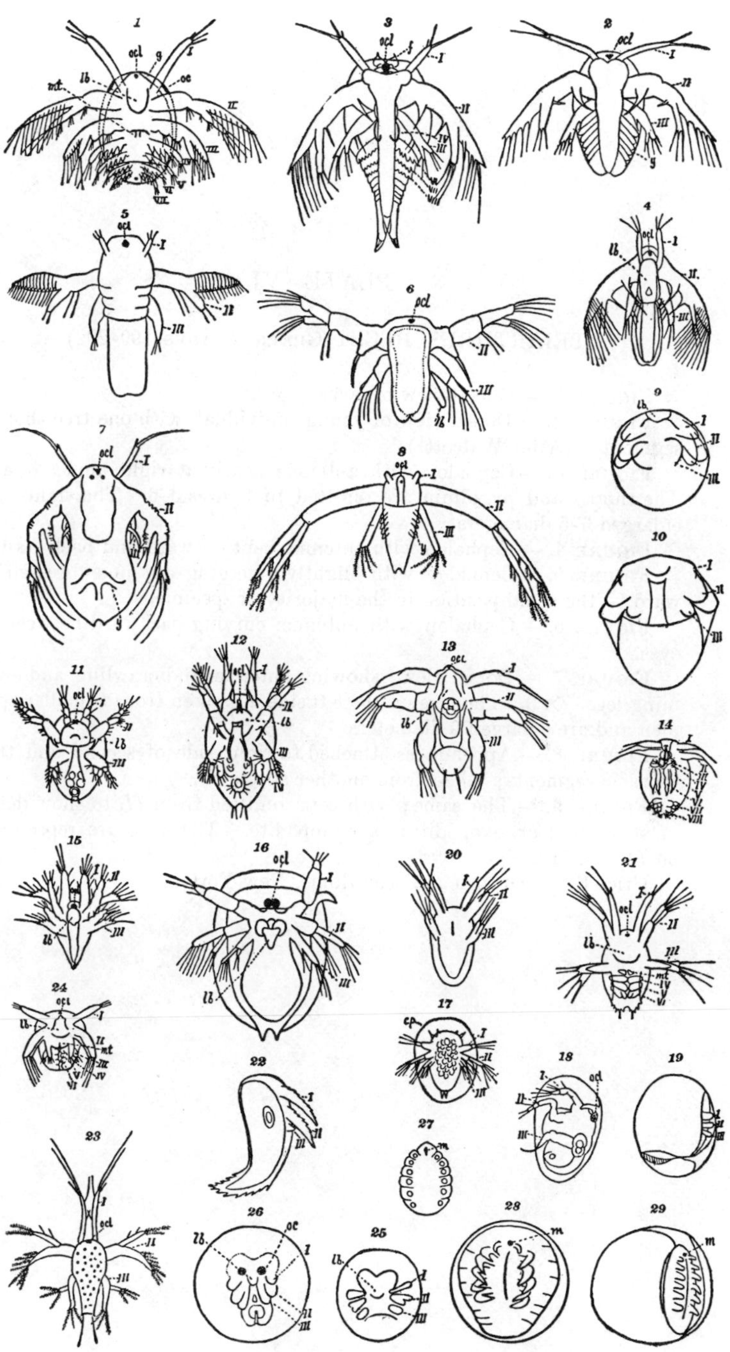

CRUSTACEAN LARVÆ

PLATE VI

TRIARTHRUS BECKI GREEN (PAGES 199–202)

FIGURE 1. — Dorsal view of larva. × 28.

FIGURE 2. — Dorsal view of young individual, with one free thoracic segment. (After Walcott.)

FIGURE 3. — Cephalon with antennæ nearly at right angles to axis. The thorax and pygidium are omitted in figures 3–6. The figures are enlarged 3–5 diameters.

FIGURE 4. — Cephalon with antennæ bent outward and backward.

FIGURE 5. — Cephalon with slightly diverging antennæ directed forward, — the usual position in the majority of specimens.

FIGURE 6. — Cephalon with antennæ curving backward between the eyes.

FIGURE 7. — Dorsal view; showing antennæ and crawling and swimming legs. × 3. The legs on the left side are taken from a smaller specimen and are enlarged 6 diameters.

FIGURE 8. — Appendages attached to right side of second and third thoracic segments; taken from another specimen.

FIGURE 9. — The same; with setæ omitted from *II*, to show details of structure; *ex*, exopodite; *en*, endopodite. The setæ are represented on *III*. × 12.

Utica Slate, Ordovician, near Rome, New York.

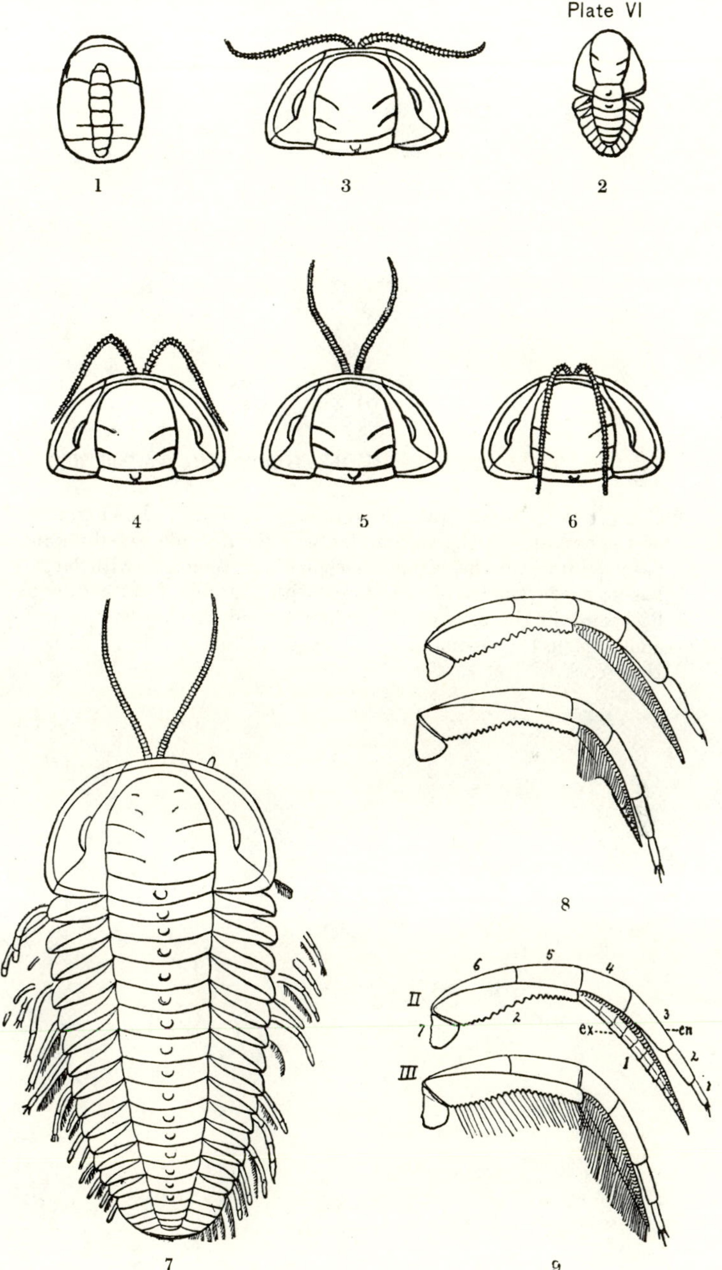

APPENDAGES OF TRIARTHRUS

PLATE VII

TRIARTHRUS BECKI Green (Pages 206-209)

Figure 1. — Ventral side of a large specimen; showing, at each side of the hypostoma, the antennules with their single flagellum and strong basal joint; also the biramous cephalic appendages with large gnathobases; the biramous thoracic limbs with gnathobases extending obliquely backward in the axis of the trilobite; and the hypostoma, metastoma, and anus, in the median line. × 4. Utica Slate, Ordovician, near Rome, New York.

1

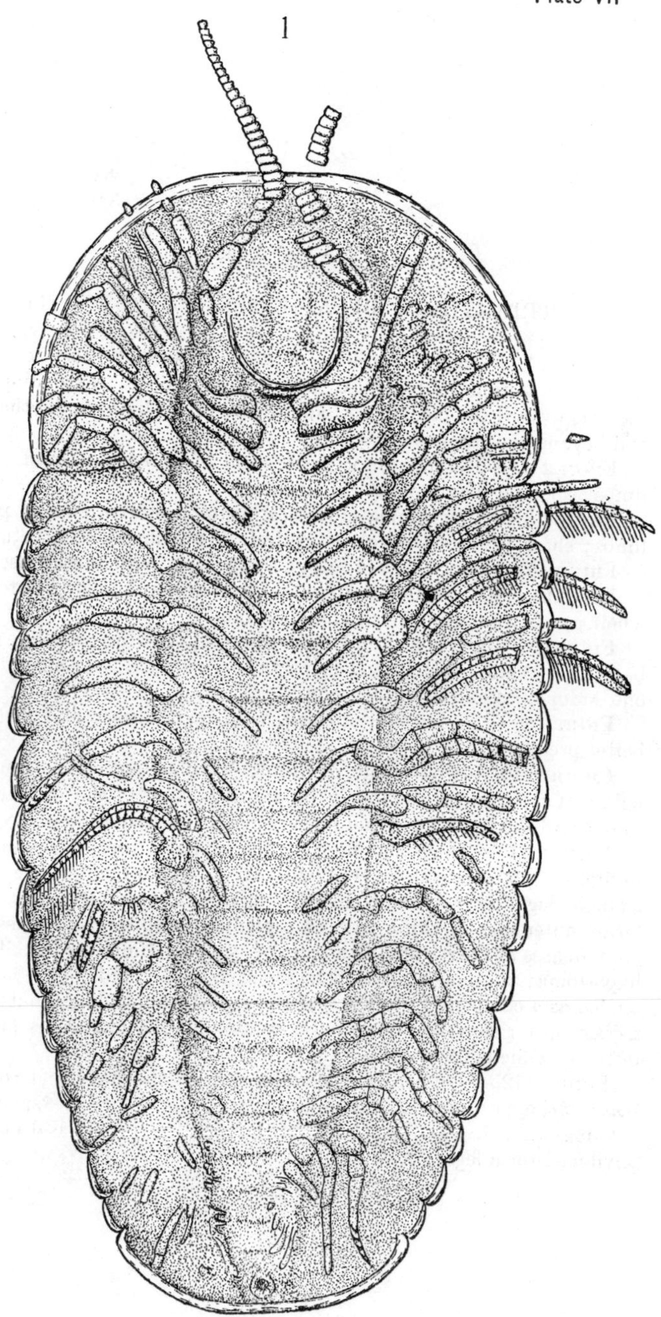

Ventral Side of Triarthrus

PLATE VIII

TRIARTHRUS BECKI GREEN (PAGES 205–211)

FIGURE 1. — Diagrammatic restoration of second thoracic limb in transverse section of trilobite; showing gnathobases extending under the axis toward the median line, and the biramous limbs under the pleura and beyond the carapace.

FIGURE 2. — Diagrammatic restoration of anterior pair of pygidial limbs; showing phyllopodiform structure.

FIGURE 3. — Diagrammatic restoration of posterior pair of pygidial limbs; showing more strongly the primitive phyllopodous structure.

FIGURE 4. — Dorsal view of second thoracic leg, with gnathobase. × 12.

FIGURES 5, 6, 7. — Dorsal views of three heads; showing antennules. Their position in figure 7 is the most common.

FIGURE 8. — Portion of under side of head; showing plate-like gnathobases of cephalic limbs, posterior part of hypostoma, the metastoma, and one endopodite on the right. × 4.

FIGURE 9. — Under side of head, with gnathobases and metastoma better preserved. × 4.

FIGURE 10. — Under side of head; showing antennules and their points of attachment, the four pairs of gnathobases of the other cephalic limbs, and the hypostoma and metastoma in the median line. × 4.

FIGURE 11. — Diagram; showing the cephalic appendages preserved in figures 5, 6, 7, and figure 1, Plate VII. *1*, shaft of antennule bearing a single flagellum ; *2*, coxopodite of first pair of biramous limbs, or posterior antennæ; *3*, third pair of cephalic limbs, or mandibles; *4*, *5*, gnathobases of fourth and fifth pairs of cephalic limbs, or maxillæ; *hy*, hypostoma; *m*, mouth ; *met,* metastoma; *s*, setæ. This figure is not intended as a complete restoration of the cephalic appendages, but only as a diagram for convenient reference, combining the characters preserved in the specimens illustrated.

FIGURE 12. — Portion of the under side of the head and thorax of *Apus*. *hy*, hypostoma ; *1*, antennule; *2*, posterior antenna ; *3*, mandibles; *4, 5*, maxillæ; *6*, maxilliped ; *7*, first thoracic leg ; *8, 9*, basal endites of phyllopodiform legs. × 3.

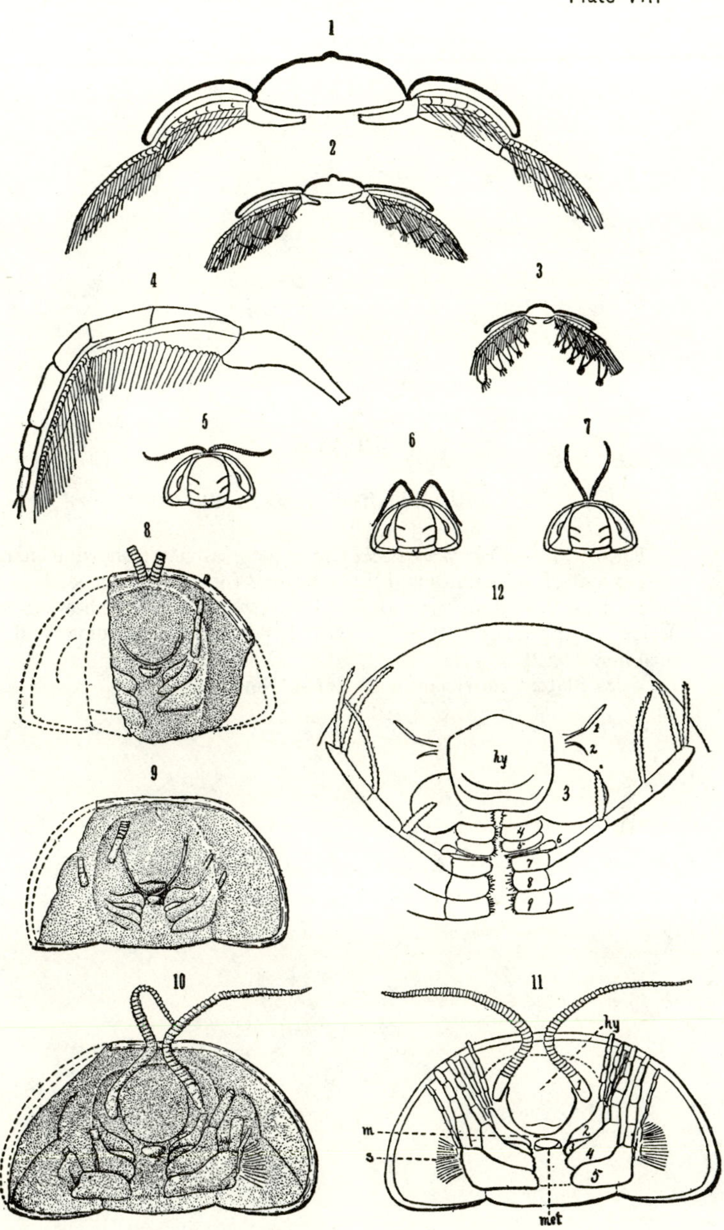

APPENDAGES OF TRIARTHRUS

PLATE IX

TRIARTHRUS (Pages 213–219)

Figure 1. — *Triarthrus Becki* Green ; dorsal view; showing character and extent of antennules and limbs beyond the carapace.   × 2½.

Figure 2. — *Triarthrus Becki* Green; ventral view; showing entire series of appendages, together with hypostoma, metastoma, and anal opening.   × 2½.

Utica Slate, Ordovician, near Rome, New York.

1

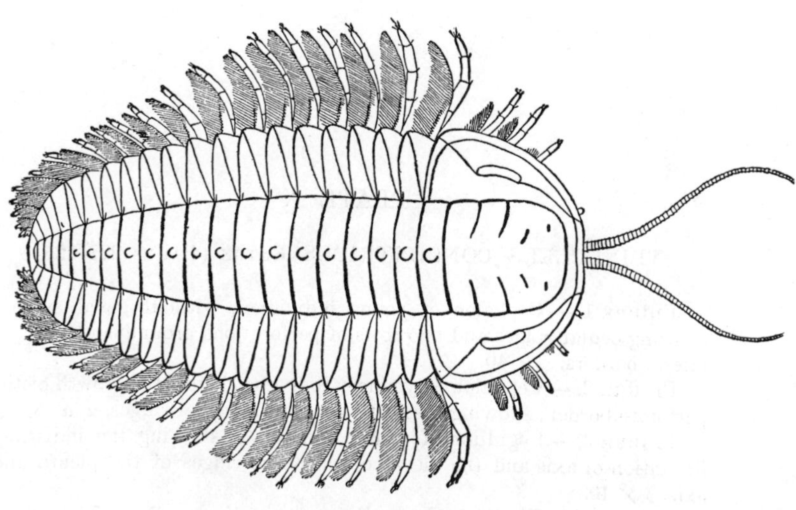

2

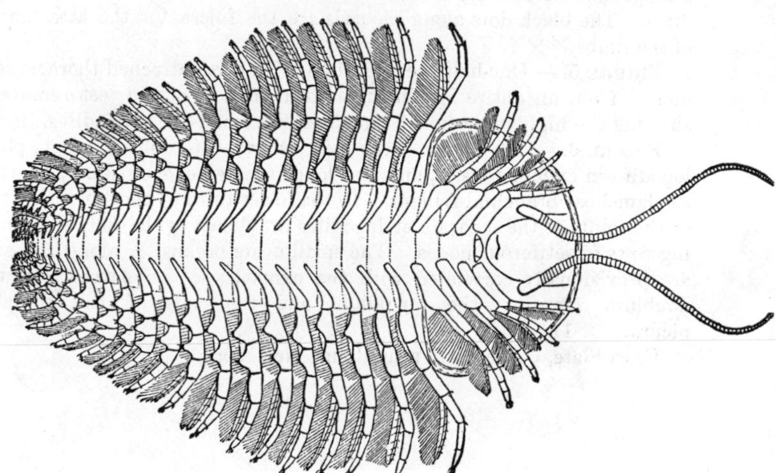

TRIARTHRUS

TRINUCLEUS CONCENTRICUS Eaton (Pages 222-225)

Figure 1. — Cephalon of young individual, without genal spines; showing ocular ridges and two rows of perforations around anterior and lateral borders.   × 40.

Figure 2. — Cephalon of younger individual before the growth of the perforate border; showing distinctly the clavate ocular ridges, *a, a*. × 40.

Figure 3. — Pygidium of young individual; showing the indistinct limitation of axis and the elevated transverse ridges of the pleura and axis.   × 40.

Figure 4. — Thorax and pygidium of an entire specimen from which the dorsal test has been removed by weathering, exposing below the fringes of the exopodites, which entirely cover the pleural portions.   The stronger lines ascending from the axis are the main stems of the exopodites.   The black dots along the axis are the fulcra for the attachment of the limbs.   × 4.

Figure 5. — One-half the pygidium with three attached thoracic segments, from an entire specimen, with a portion of the test removed; showing the highly developed, lamellose fringes of the exopodites. × 11.

Figure 6. — The same; lower side; showing the short, stout, phyllopodiform endopodites, *a*, and the long, slender exopodites, *b*, bearing the lamellose branchial fringes.   In the lower third of the figure the ends of the joints of the separate endopodites are shown by the oblique ascending rows of setiferous nodes.   The small ovate organs, *c*, along the side are provisionally correlated with the exopodites.   A narrow striated doublure margins the pygidium and the ends of the thoracic pleura.   × 11.

Utica Slate, Ordovician, near Rome, New York.

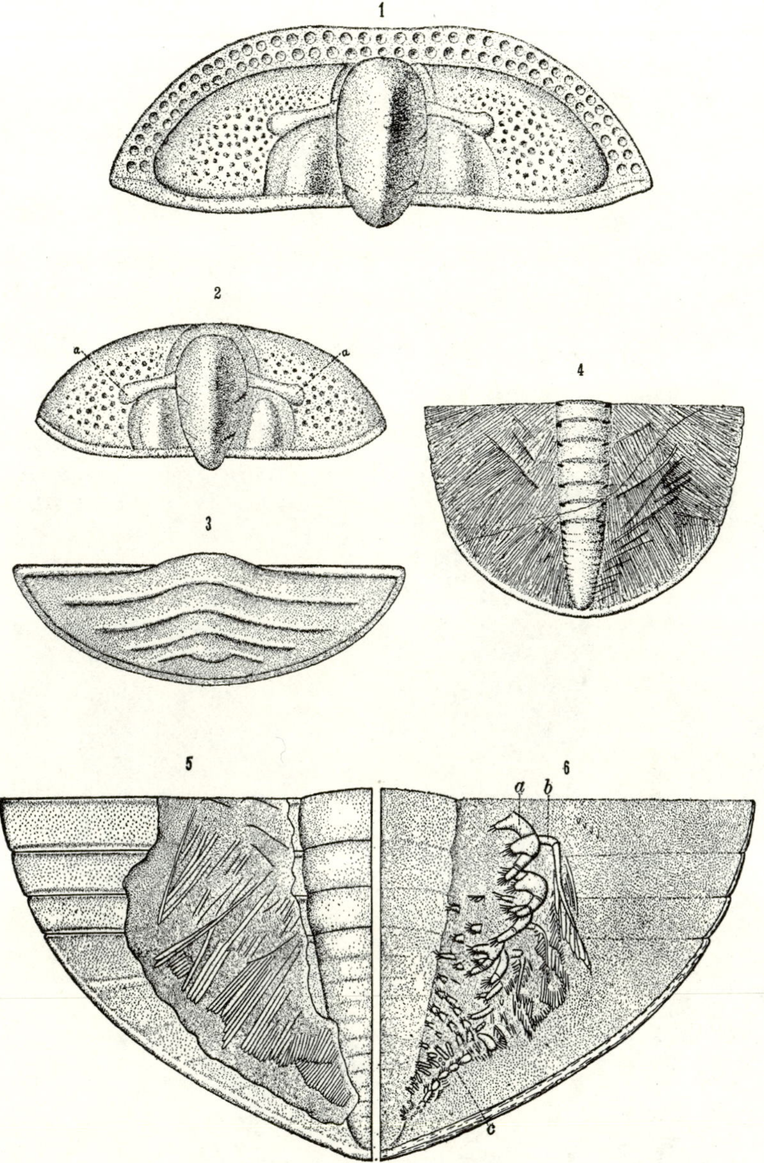

APPENDAGES OF TRINUCLEUS

PLATE XI

ATREMATA (Page 243)

Figure 1. — Dorsal valve of *Iphidea labradorica swantonensis* Billings. × 3.

Figure 2. — Ventral valve of young specimen. × 3.

Cambrian, near Georgia, Vermont.

Figure 3. — Apex of ventral valve of *Glottidia pyramidata* Stimpson. × 25.

Figure 4. — The same; dorsal valve; showing more distinctly terminal protegulum. × 25.

Recent, Beaufort, North Carolina.

NEOTREMATA (Page 244)

Figure 5. — Upper valve of nepionic *Orbiculoidea minuta* Hall; representing protegulum, *p*, and *Paterina* stage. × 25.

Figure 6. — More advanced condition; showing acquisition of discinoid characters. × 25.

Figure 7. — Lower valve of young specimen; showing protegulum and open pedicle-notch. × 25.

Devonian, Marcellus Shale, Avon, New York.

Figure 8. — Accelerated, discinoid, dorsal protegulum of *Discinisca lævis* Sowerby ; corresponding to neanic stage of *Orbiculoidea minuta*, figure 6. × 25.

Figure 9. — Ventral protegulum of same species similarly modified ; agreeing with figure 7. × 25.

Figure 10. — Lower valve of same species; showing sub-marginal position of pedicle-opening. Natural size.

Recent, Callao, Peru.

Figure 11. — Lower valve of *Schizotreta tenuilamellata* Hall; showing centripetal tendency of pedicle-opening. Natural size.

Niagara Group, Hamilton, Ontario. (*Pal. N. Y.* Extract from vol. viii, pl. iv E, fig. 10, 1890.)

Figure 12. — Lower valve of *Acrothele subsidua* (after Linnarsson) ; showing sub-central position of pedicle-opening. Natural size. Cambrian, Sweden.

PROTREMATA (Page 245)

Figure 13. — Dorsal protegulum and early nepionic growth-lines of *Plectambonites segmentina* Angelin. × 80. Upper Silurian, Gotland, Sweden.

Figure 14. — Dorsal protegulum of *Chonetes scitulus* Hall. × 80. Hamilton Group, Thedford, Ontario.

Figure 15. — Ventral protegulum of *Chonetes granuliferus* Owen; showing pedicle-notch. × 80. Coal Measures, Manhattan, Kansas.

Figure 16. — Nepionic stages of ventral valve of *Orthothetes elegans* Bouchard. × 25. (Compare with figure 12 of *Acrothele*.) Devonian, Ferques, France.

Figure 17. — Nepionic stages of *Stropheodonta perplana* Conrad; showing pedicle perforation, deltidium, and hinge-area. × 25. Hamilton Group, Falls of the Ohio.

Figure 18. — Ventral nepionic stage of *Leptœna rhomboidalis* Wilckens. × 25.

Figure 19. — Profile of the same. × 25.

Lower Helderberg Group, Albany county, New York.

Figure 20. — Hinge of a specimen 2 mm. in length; showing deltidium and hinge-area.

Figure 21. — Ventral view of specimen having same dimensions; showing nepionic and neanic stages, and relative proportions of pedicle-opening and shell at this stage. Niagara Group, Waldron, Indiana.

Figures 20 and 21 are taken from "Development of Some Silurian Brachiopoda," *Mem. N. Y. State Museum*, vol. i, no. 1, pl. ii, figures 2, 12, 1889.

TELOTREMATA (Pages 245, 246)

Figure 22. — Ventral view of young *Kraussina (Megerlina) Lamarckiana* Davidson; showing protegulum and early nepionic stages. × 80.

Figure 23. — Dorsal view of same; showing dorsal protegulum and pedicle-opening in ventral valve. × 80.

Recent, Port Jackson, Australia.

Figure 24. — Dorsal view of beaks of young *Terebratulina septentrionalis* Couthouy; showing dorsal protegulum and pedicle-opening in ventral valve. × 80. Recent, Eastport, Maine.

Figures 25–28. — Diagrammatic representation of ventral areas; showing progressive development of deltidial plates.

Figure 25 is without plates, as in ventral area of figure 23. Figure 26 shows two triangular plates, which unite by symphysis in figure 27, making an elongate pedicle-opening. In figure 28 the pedicle perforation is sub-circular and truncates ventral beak. This series corresponds essentially with that shown in *Rhynchotreta cuneata* Dalman, in "Development of Some Silurian Brachiopoda," *loc. cit.*, pl. iv, figures 16–22.

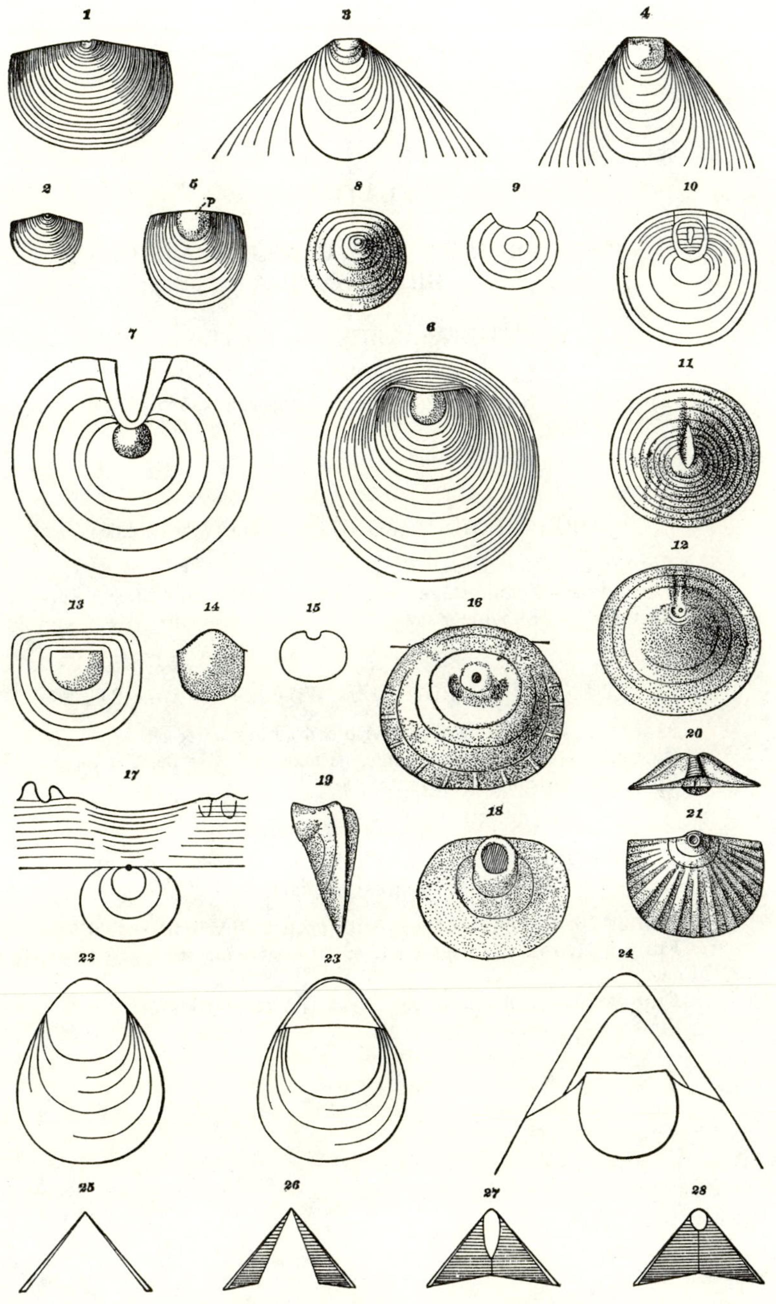

STAGES OF GROWTH OF BRACHIOPODA

PLATE XII

DEVELOPMENT AND CLASSIFICATION OF BRACHIOPODA.

GLOTTIDIA ALBIDA Hinds (Page 269)

Figure 1. — Nepionic shell; *Obolella* stage.   × 36.

Figure 2. — Neanic stage; showing anterior growth producing *Lingula*-like form.   × 16.

Figure 3. — Ephebic stage.   × $\frac{3}{2}$.

ORBICULOIDEA MINUTA Hall (Page 269)

Figure 4. — Nepionic shell; *Paterina* stage.   × 36.

Figure 5. — Neanic stage; first holoperipheral growth.   × 16.

Figure 6. — Ephebic stage.   × 10.

LEPTÆNA RHOMBOIDALIS Wilckens (Pages 268, 269)

Figure 7. — Nepionic stage, with short hinge.   × 36.

Figure 8. — Early neanic stage, with radiating striæ.   × 10.

Figure 9. — Ephebic stage.   × $\frac{2}{3}$.

TEREBRATULINA SEPTENTRIONALIS Couthouy (Pages 268, 269)

Figure 10. — Nepionic stage, with open delthyrium.   × 26.

Figure 11. — Early neanic stage, with radiating striæ.   × 16. (After Morse.)

Figure 12. — Ephebic stage.   × $\frac{2}{3}$.   (After Davidson.)

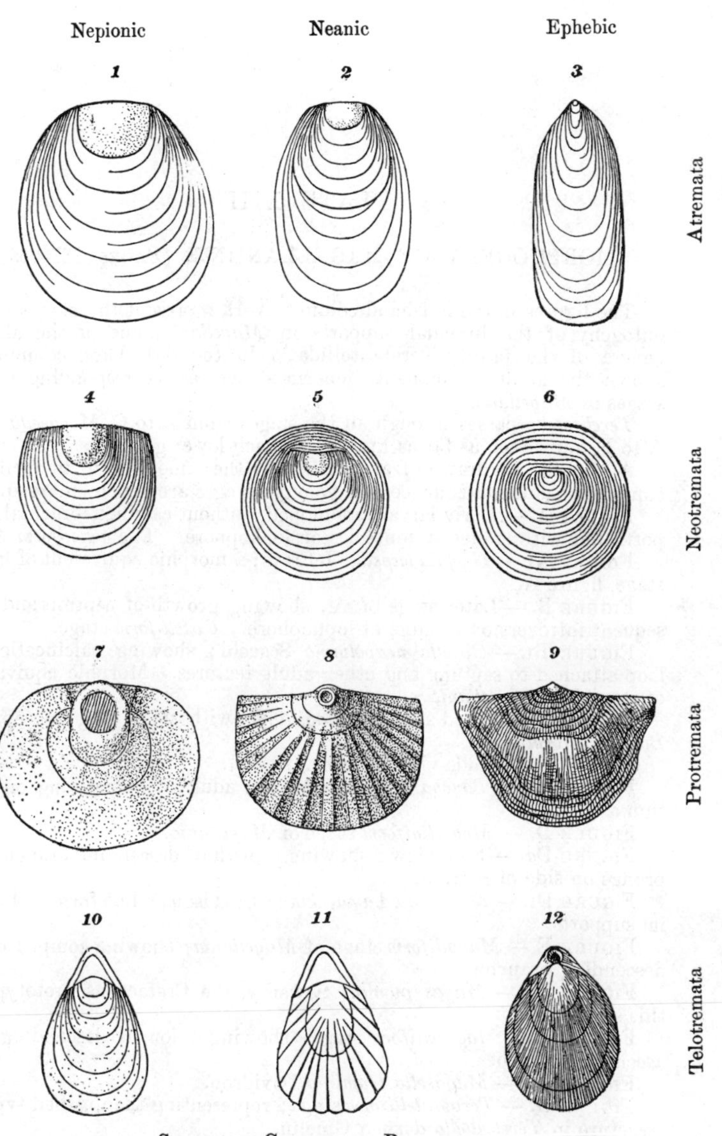

STAGES OF GROWTH OF BRACHIOPODA

PLATE XIII

MORPHOGENY OF MAGELLANIINÆ (Pages 287–289)

The figures in the left-hand column, A–H, represent the stages in the
ontogeny of the brachial supports in *Magellania*, one of the highest
genera of the family Terebratellidæ.   In the right-hand column are
shown the adult, permanent, generic structures, corresponding to the
stages of *Magellania*.

Terebratella passes through all the stages from A to G, *Magasella* from
A to F, and so on, as far as known, for each lower genus.

All figures are drawn of approximately the same length, to facilitate
comparison; in consequence, the younger stages are much enlarged.

Figure A. — Early larval brachiopod, without calcified brachial supports, but with circlet of tentacles on lophophore.   The *gwyniform* stage.

Figure A1. — *Gwynia capsula* Jeffreys; a morphic equivalent of larval
stage, figure A.

Figure B. — Later stage of A ; showing growth of septum and consequent introversion of edge of lophophore.   *Cistelliform* stage.

Figure B1. — *Cistella neapolitana* Scacchi; showing calcification of
loop attached to septum, and other adult features.   Morphic equivalent
of stage B of *Magellania*.

Figure C. — Third stage of *Magellania*, with small ring on septum.
Bouchardiform stage.

Figure Ca. — Side view of same.

Figure C1. — *Bouchardia rosea* Mawe; adult; showing ring on septum as in C.

Figure D. — *Megerliniform* stage of *Magellania*.

Figure Da. — Side view ; showing growth of descending branches as
prongs on side of septum.

Figure D1. — *Megerlina Lamarckiana* Davidson; adult form of brachial supports.

Figure E. — *Magadiform* stage of *Magellania;* showing completion of
descending branches.

Figure E1. — *Magas pumilus* Sowerby, the Cretaceous prototype of
this structure.

Figure F. — *Magaselliform* stage ; showing union of descending and
ascending branches.

Figure F1. — *Magasella Cumingi* Davidson.

Figure G. — *Terabratelliform* stage ; representing the finished type of
structure in *Terebratella dorsata* Gmelin.

Figure G1. — *Terebratella rubicunda* Sowerby.   Morphically equivalent to G, but showing more mature features.

Figure H. — Final stage of *Magellania venosa* Solander, produced by
resorption of the septum and connecting bands of the *terebratelliform* stage.

Figure H1. — *Magellania flavescens* Lamarck.

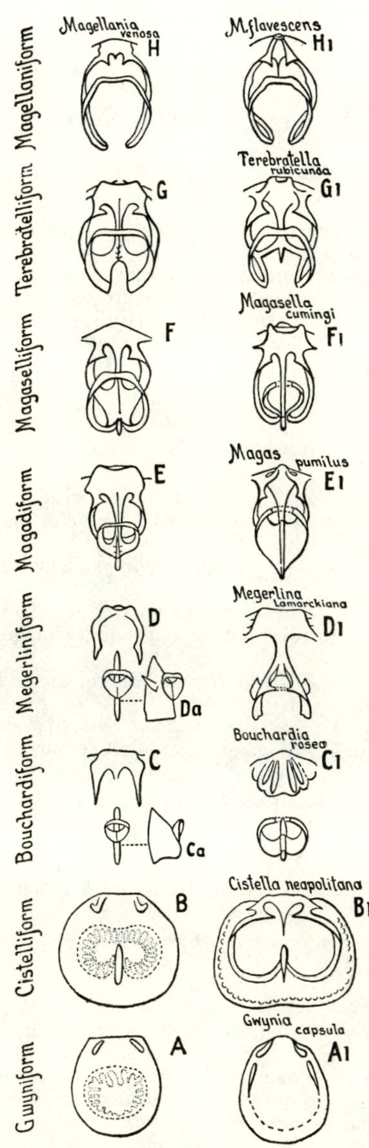

PARALLELISM IN BRACHIOPODA (*Magellania* Series)

PLATE XIV

ONTOGENY AND PHYLOGENY OF THE TEREBRATELLIDÆ

Figures represent brachial supports in various stages of growth in different genera and species. All are drawn of approximately the same length to facilitate comparison, so that in general the younger stages are much enlarged. Vertical rows of figures connected by dotted lines indicate ontogeny as far as known. Horizontal rows indicate the same growth-stages of higher forms and the adult structure in genera representing these stages. All figures are of recent species unless otherwise stated.

MORPHOGENY OF MEGATHYRINÆ (Page 300)

FIGURE A. — Early larval brachiopod without calcified brachial supports, but with circlet of centripetal tentacles on the lophophore. The *gwyniform* stage.

FIGURE A*a*. — *Gwynia capsula* Jeffreys ; a morphic equivalent of larval stage, figure A.

FIGURE B. — Later immature stage of A ; showing growth of septum, and consequent introversion of edge of lophophore. Early *cistelliform* stage.

FIGURE B*a*. — *Zellania liasina* Moore, from the Lias; a morphic equivalent of figure B.

FIGURE B1. — *Cistella neapolitana* Scacchi; showing calcification of loop attached to septum, and other adult features.

FIGURE B2. — *Megathyris decollata* Chemnitz; adult shell; showing advance over *Cistella* in the two lateral septa, thus increasing the length of the loop.

MORPHOGENY OF DALLININÆ (Pages 295–299)

Ontogeny of *Macandrevia cranium* Müller

FIGURE A. — *Gwyniform* stage.

FIGURE B. — *Cistelliform* stage.

FIGURE C1. — *Platidiform* stage; showing union of descending lamellæ with dorsal septum.

FIGURE C1′. — Side view of septum of preceding.

FIGURE D1. — More advanced *platidiform* stage; showing growth of ascending branches, or secondary loop.

FIGURE D1′. — Side view of same.

FIGURE E1. — *Ismeniform* stage, with ascending branches still attached to septum.

FIGURE F1. — *Mühlfeldtiform* stage; showing holes in ascending lamellæ as in *Mühlfeldtia*, figure F3.

Figure G1. — *Terebrataliform* stage, with ascending branches attached to ends of descending branches.

Figure H1. — Final adult condition of *Macandrevia cranium* Müller; showing absorption of septum and connecting bands from descending branches.

Figure H2. — Adult stage of *Macandrevia tenera* Jeffreys.

Ontogeny of *Dallina septigera* Lovén

Figures A, B, C1. — Stages of growth common to all the genera of Dallininæ.

Figure D2. — Late *platidiform* stage.

Figure E2. — *Ismeniform* stage.

Figure F2. — *Mühlfeldtiform* stage.

Figure G2. — *Terebrataliform* stage.

Figure H3. — Adult *Dallina septigera* Lovén.

Figure H4. — *Dallina Raphaelis* Dall.

Figure H5. — *Dallina floridana* Pourtales.

Figure H6. — *Dallina Grayi* Davidson.

Figures C2, D3, E3, F3.—Ontogeny of *Mühlfeldtia sanguinea* Chemnitz.

Figure G3. — *Terebratalia coreanica* Adams and Reeve; adult structure.

Figure F4. — *Mühlfeldtiform* stage of *Laqueus californicus* Koch.

Figure G4. — Adult *Laqueus californicus* Koch; showing connecting bands from ascending and descending lamellæ.

Figure G5. — *Laqueus pictus* Chemnitz.

Figure G6. — *Trigonosemus elegans* Koenig; a Cretaceous representative of the *Terebratalia* structure.

Figure E4. — *Ismenia furcata* Sowerby; Jurassic.

Figure E5. — *Ismenia Buckmani* Moore; Jurassic.

Figure C3. — *Platidiform* stage of *Dallina floridana* Pourtales?

Figure C4. — Adult *Platidia anomioides* Scacchi.

Morphogeny of Magellaniinæ (Pages 293–295)

Ontogeny of *Terebratella dorsata* Gmelin, and *Magellania venosa* Solander

Figure A. — *Gwyniform* stage.

Figure B. — *Cistelliform* stage.

Figure Ca. — *Bouchardiform* stage, with small ring on septum.

Figure Ca'. — Side view of same.

Figure Da. — *Megerliniform* stage; showing growth of ring, or ascending branches.

Figure Da'. — Side view; showing growth of descending branches as prongs on side of septum.

Figure Ea. — *Magadiform* stage; showing completion of descending branches.

Figure Fa. — *Magaselliform* stage; showing union of descending and ascending branches.

Figure Ga. — *Terebratelliform* stage; representing the finished type of structure in *Terebratella dorsata* Gmelin.

Figure Ha. — Final stage of *Magellania venosa* Solander; produced by resorption of the septum and connecting bands of the *terebratelliform* stage.

Figure H*b*. — *Magellania Wyvillii* Davidson.

Figure H*d*. — *Magellania flavescens* Lamarck.

Figure H*e*. — *Magellania kerguelenensis* Davidson.

Figures E*b*, F*b*, G*b*, H*c*. — Ontogeny of *Magellania* (*Neothyris*) *lenticularis* Deshayes, from the *magadiform* to the *magellaniform* stages.

Figures E*c*, F*c*, G*c*. — The *magadiform, magaselliform,* and *terebratelliform* stages of *Terebratella cruenta* Dillwyn.

Figures E*d*, F*d*, G*d*. — The same stages in *Terebratella rubicunda* G. B. Sowerby.

Figure E. — *Magas pumilus* Sowerby; the Cretaceous prototype of this structure.

Figures E*f*, F*e*. — The *magadiform* and *magaselliform* stages of *Magasella Cumingi* Davidson.

Figure D*b*. — *Megerlina Lamarckiana* Davidson; adult form of brachial supports.

Figure C*b*. — *Kraussina rubra* Pallas; adult; showing septum and two branches.

Figure C*c*. — *Kraussina pisum* Valenciennes apud Lamarck.

Figure C*d*. — *Bouchardia rosea* Mawe; adult; showing thickened cardinal process, hinge-plate, and ring on the septum as in C*a*.

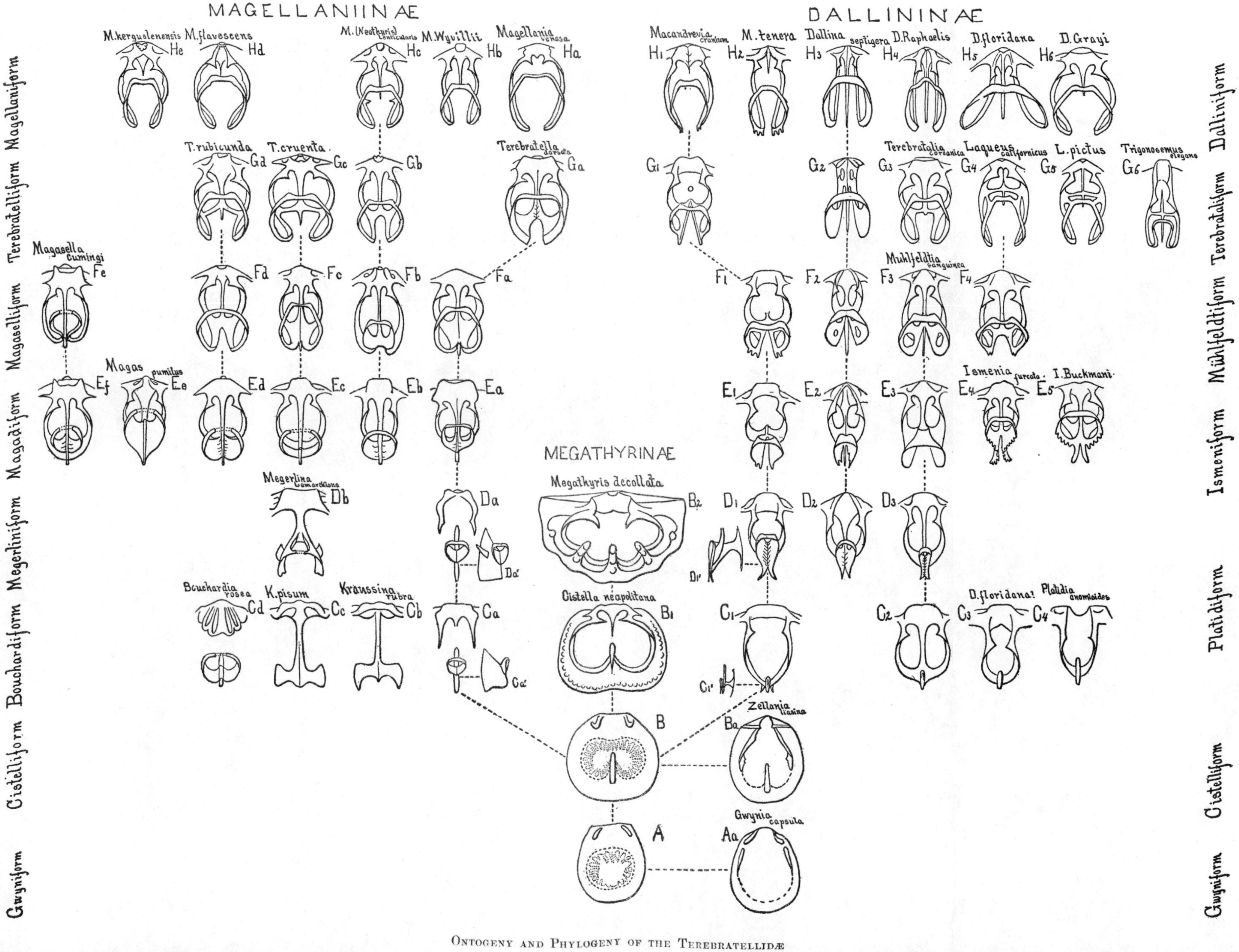

MAGELLANIINÆ

DALLININÆ

M.kerguelenensis   M.flavescens   M.(Neothyris)lenticularis   M.Wyvillii   Magellania venosa
He   Hd   Hc   Hb   Ha

Macandrevia cranium   M.tenera   Dallina septigera   D.Raphaelis   D.floridana   D.Grayi
H₁   H₂   H₃   H₄   H₅   H₆

T.rubicunda   T.cruenta   Terebratella dorsata
Gd   Gc   Gb   Ga

Terebratalia coreanica   Laqueus californicus   L.pictus   Trigonosemus elegans
G₁   G₂   G₃   G₄   G₅   G₆

Magasella Cumingi
Fe   Fd   Fc   Fb   Fa

Mühlfeldtia sanguinea
F₁   F₂   F₃   F₄

Magas pumilus
Ef   Ee   Ed   Ec   Eb   Ea

Ismenia furcata   I.Buckmani
E₁   E₂   E₃   E₄   E₅

MEGATHYRINÆ

Megerlina Lamarckiana
Db

Megathyris decollata
B₂

D₁   D₂   D₃

Da

Dr

Bouchardia rosea   K.pisum   Kraussina rubra
Cd   Cc   Cb

Ca

Cistella neapolitana
B₁

Da'

D.floridana?   Platidia anomioides
C₂   C₃   C₄

C₁

Cr

Ca'

B   Ba

Zellania liasina

A   Aa

Gwynia capsula

Magellaniform
Terebratelliform
Magaselliform
Magadiform
Megerliniform
Bouchardiform
Cistelliform
Gwyniform

Dalliniform
Terebrataliform
Mühlfeldtiform
Ismeniform
Platidiform
Cistelliform
Gwyniform

ONTOGENY AND PHYLOGENY OF THE TEREBRATELLIDÆ

PLATE XV

CRANIA SILURIANA Hall (Page 317)

Figure 1. — The youngest individual observed ; having a height of 1 mm. and a width across the base of 1.5 mm. The elevation of the shell is in strong contrast to that of the mature form.

Figure 2. — A mature individual attached to a shell of *Platystoma*. (28*th Rept. N. Y. State Mus. Nat. Hist.*, pl. 21, fig. 5.)

DALMANELLA ELEGANTULA Dalman (Pages 317–321)

See Plate XXII

Figure 3. — Dorsal view of the youngest embryo observed, its length being .5 mm., its width, .75 mm. The median sinus has already formed, and three pairs of plications have appeared, of which the middle pair is evidently the oldest.

Figure 3*a*. — Outline profile of the same; showing the slightly greater convexity of the ventral valve.

Figure 4. — A larger example viewed from the dorsal side, its length being 1 mm., its width 1.5 mm. The plications have now increased to the number of six pairs, one of which has appeared between the median plications seen in figure 3.

Figure 5. — Ventral view of an individual in about the same stage of development ; showing a strong median plication corresponding to the dorsal sinus, and five pairs of lateral plications.

Figure 6. — Cardinal view of an individual in the growth-stage represented by figure 5. The valves have nearly the same convexity, while the width of the cardinal area and the size of the pedicle-passage are the same for each. The latter is seen to be quite unobstructed and without further differentiation than a slight thickening of the margins.

Figure 6*a*. — Outline profile of the same.

Figure 7. — Cardinal view of an individual which has reached a size of 3 × 3 mm. Here a change is apparent in the development of the cardinal area and foramen of the valves. The primary indication of the callosity or cardinal process is in the apex of the dorsal opening. The difference in the convexity of the valves has also noticeably increased.

Figure 7*a*. — Outline profile of the same.

Figure 8. — Cardinal view at a size of 5 × 5 mm. The ventral beak has become strongly incurved, and the cardinal process is now sub-divided into three parts. The cross-lines, representing the natural size of the specimen, are too short.

FIGURE 8a. — Outline profile of the same.

FIGURE 9. — Cardinal view at a size of 12 × 11 mm. The ventral valve and area have become greatly curved, and the dorsal aperture is quite filled by the tripartite cardinal process.

FIGURE 9a. — Outline profile of the same.

FIGURE 10. — Cardinal view of a large adult; size 18 × 18.5 mm. The areas are closely appressed, and the dorsal aperture is wholly filled by the cardinal process, the central portion of which extends into the aperture of the other valve.

FIGURE 10a. — Outline profile of the same.

FIGURE 11. — Cardinal view of a normal adult; natural size.

FIGURE 11a. — Profile of the same. (*op. cit.*, pl. 21, figs. 17, 14.)

FIGURE 12. — Dorsal view of a small adult.

FIGURE 12a. — Ventral view of the same. (*op. cit.*, figs. 12, 11.)

RHIPIDOMELLA HYBRIDA SOWERBY (PAGES 321, 322)

FIGURE 13. — Cardinal view of a very young individual having a length of 1 mm. and a width of 1.5 mm. The valves are nearly equi-convex, the area and apertures as in the earlier stages of *D. elegantula.*

FIGURE 13a. — Outline profile of the same.

FIGURE 14. — Cardinal view of a somewhat gibbous example measuring 10 × 8 mm. The relatively short areas are about equally developed, and the cardinal callosity of the dorsal valve has already filled the dorsal aperture.

FIGURE 14a. — Outline profile of the same.

FIGURE 15. — Cardinal view of a normal adult 12 × 10 mm. ; showing the short area and the projection of the cardinal process into the ventral aperture.

FIGURE 15a. — Outline profile ; showing the incurvature of the areas.

FIGURE 16. — Dorsal view of a large adult. (*op. cit.*, fig. 20.)

FIGURE 17. — Profile of a normal adult. (*op. cit.*, fig. 21.)

FIGURE 18. — Dorsal view of a small, gibbous example ; showing strong varices. Enlarged to two diameters.

FIGURE 18a. — Profile of the same.

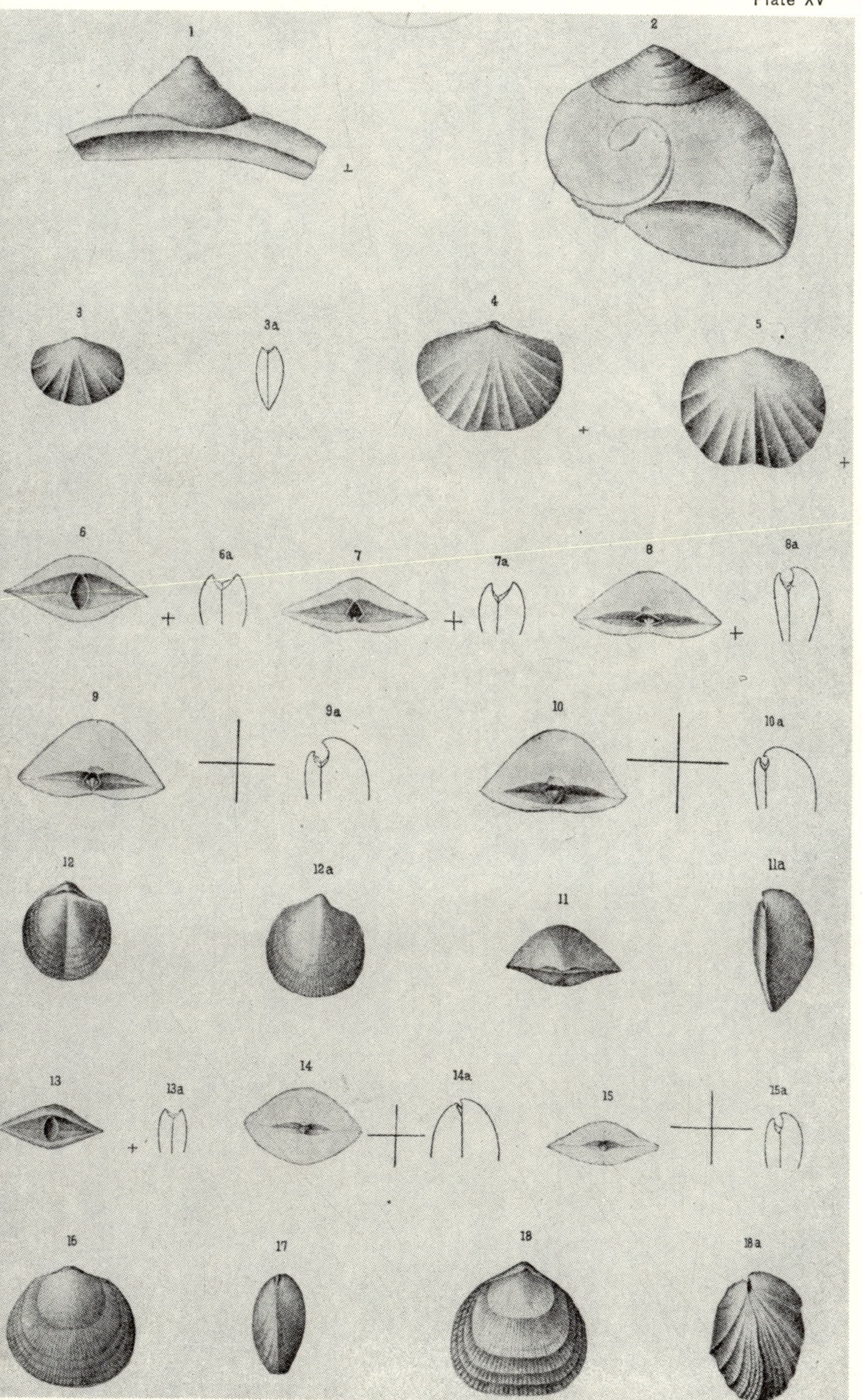

Stages of Growth in Silurian Brachiopoda

PLATE XVI

Leptæna Rhomboidalis Wilckens (Pages 322–327)

Figure 1. — Ventral view of the youngest shell observed, its length being 1.25 mm. The aperture of the embryonal pedicle-sheath is very conspicuous, and its margins are very thick. The surface shows a faint median depression, indications of two concentric growth-lines, and outside the latter of these, obscure traces of plications.

Figure 1a. — Outline profile of the same; showing the prominence of the sheath.

Figure 2. — Ventral view of an individual with a length of 2 mm. The aperture of the pedicle-sheath is relatively somewhat diminished in size, its margins have become thinner, and the radiating plications numerous and sharply defined.

Figure 2a. — Outline profile of the same; indicating diminution in the prominence of the sheath.

Figure 3. — Ventral view of an individual having a length of 4 mm.; showing the increase in the number of plications, the appearance of numerous concentric undulations and striæ, and the narrowing pedicle-aperture.

Figure 3a. — Outline profile of the same; showing the concentric undulations and the diminishing pedicle-sheath.

Figure 4. — Ventral view of a normal adult, having a length of 28 mm.; showing the characters of maturity. Natural size.

Figure 4a. — Profile of another individual of full growth; showing the anterior geniculation and the length of the anterior slope, or *curtain*. (28*th Rept. N. Y. State Mus. Nat. Hist.*, pl. 22, figs. 6, 7.)

Figure 5. — Cardinal view of the specimen represented by figure 1.

Figure 6. — Similar view of the specimen represented by figure 2.

Figure 7. — Similar view of an individual 2.65 mm. in length.

Figure 8. — Similar view of the specimen represented by figure 3.

Figure 9. — Similar view of an individual 9 mm. in length.

These cardinal views are drawn with the same degree of enlargement, and show the gradual diminution in height and in diameter of aperture in the pedicle-sheath, and the increasing development of the grooved callosity on the dorsal valve.

Figure 10. — Cardinal view of a normal adult; showing the great size of the grooved callosity, and the cæcal opening, representing the atrophied pedicle-sheath. Natural size. (*op. cit.*, fig. 10.)

Figure 11. — The cardinal area represented in figure 5 (length 1.25 mm.), still further enlarged; showing the broad, prominent, exsert sheath, embracing, at its base, the faint, grooved dorsal callosity.

Figure 12. — The cardinal area shown in figure 6 (length 2 mm.), enlarged to the size of figure 11; showing the depression of the sheath, the narrowing of the cardinal area, and the increasing aperture between the sheath and the callosity.

Figure 13. — The pedicle-area of a mature individual. The sheath is now wholly absorbed, the sole trace of it being seen in the cæcal foramen, surrounded by the umbonal portion of the shell. The callosity is strongly developed, but not sufficiently to close the gap between it and the opposite valve, thus leaving a passage between the valves and along the dorsal groove.   × 2.

ORTHOTHETES SUBPLANUS Conrad (Pages 327–330)

See Plate XXII

Figure 14. — Ventral view of the smallest individual observed; having a length of 2.25 mm. Both primary and secondary plications and concentric growth-lines have already appeared; indicating the very early assumption of these characters.

Figure 14a. — Outline profile of the same; showing the convexity of the valves.

Figure 15. — A normal adult; dorsal view.

Figure 15a. — The same in profile. (*op. cit.*, pl. 21, figs. 30, 31.)

Figure 16. — Cardinal view of specimen somewhat larger than that represented in figure 14. The ventral valve bears a small pedicle-sheath, the dorsal, the inception of a cardinal process or callosity, while between the two is a broad opening which serves to indicate that at this early age the pedicle-sheath had ceased its function.

Figure 17. — Cardinal view of an individual slightly below normal full growth, but with essentially mature characters.

Figure 18. — The pedicle-area of the specimen represented in figure 16.

Figure 19. — Pedicle-area of a shell having a length of 4 mm. At this stage of growth the sheath has relatively diminished in size, while the dorsal callosity has increased and shows a median groove on its inner edge. Deltidial plates have also begun to develop along the margins of the ventral aperture.

Figure 20. — Pedicle-area of the specimen represented in figure 17. The sheath is now atrophied and altogether obsolete, the dorsal callosity is very large, nearly filling the aperture between the valves, and the deltidial plates have attained the maximum development observed in the Strophomenidæ.

The last three figures have the same degree of enlargement.

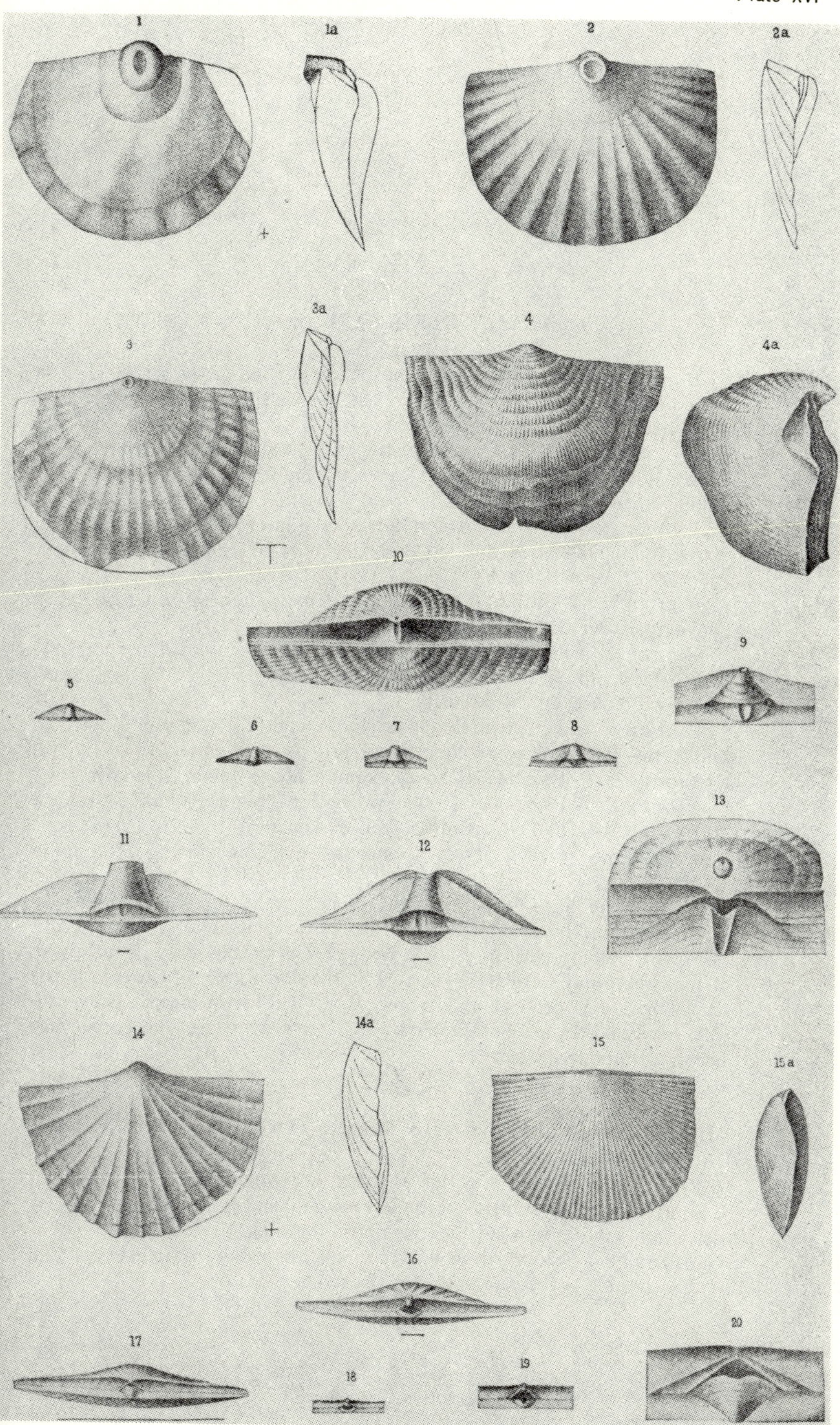

STAGES OF GROWTH IN SILURIAN BRACHIOPODA

PLATE XVII

STROPHONELLA STRIATA Hall (Pages 330–334)

Figure 1. — Ventral view of the incipient shell of the series ; length
2.25 mm. ; showing the opening of the pedicle-sheath, and the primary
surface plications.

Figure 1a. — Outline profile of the same; showing the complete convexity of the ventral valve, and essentially conformable concavity of the
dorsal valve.

Figure 2. — Ventral valve of a normal adult; showing the umbonal
convexity of the shell and general concavity over the pallial region.
(28th Rept. N. Y. State Mus. Nat. Hist., pl. 23, fig. 1.)

Figure 2a. — Outline profile of the same; showing the reversal in
convexity from the embryonic condition.

Figure 3. — Pedicle-area of the specimen represented in figure 1;
length 2.25 mm. ; showing the well-developed, slightly exsert sheath
and the obscure dorsal callosity.

Figure 4. — Pedicle-area of an individual 2.5 mm. in length, in
which the sheath is extravagantly exsert.

Figure 5. — Pedicle-area in an example about 6 mm. in length.

Figure 6. — Pedicle-area, when a length of 8 mm. has been attained.

Figure 7. — Pedicle-area in a shell measuring 13 mm. in length.

Figure 8. — Pedicle-area in a normal adult measuring 17 mm. in
length.

Figures 3–8 have been drawn to the same scale, and show the successive phases in the development of the pedicle characters. The sheath
ceases its function as a pedicle-passage before maturity is attained,
though retaining its relative size ; while the dorsal callosity, which in the
earlier stages is grooved and largely enveloped by the sheath, is eventually separated from the sheath by a narrow aperture, and its surface
becomes uninterrupted.

MIMULUS WALDRONENSIS Miller and Dyer (Pages 334, 335)

Figure 9. — Dorsal view of a young individual having a length of
3 mm. The shell is nearly symmetrical, and shows an open triangular
deltidium, ending in a sub-circular apical foramen.

Figure 9a. — Ventral view of the same ; showing the apical foramen.

Figure 9b. — Outline profile of the same.

Figure 10. — Cardinal view of an adult specimen; showing the asymmetrical shell.

Figure 10a. — Dorsal view of the same.

Figure 10b. — Ventral view of the same. (11th Ann. Rept. State Geol. Ind., pl. 27, figs. 21, 19, 20.)

DICTYONELLA RETICULATA Hall (Pages 335–337)

Figure 11. — A young individual having a length and width of 3 mm.; showing the sub-circular outline and undefined median fold.

Figure 12. — Axial section of a larger, but immature form; indicating the character of the articulation, and showing the internal ventral plate and dorsal septum.

Figure 13. — Dorsal view of an adult shell. × 2.

Figure 13 a. — Cardinal view of the same; showing the bare umbonal area and the lines of lateral attachment of the internal plate. × 2.

(Figures 13, 13a, from the 28th Rept. N. Y. State Mus. Nat. Hist., pl. 26, figs. 53, 54.)

ANASTROPHIA INTERNASCENS Hall (Pages 337–339)

Figure 14. — Ventral view of the youngest specimen observed; length 2 mm.

Figure 14a. — Outline profile of the same; showing the elevation of the ventral beak and cardinal area.

Figure 15. — Dorsal view of a large adult.

Figure 16. — Ventral view of an average adult; showing the overlapping dorsal valve.

Figure 16a. — Profile of the same; showing the relative convexity of the valves.

(Figures 15–16a from the 28th Rept. N. Y. State Mus. Nat. Hist., pl. 26, figs. 49, 45, 44.)

CAMAROTŒCHIA INDIANENSIS Hall (Pages 346–351)

Figure 17. — The earliest stage of growth observed, the shell having a length of .65 mm. The characters are essentially primitive; the surface is without plications, the foramen triangular and devoid of plates or even marginal thickening.

Figure 17a. — Outline profile of the same; showing the elevation of the ventral umbo.

Figure 18. — A later stage of growth in which the shell has a length of 1.5 mm. With the formation of the first growth-line, a number of faint plications have appeared, and the margins of the foramen have become thickened.

Figure 18a. — Outline profile of the same.

FIGURE 19. — A young example with a length of 6 mm.; showing the inception of the median fold. (11*th Ann. Rep. Geol. Surv. Ind.*, pl. 27, fig. 6.)

FIGURE 20. — Dorsal view of a larger example, having two plications on the fold, and abnormal in the absence of all lateral plications. Natural size. (*op. cit.*, fig. 5.)

FIGURE 21. — Dorsal view of a small, essentially mature example, with two plications on the fold. (*op. cit.*, fig. 4.)

FIGURE 22. — Dorsal view of an adult with three plications on the fold. (28*th Rept. N. Y. State Mus. Nat. Hist.*, pl. 26, fig. 13.)

FIGURE 23.— Front view of an adult with four plications on the fold. (*op. cit.*, fig. 22.)

FIGURE 24. — Dorsal view of a similar individual.

FIGURE 24*a*. — Profile of the same. (*op. cit.*, figs. 15, 19.)

FIGURE 25. — An enlargement of the cardinal area in an individual with a length of 1.5 mm.; showing essentially the primitive characters seen in figure 17.

FIGURE 25*a*. — Outline profile of the same.

FIGURE 26. — The condition of the ventral cardinal area and foramen in an individual 3.5 mm. in length. The margins of the foramen are thickened by the inception of the deltidial plates, and the aperture is seen to encroach upon the apical portion of the shell.

FIGURE 26*a*. — Outline profile of the same.

FIGURE 27. — The cardinal features in an individual of 7 mm. length. The advance upon the last stage is chiefly in rapid development of the deltidial plates, which have narrowed the opening, slightly constricting it near the now incurved umbo.

FIGURE 27*a*. — Outline profile of the same.

FIGURE 28. — The cardinal features in an adult, having a length of 10 mm. The deltidial plates have united at their base, forming an elongate oval aperture, encroaching upon the umbo. The species does not pass this stage of development.

FIGURE 28*a*. — Outline profile of the same.

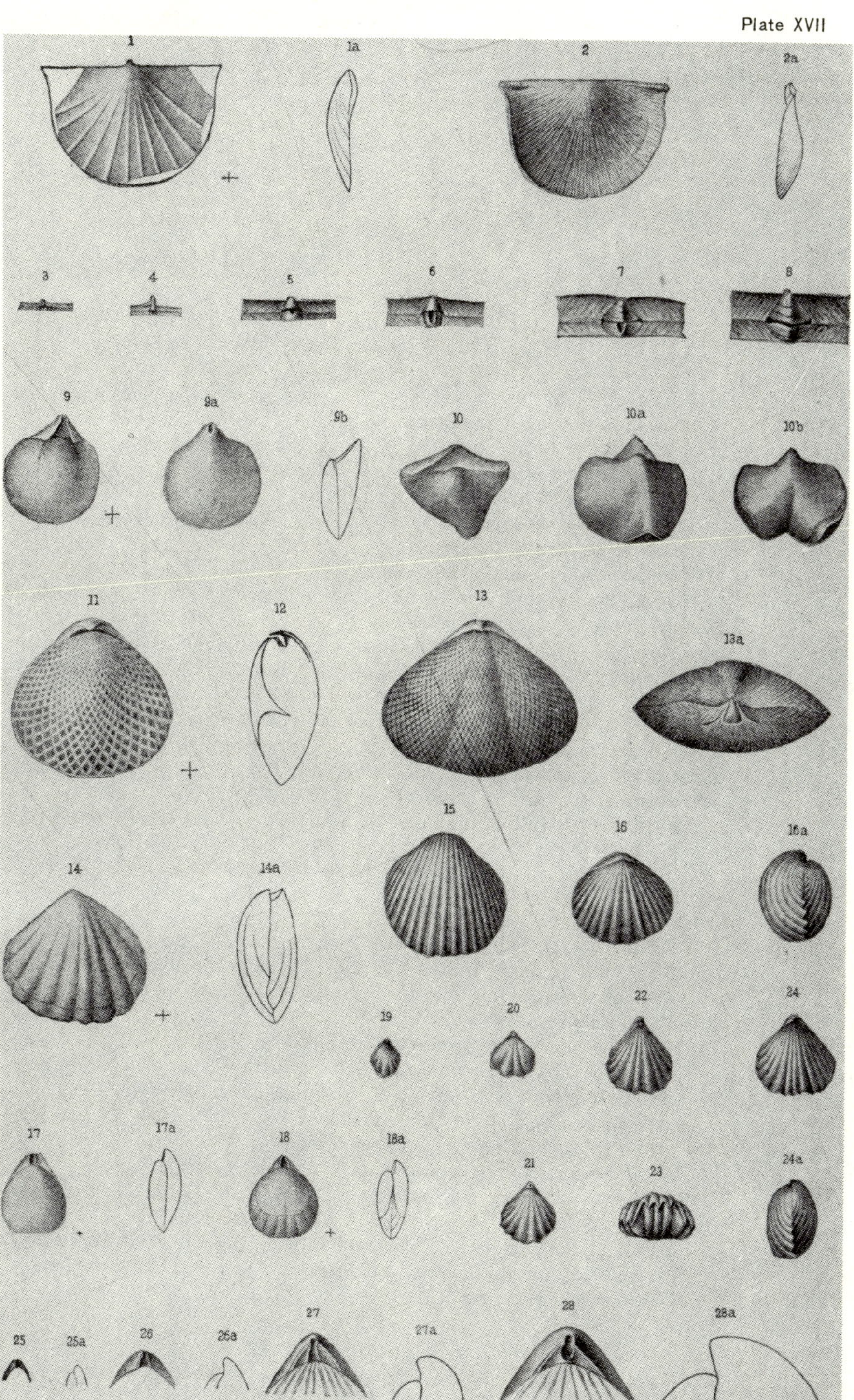

Stages of Growth in Silurian Brachiopoda

PLATE XVIII

CAMAROTŒCHIA WHITII Hall (Pages 344–346)

Figure 1. — The earliest observed stage of growth, the shell measuring 2.75 mm. in length by 2 mm. in width. The deltidial plates have already begun to form along the edges of the triangular foramen, and four pairs of plications are visible on the dorsal valve.

Figure 1a. — Outline profile of the same.

Figure 2. — Dorsal view of a normal adult having two plications on the fold. Natural size.

Figure 2a. — Profile of the same.

Figure 2b. — Front view of the same. (28th Rept. N. Y. State Mus. Nat. Hist., pl. 26, figs. 23, 26, 25.)

Figure 4. — The cardinal features of a young example with a length of 3.25 mm.; showing the inception of the deltidial plates.

Figure 4a. — Profile of the same.

Figure 5. — Cardinal features of an individual which has attained a length of 6 mm. The umbo has become incurved and the development of the plates is well advanced, but it is arrested at this stage, the foramen not becoming enclosed at maturity.

Figure 5a. — Outline profile of the same.

CAMAROTŒCHIA NEGLECTA Hall (Page 341–344)

Figure 6. — Dorsal view of the youngest shell observed; having a length of .75 mm., and a width of .5 mm. Though the foramen is open and without evidence of thickened margins, four fine plications have already appeared on the dorsal valve.

Figure 6a. — Outline profile of the same.

Figure 7. — Dorsal view of an individual with a length of 2.25 mm. The deltidial plates are in an incipient condition and the plications of the surface have considerably increased.

Figure 7a. — Outline profile of the same.

Figure 8. — Dorsal view of a mature example. Natural size.

Figure 8a. — The same in profile. (28th Rept. N. Y. State Mus. Nat. Hist., pl. 26, figs. 3, 5.)

Figure 3. — A somewhat abnormal adult, having three plications on the fold and the lateral plications obsolescent. × 2. (11th Ann. Rept. Geol. Surv. Ind., pl. 27, fig. 3.)

CAMAROTŒCHIA ACINUS Hall (Pages 339-341)

Figure 9. — Ventral view of a young shell in a secondary stage of growth; showing a single growth-line and the embryonal median ridge.

Figure 9a. — Dorsal view of the same. The foramen is slightly narrowed, but without evidence of deltidial plates; the embryonal sinus is well defined.

Figure 9b. — Outline profile of the same.

Figure 10. — Ventral view of a specimen 3.25 mm. in length.

Figure 10a. — Dorsal view of the same. The mature fold and sinus have not yet begun to develop; the foramen shows increased constriction at its base, and slightly thickened margins.

Figure 10b. — Outline profile of the same.

Figure 11. — Dorsal view of an average adult. × 4.

Figure 11a. — Profile of the same.

Figure 11b. — Front view of the same; showing the elevation of the median fold.

RHYNCHOTRETA CUNEATA Dalman, var. AMERICANA
Hall (Pages 351-356)

See Plate XXII

Figure 12. — Dorsal view of a shell representing the earliest stage of growth observed; length 1.5 mm. The foramen is widely triangular and unobstructed, the surface is covered with all the plications of maturity, and the sub-circular valve bears a broad median depression.

Figure 12a. — Outline profile of the same.

Figure 13. — Similar view of a shell 3 mm. in length. The outline has become elongate, the foramen narrowed, and its margins thickened.

Figure 13a. — Outline profile of the same.

Figure 14. — Dorsal view of a normal adult individual; showing the elevated beak and broad plications. Natural size.

Figure 14a. — Profile of the same; showing the elevation of the dorsal fold.

Figure 14b. — Front view of the same. (*28th Rept. N. Y. State Mus. Nat. Hist.*, pl. 25, figs. 31, 34, 36.)

Figure 15. — A series of outlines of the anterior margin; showing the embryonal sinus or depression in the dorsal valve in young forms, reaching its maximum in individuals having a length of 4.5, and becoming obsolete, or the line of junction straight, in specimens 7 mm. in length. Outline 6 shows the inception of the dorsal fold. In 8 the four median plications are distinctly elevated; 9 is from a normal full-grown individual, and 10 represents the maximum elevation of the fold. The outlines, except in 9 and 10, are enlarged to the diameter of a fully developed specimen.

Figure 16. — Cardinal area of figure 12.

Figure 16a. — Outline profile of the same.

FIGURE 17. — Cardinal area of figure 13.

FIGURE 17a. — Outline profile of the same.

FIGURE 18. — Cardinal area in an individual 4.5 mm. in length; showing the considerably advanced development of the deltidial plates and the encroachment of the foramen on the apex of the valve.

FIGURE 18a. — Outline profile of the same.

FIGURE 19. — Cardinal area at a length of 5 mm.

FIGURE 19a. — Outline profile of the same.

FIGURE 20. — Cardinal area at a length of 7 mm.

FIGURE 20a. — Outline profile of the same.

FIGURE 21. — Cardinal area in an individual which has attained a length of 12 mm. The plates are united for nearly one-half their length, and slightly bent outward about the base of the foramen.

FIGURE 21a. — Outline profile of the same.

FIGURE 22. — The cardinal area at maturity; showing a stronger flexion of the plates below the foramen, and a sub-circular foramen encroaching upon the apex of the shell.

FIGURE 22a. — Outline profile of the same.

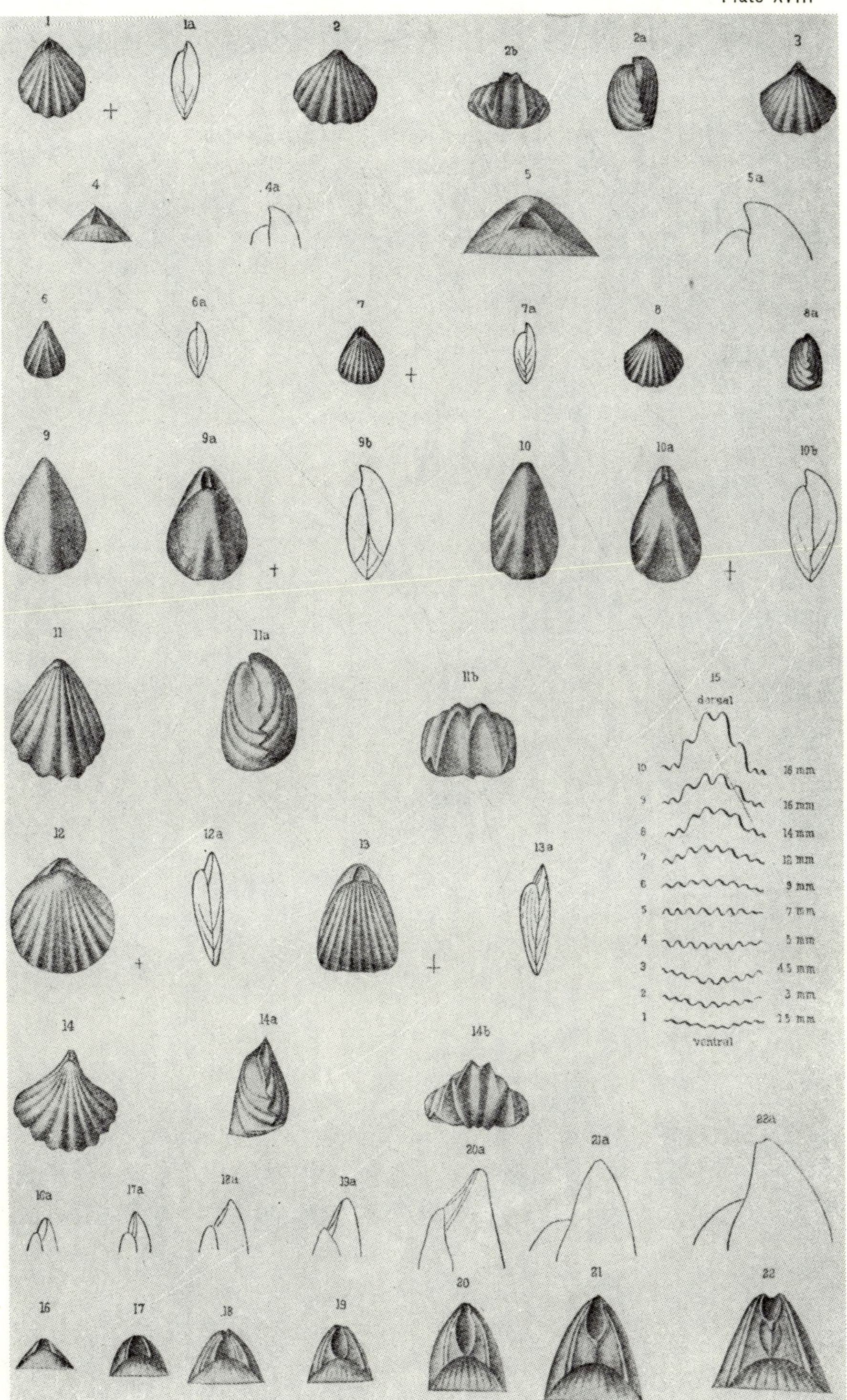

STAGES OF GROWTH IN SILURIAN BRACHIOPODA

PLATE XIX

HOMŒOSPIRA EVAX Hall (Pages 360–365)

See Plate XXII

Figure 1. — Dorsal view of the youngest individual observed; having a length of 1 mm. and a width of .8 mm. The ventral umbo is erect, the foramen triangular and without deltidial plates; the surface bears two faint lines of growth, and outside the second of these are three fine plications on each side of a median sinus.

Figure 1a. — Outline profile of the same; showing the very shallow valves.

Figure 2. — Dorsal view of an average adult; showing the characters of advanced growth.

Figure 2a. — Profile of the same. (*28th Rept. N. Y. State Mus. Nat. Hist.*, pl. 25, figs. 14, 18.)

Figure 3. — Dorsal view of an immature individual 5 mm. in length, in which the plications on the earlier portion of the shell end abruptly at a growth-line, from there outward the surface characters being altogether primitive.

Figure 3a. — Outline profile of the same.

Figure 4. — The cardinal area in its earliest observed condition; enlarged from figure 1.

Figure 5. — A later stage of growth in the cardinal area, the deltidial plates having a considerable development.

Figure 6. — A still later condition of this area, in which the plates have united, enclosing the foramen.

Figure 7. — The cardinal portions of an individual with an unusually elevated ventral umbo; showing also an advance in growth from the condition represented in figure 6.

Figure 8. — The character of the cardinal area in a normal adult of about the size represented in figure 2. The foramen has become circular, and the ventral umbo so incurved as to conceal the deltidial plates.

Figure 9. — A small obese example in which the foramen is almost wholly concealed.

HOMŒOSPIRA SOBRINA sp. nov. (Pages 366–369)

Figure 10. — The youngest shell observed; having a length of 2 mm. and a width of 1.6 mm. The shell already bears two plications on each side of the median sinus, and two much fainter elevated striæ in the sinus itself.

Figure 10a. — Outline profile of the same.

Figure 11. — Dorsal view of an adult specimen, somewhat above the average size, and enlarged to 1½ diameters; showing the normal features of maturity.

Figure 11a. — Profile of the same.

Figure 11b. — Ventral view of the same individual.   Natural size.

Figure 12. — Transverse section of an individual; showing the spiral ribbon and the number of volutions.

Figure 13. — The character of the pedicle-passage in the youngest example (figure 10).   The deltidial plates are absent, but the margins of the foramen are thickened.

Figure 13a. — Outline profile of the same.

Figure 14. — A later stage of growth, in which the plates are considerably developed.

Figure 14a. — Outline profile of the same.

Figure 15. — A stage of growth in which the foramen has become circular and apical.   The plates are slightly flexed along two oblique lines which converge toward the base of the area.

Figure 15a. — Outline profile of the same.

Figure 16. — The mature condition of the pedicle-area, the deltidial plates being somewhat concealed by the incurvature of the ventral umbo.

Figure 16a. — Outline profile of the same.

Figures 13–16a are drawn to a scale.

ATRYPINA DISPARILIS Hall (Pages 369–372)

Figure 17. — A young individual 2.5 mm. in length; showing surface and pedicle characters essentially as at maturity.   The sinal plications apparent at this early stage appear, in the usual adult form, as the broad median fold.

Figure 17a. — Outline profile of the same; showing the erect beak and shallow valves.

Figure 18. — A young individual; showing asymmetry in the development of the median fold, and a greater number of lateral plications than in the normal adult (enlarged).

Figure 19. — Dorsal view of a normal adult.

Figure 19a. — Ventral view of the same.

Figure 19b. — Profile of the same.   All are enlarged to two diameters. (28th Rept. N. Y. State Mus. Nat. Hist., pl. 25, figs. 39–41.)

Figures 20–23. — Enlarged views of the umbonal region of the ventral valve; showing the variations of form and position in the foramen as observed among specimens which differ but little in actual size.

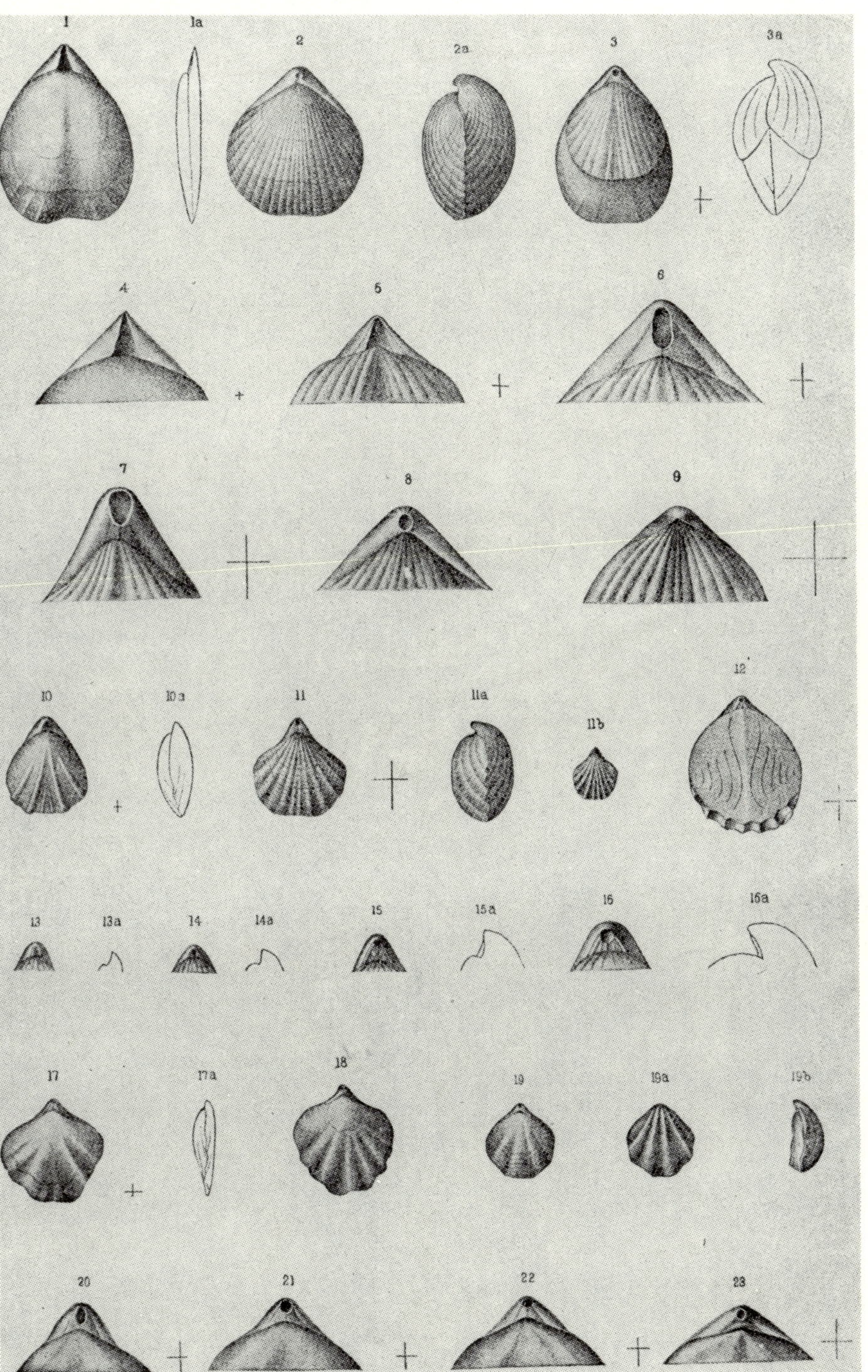

STAGES OF GROWTH IN SILURIAN BRACHIOPODA

PLATE XX

RETICULARIA BICOSTATA, var. PETILA Hall (Pages 380–382)

Figures 1, 1a. — The youngest example observed, and outline profile of the same.

Figure 2. — Dorsal view of a large specimen.

Figure 3. — Cardinal view of the same. × 3. (11*th Rept. State Geol. Ind.*, pl. 27, figs. 8, 9.)

SPIRIFER CIRSPUS, var. SIMPLEX Hall (Pages 380–382)

Figure 4. — Dorsal view of the youngest growth-stage observed. Within the single growth-line, the smooth sub-circular initial shell is visible, but outside this the shell has developed the median fold and the surface fimbriæ.

Figure 4a. — Outline profile of the same.

Figure 5. — Dorsal view of a mature individual. × 2.

Figure 5a. — Profile of the same. (28*th Ann. Rept. N. Y. State Mus. Nat. Hist.*, pl. 24, figs. 1, 4.)

SPIRIFER CRISPUS Hisinger (Pages 380–382)

Figure 6. — Dorsal view of a young form, in which the deltidial plates are in an incipient condition.

Figure 6a. — Outline profile of the same.

Figure 7. — Dorsal view of a mature specimen.

Figure 7a. — Profile of the same. (28*th Ann. Rept. N. Y. State Mus. Nat. Hist.*, pl. 24, figs. 8, 12.)

SPIRIFER NIAGARENSIS Conrad (Page 385)

Figure 8. — The cardinal area of a ventral valve, from Lockport, New York. The deltidial plates are in the third stage of development; the foramen not yet enclosed, but confluent with the hiatus below the basal edges of the plates. Natural size.

SPIRIFER RADIATUS Sowerby (Pages 382–386)

Figure 9. — A cardinal view; showing a similar stage of development of the deltidium; from an individual which has not attained normal full-growth.

Figure 10. — The youngest growth-stage observed.

Figure 10a. — Outline profile of the same.

Figure 11. — Dorsal view of a normal adult.

Figure 11a. — Profile of the same. (*28th Ann. Rept. N. Y. State Mus. Nat. Hist.*, pl. 24, figs. 23, 25.)

ATRYPA RETICULARIS Linnæus (Pages 356–360)

Figure 12. — The incipient shell; having a length of 2.25 mm. The deltidial plates have already begun to form, and the shell has developed two growth-lines.

Figure 12a. — Outline profile of the same.

Figure 13. — Ventral view of a full-grown individual.

Figure 13a. — Profile of the same. (*28th Ann. Rept. N. Y. State Mus. Nat. Hist.*, pl. 25, figs. 46, 47.)

Figure 14. — A series of outlines of the anterior margin of the valves; showing the rapid increase in plications and the reversion from the embryonic fold and sinus to the sinus and fold of maturity.

Figure 15. — The deltidial characters of the specimen represented in figure 12.

Figure 15a. — Outline profile of the same.

Figure 16. — Similar view of a specimen measuring 3 × 3 mm.

Figure 16a. — Outline profile of the same.

Figure 17. — Similar view of a specimen measuring 3.5 × 4 mm.

Figure 17a. — Outline profile of the same.

Figure 18. — Similar view of a specimen measuring 8 × 7 mm.

Figure 18a. — Outline profile of the same.

Figure 19. — Similar view of a specimen measuring 21 × 20 mm.

Figure 19a. — Outline profile of the same.

Figure 20. — Similar view of a full-grown specimen measuring 24 × 22 mm.

Figure 20a. — Outline profile of the same.

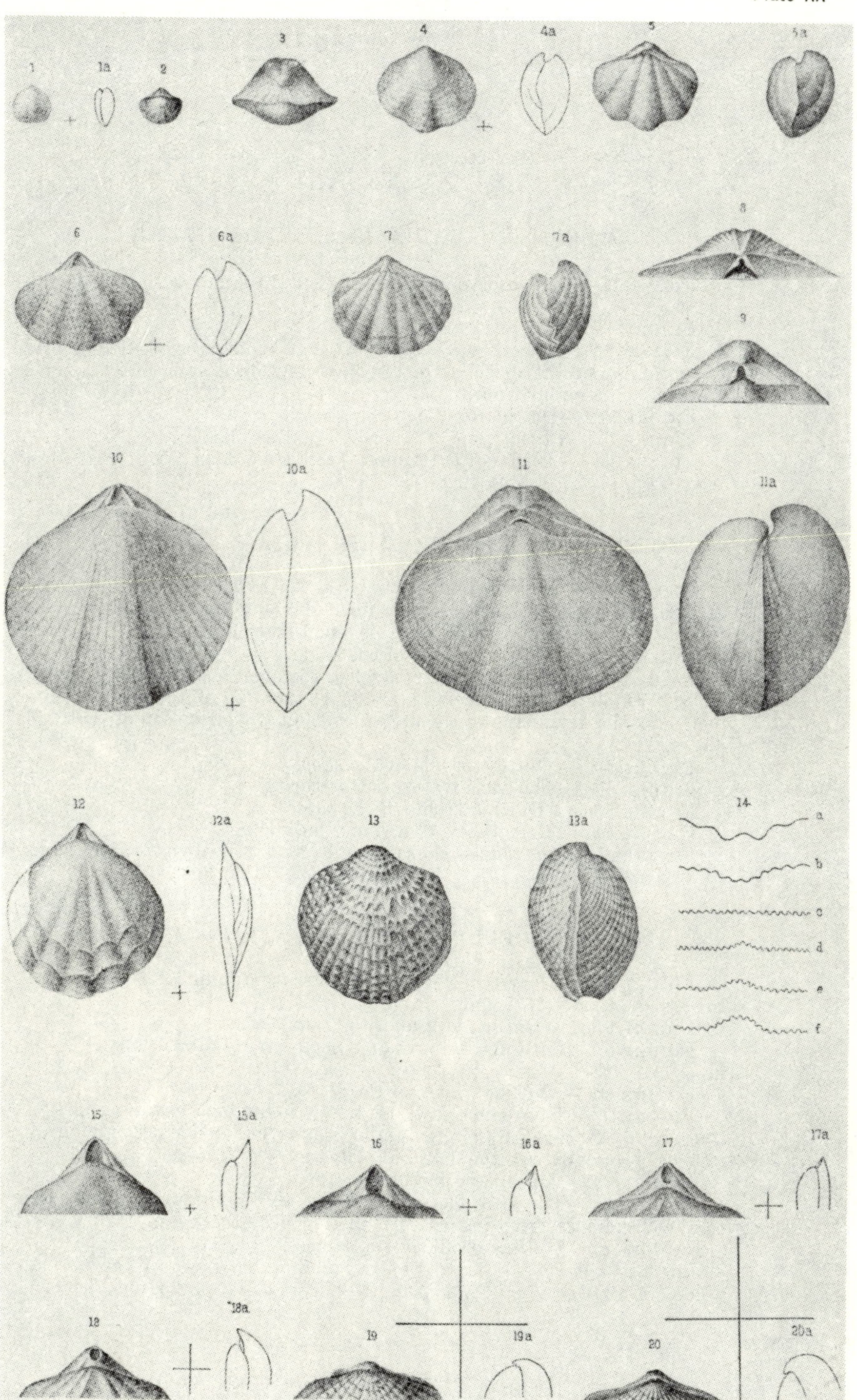

STAGES OF GROWTH IN SILURIAN BRACHIOPODA

PLATE XXI

MERISTINA MARIA Hall (Pages 377–380)

Figure 1. — An embryo measuring .75 × .75 mm.; without deltidial plates.

Figure 1a. — Outline profile of the same.

Figure 2. — A later stage of growth, at which the shell measures 5 × 5 mm., and the deltidial plates have developed sufficiently to give the foramen a circular outline.

Figure 2a. — Outline profile of the same.

Figure 3. — Dorsal view of a normal, mature shell.

Figure 3a. — Profile of the same. (28*th Ann. Rept. N. Y. State Mus. Nat. Hist.*, pl. 25, figs. 9, 12.)

MERISTINA RECTIROSTRIS Hall (Pages 372–374)

Figure 4. — The cardinal features at an extremely early stage of growth, the shell measuring 3 × 2 mm.; showing the broad, triangular foraminal opening and its faintly thickened margins.

Figure 4a. — Outline profile of the same.

Figure 5. — The same features when the shell has attained a size of 4.5 × 3.5 mm.; showing the gradual approximation of the sides of foramen and the narrowing of the umbo, without the formation of deltidial plates.

Figure 5a. — Outline profile of the same.

Figure 13. — The same features at maturity.

Figure 13a. — Outline profile of the same.

Figure 11. — Dorsal view of a young individual.

Figures 12, 12a, 12b. — Dorsal, profile, and ventral views of an adult specimen. (11*th Ann. Rept. Geol. Surv. Ind.*, pl. 27, figs. 1, 10–12.)

WHITFIELDELLA NITIDA Hall (Pages 374–377)

Figure 10. — Dorsal view of the youngest individual observed; having a size of 2.5 × 1.75 mm.

Figure 10a. — Outline profile of the same.

Figure 6. — Cardinal features of a shell having dimensions of 3 × 2 mm.

Figure 6a. — Outline profile of the same.

Figure 7. — The same features in an individual having a size of 6 × 4 mm. In this example the umbo is abnormally elongate, giving an unusual prominence to the deltidial plates.

Figure 7a. — Outline profile of the same.

Figure 8. — The same features at a size of 5 × 4 mm.; showing the total concealment, from incurvature, of the deltidial plates.

Figure 8a. — Outline profile of the same.

Figures 9, 9a, 9b. — Dorsal, ventral, and profile views of the normal adult. (28*th Ann. Rept. N. Y. State Mus. Nat. Hist.*, pl. 25, figs. 3, 4, 5.)

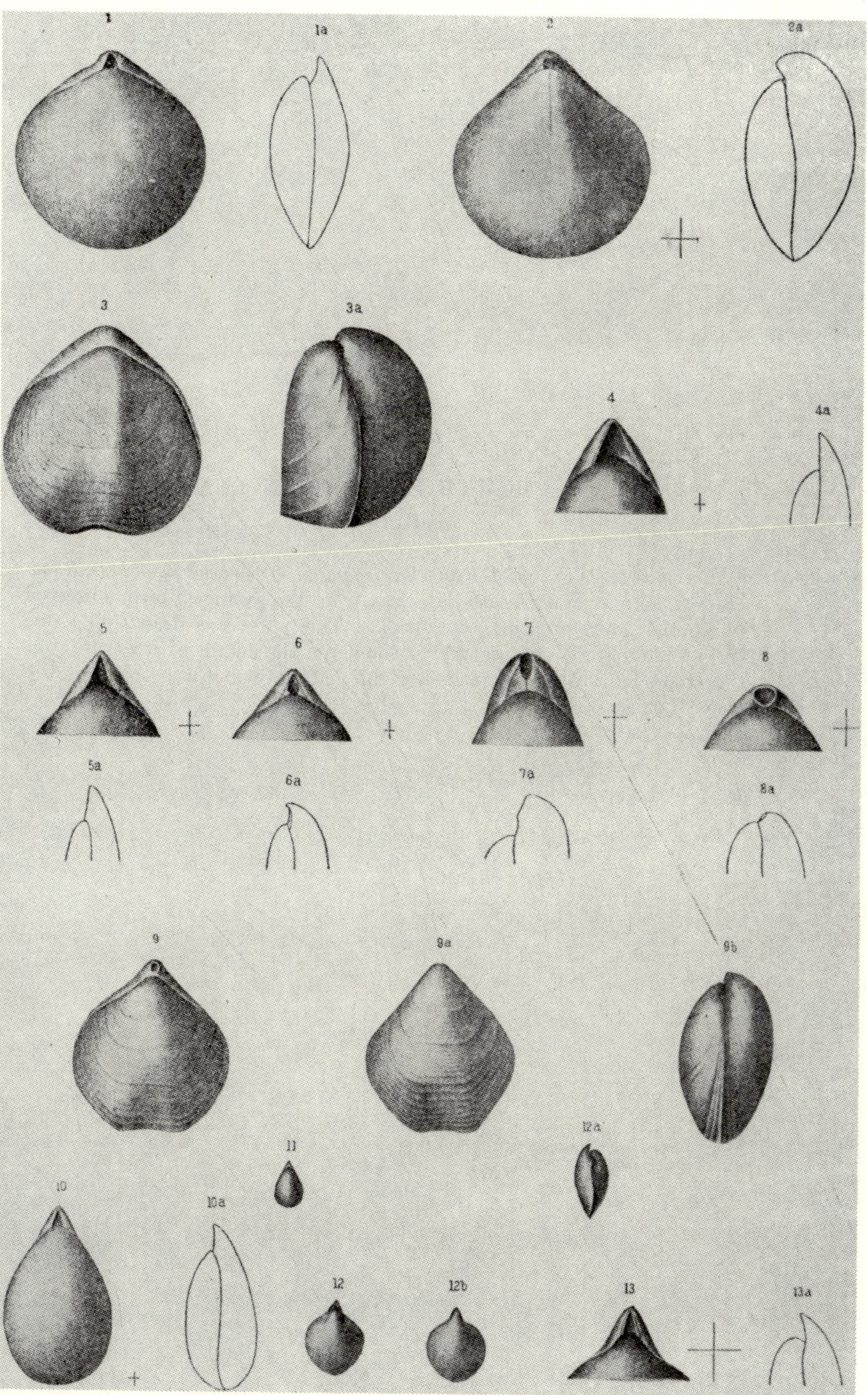

Stages of Growth in Silurian Brachiopoda

PLATE XXII

STAGES OF GROWTH IN SILURIAN BRACHIOPODA
(Page 312)

Series of the shells of *Dalmanella elegantula, Orthothetes subplanus, Rhynchotreta cuneata*, and *Homœospira evax*, from the youngest form observed up to the adult or senile condition. The plate was drawn on stone from a photograph, and serves to show, not the details of structure, but the character and completeness of the material which has served as the basis of this work.

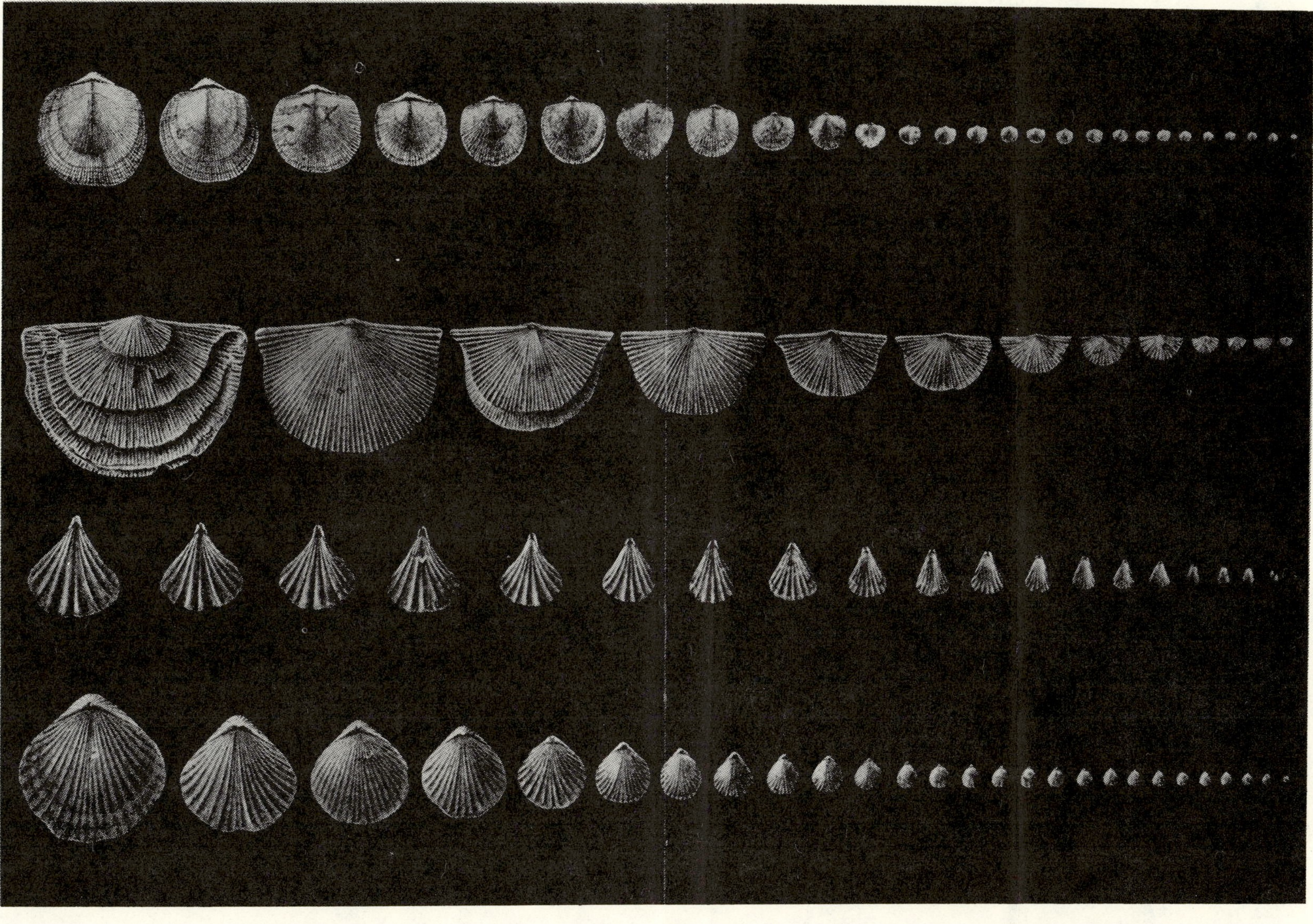

Stages of Growth in Silurian Brachiopoda

PLATE XXIII

BILOBITES ACUTILOBUS Ringueberg (Page 400)

Figure 1. — Outline of specimen from Niagara Group, Lockport, New York. × 4.

BILOBITES VERNEUILIANUS Lindström (Page 400)

Figure 2. — Common elongate form from Upper Silurian, Gotland, Sweden. × 4.

BILOBITES VARICUS Conrad (Pages 402–405)

Figure 3. — Dorsal view of youngest individual observed; showing inception of radiating striæ and concealment of hinge-areas. × 18.

Figure 4. — Profile of same; showing depth and extent of both valves. × 18.

Figure 5. — Hinge view of preceding. × 18.

Figure 6. — Dorsal side of specimen; showing beginning of anterior marginal sinus. × 18.

Figure 7. — Profile of same. × 18.

Figure 8. — Posterior view of same. × 18.

Figure 9. — Dorsal view of specimen, figure 15 ; showing concealment of ventral area. × 9.

Figure 10. — Ventral view of same; showing dorsal area. × 9. Compare this with dorsal view of larger specimen, figure 21, in series.

Figures 11–26. — Series of specimens; seen from dorsal side; exhibiting observed stages of growth, variation and development of hinge, hinge-area, and marginal sinus. × 4.

Figure 27. — Interior of ventral valve; showing teeth, muscular impressions, minute concave plate in apex of fissure, and arrangement of punctæ between nodes and ribs. × 6.

Lower Helderberg Group, Albany county, New York.

BILOBITES BILOBUS Linné (Page 400)

Figure 28. — Outline; showing characteristic form of this species as occurring in Upper Silurian of Gotland, Sweden.

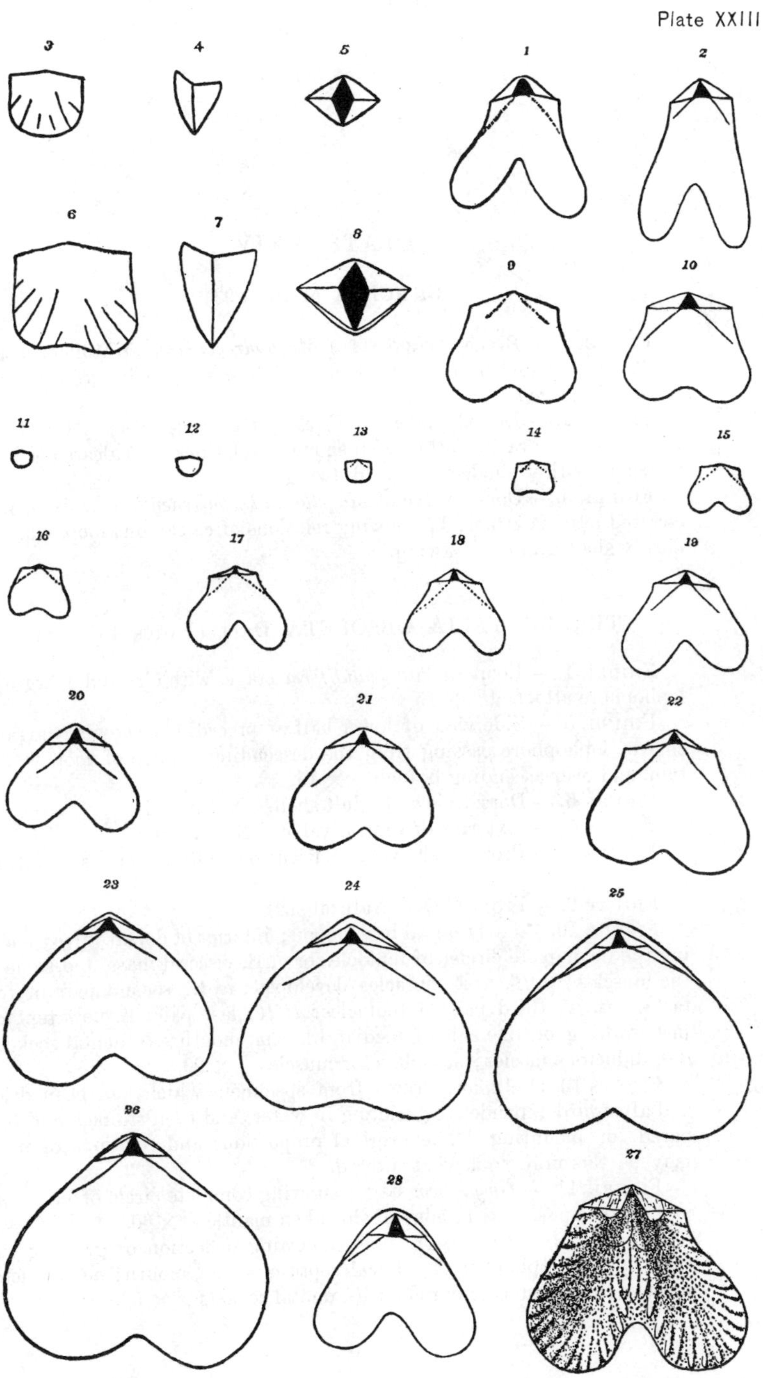

DEVELOPMENT OF BILOBITES

PLATE XXIV

BRACHIA (Page 293)

FIGURE 1. — Brachial supports of *Macandrevia cranium* Müller; nearly adult; *p, p,* points of former attachment to a septum in *terebrataliform* stage. Enlarged.

FIGURE 2. — Dorsal valve of *Terebratulina septentrionalis* Couthouy, with cirrated brachia attached; showing relations of calcareous loop, which is darkly shaded in the drawing.

FIGURE 3. — Dorsal valve of *Magellania kerguelenensis* Davidson, with cirrated brachia attached; showing relations of calcareous loop, which is darkly shaded in the drawing.

TEREBRATALIA OBSOLETA Dall (Pages 406–408)

FIGURE 4. — Loop in late *platidiform* stage, with cirrated margin of lophophore attached. × 15.

FIGURE 5. — Side view of lower half of preceding; showing cirrated edge of lophophore passing from the descending branch of loop to septum, and over ascending branch. × 15.

FIGURE 6. — Dorsal view of adult shell. Natural size.

FIGURE 7. — Exterior of ventral valve. Natural size.

FIGURE 8. — Profile; showing relative convexity of valves. Natural size.

FIGURE 9. — Front view. Natural size.

FIGURE 10. —Early larval brachiopod; interior of dorsal valve; showing the incomplete circlet of tentacles or cirri, visceral mass, and some of the muscles; *t*1, *t*1′, first tentacles developed; *t*2, *t*2′, second pair of tentacles; *t*3, *t*3′, third pair of tentacles; *t*4, *t*4′, last pair; *t*5, new tentacle just growing on one side of median line; *m*, mouth; *ds*, dental sockets; *did*, diductor muscles; *ad*, adductor muscles. × 90.

Figures 10, 11, 12, are drawn from specimens which had been dried and afterward expanded by soaking in water, and then stained and prepared for mounting. The original proportions and relations of parts may be therefore somewhat altered.

FIGURE 11. — *Gwyniform* stage; showing complete circle of cirri; *ps*, pallial sinus; *se*, setæ in edge of shrunken mantle. × 60.

FIGURE 12. —*Cistelliform* stage; showing deflection of growing cirrated edge of lophophore by dorsal septum (*s*); *m*, mouth; *ad*, adductor muscles; *did*, diductor muscles; *ds*, dental sockets. × 60.

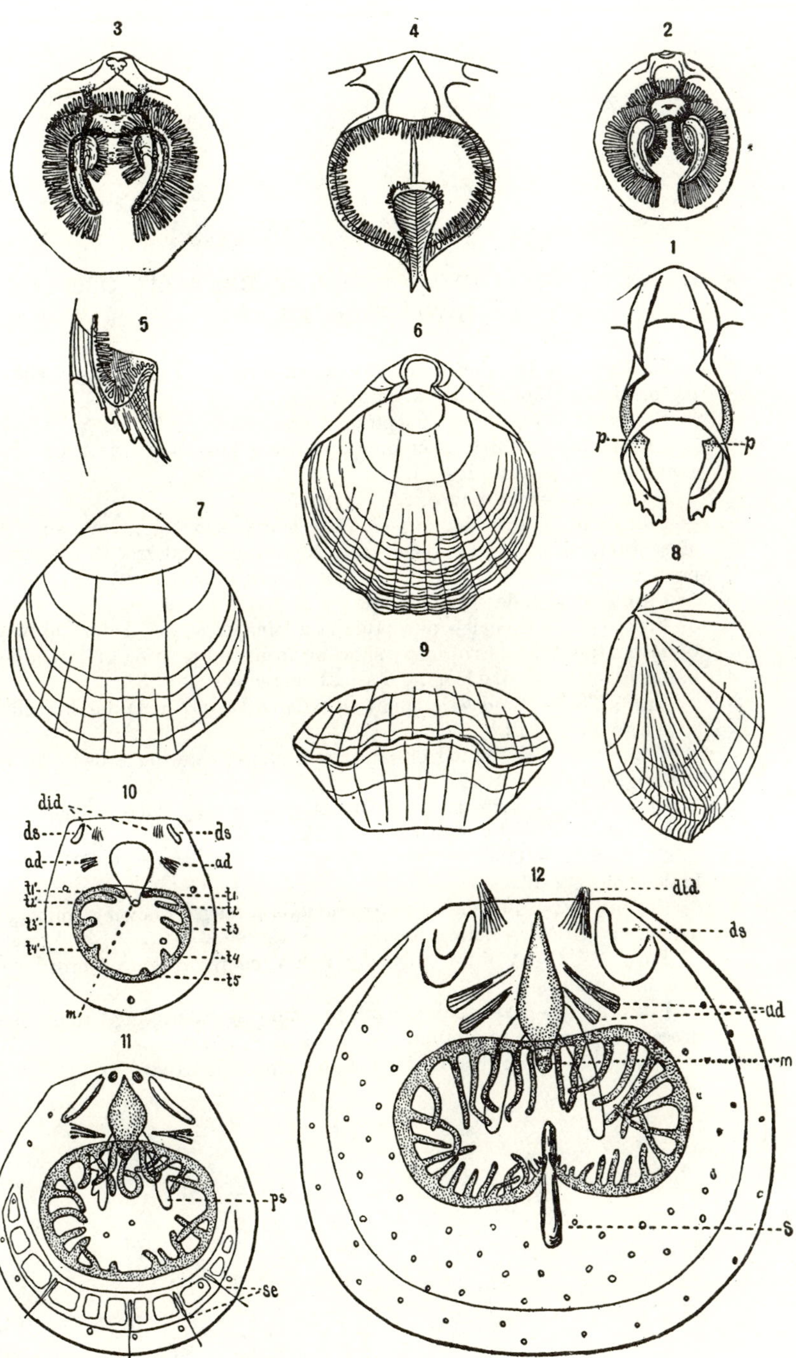

DEVELOPMENT OF TEREBRATALIA

PLATE XXV

ONTOGENY OF THE LOOP IN TEREBRATALIA OBSOLETA
DALL (PAGES 408, 409)

FIGURE 1. — Interior of dorsal valve in early *cistelliform* stage; showing median septum and dental sockets.   × 25.

FIGURE 2. — Side view of septum of preceding.   × 25.

FIGURE 3. — A little later stage; showing growth of two transverse expansions on septum.   × 25.

FIGURE 4. — Side view of same.   × 25.

FIGURE 5. — Beginning of *platidiform* stage; showing groove on edge of septum, with arched covering at posterior end, and growth of crural processes.   × 25.

FIGURE 6. — Side view of same.   × 25.

FIGURE 7. — Side view of septum at a later stage, just before growth of descending branches of loop; showing form and position of ascending branches, or secondary loop, and septal characters.   × 25.

FIGURE 8. — *Platidiform* stage; showing union of descending branches with septum.   × 12.

FIGURE 9. — Beginning of *ismeniform* stage; showing growth of ascending branches.   × 12.

FIGURE 10. — Early *mühlfeldtiform* stage; showing holes developed in ascending branches by resorption.   × 12.

FIGURE 11. — Oblique view of septum, with ascending branches broken off.   × 12.

FIGURE 12. — Beginning of *terebrataliform* stage; showing union of ascending with descending branches.   × 12.

FIGURE 13. — A little later stage; showing resorption of broad descending branches.   × 7.

FIGURE 14. — Late adolescent stage; showing connecting bands from descending lamellæ to septum.   × $2\frac{1}{2}$.

FIGURE 15. — Fully adult loop, with connecting bands.   × $2\frac{1}{2}$.

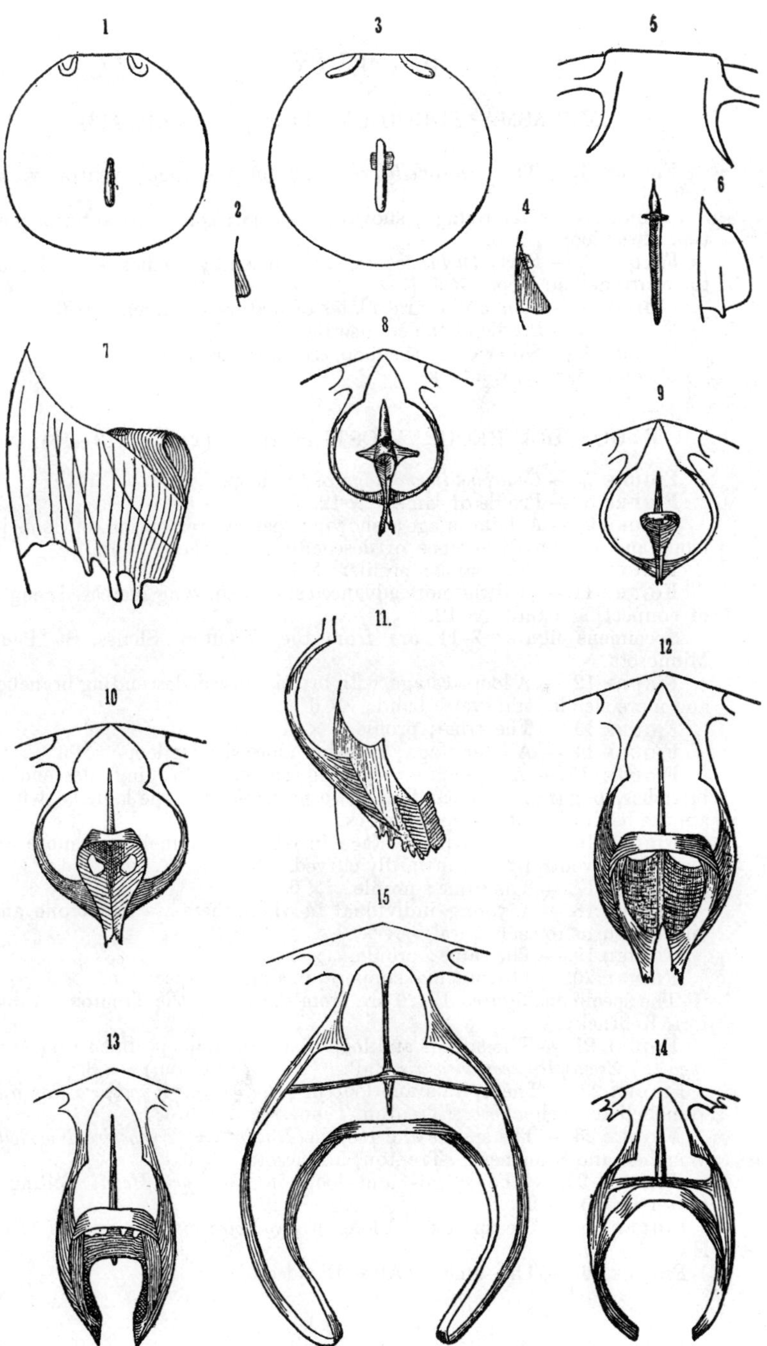

DEVELOPMENT OF TEREBRATALIA

PLATE XXVI

DIELASMA TURGIDUM Hall (Pages 412, 413)

Figure 1. — The *centronelliform* stage of the loop; ventral view. × 9.

Figure 2. — A later stage; showing the resorption of the anterior portion of the loop. × 6.

Figure 3. — Early *Dielasma* stage, produced by further resorption of the centronelloid loop. × 6.

Figure 4. — Loop and crural plates of mature specimen. × 6.

Figure 5. — Profile of the connecting band. × 6.

Figure 6. — Side view of the loop, crura, and septum. × 6.
St. Louis Group, Kentucky.

ZYGOSPIRA RECURVIROSTRIS Hall (Pages 413–416)

Figure 7. — *Centronelliform* stage of the loop. × 12.

Figure 8. — Profile of same. × 12.

Figure 9. — A later stage; showing partial resorption of loop in front and greater divergence of descending branches. × 12.

Figure 10. — The same; profile. × 12.

Figure 11. — A little more advanced stage; showing increased length of connecting band. × 12.
Specimens figures 7–11 are from the Trenton Shales, St. Paul, Minnesota.

Figure 12. — A looped stage, with broad, curved, descending branches and more slender transverse band. × 6.

Figure 13. — The same; profile. × 6.

Figure 14. — A later stage; showing more slender loop. × 6.

Figure 15. — A specimen showing curved, diverging, descending branches, long transverse band, and two projections of the lamellæ, which are the beginning of the spiral cones. × 6.

Figure 16. — A subsequent stage in which the lamellæ are more extended and ventrally and inwardly curved. × 6.

Figure 17. — The same; profile. × 6.

Figure 18. — A young individual in which there are about one and one-half turns to each spiral. × 6.

Figure 19. — The same; profile. × 6.

Figure 20. — The brachial supports in a mature specimen. × 6.
The specimens figures 12–20 are from the top of the Trenton, Frankfort, Kentucky.

Figure 21. — The spirals and loop in a Canadian specimen of (*Anazyga =*) *Zygospira recurvirostris* Hall. (After Davidson.) × 3.

Figure 22. — The spirals and loop in (*Hallina =*) *Zygospira Saffordi* Winchell and Schuchert. Trenton, Tennessee. × 6.

Figure 23. — The spirals and loop in (*Hallina =*) *Zygospira Nicolletti* Winchell and Schuchert. Trenton, Minnesota. × 6.

Figure 24. — The spirals and loop in *Catazyga Headi* Billings. (After Hall.) × 2.

Figure 25. — The spirals and loop in *Zygospira modesta* Say. (After Hall.) × 2¼.

Figure 26. — The same. (After Davidson.) × 3½.

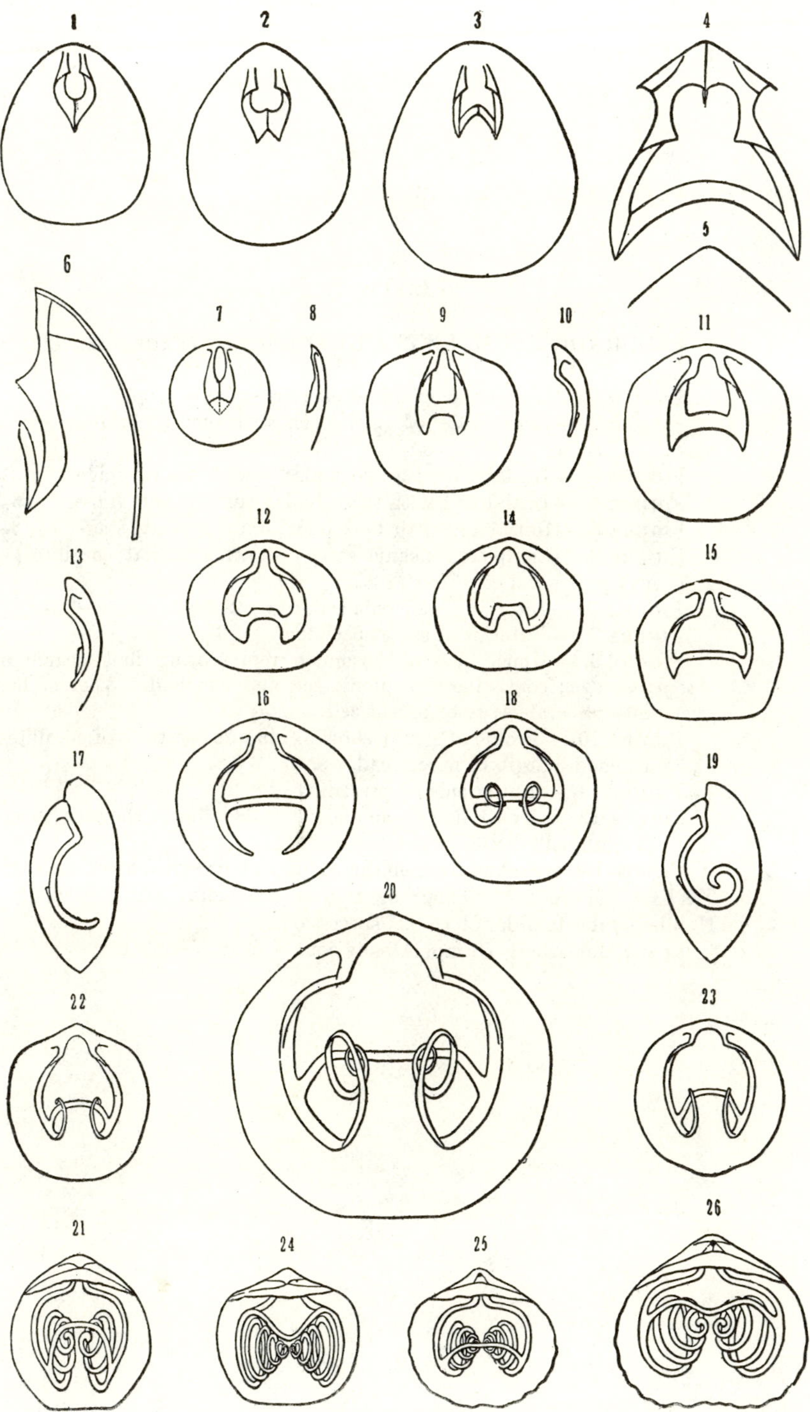

Plate XXVI

BRACHIAL SUPPORTS IN DIELASMA AND ZYGOSPIRA

PLATE XXVII

PLEURODICTYUM LENTICULARE Hall (Pages 422–425)

Figure 1. — Lower side of initial cell, or nepionic stage. × 3½.

Figure 2. — Lower side of initial cell with one bud ; first neanic stage. × 3½.

Figure 3. — Initial cell with two buds ; second neanic stage. × 3½.

Figure 4. — Initial cell with three buds ; third neanic stage. × 3½.

Figure 5. — Initial cell with four buds ; fourth neanic stage. × 3½.

Figure 6. — Completed neanic stage ; showing initial corallite (1) and seven peripheral corallites (2–8). × 3½.

Figure 7. — Interior of nepionic corallite. × 3½.

Figure 8. — Exterior of same specimen. × 3½.

Figure 9. — Upper side of specimen representing first neanic or *Aulopora* stage ; consisting of nepionic cell and one bud. Apex of bud opens into visceral cavity of parent cell. × 3½.

Figure 10. — Profile of same ; showing oblique apertures of corallites, and thickened margin of parent cell. × 3½.

Figure 11. — Lower side of preceding. × 3½.

Figure 12. — Upper side of completed neanic stage ; showing inception of eighth cell. × 3½.

Figures 1–6 are taken from epithecal lines of growth shown in Plate XXVIII, figure 2. Remaining figures are from actual specimens. Numbers refer to order of calical succession.

Lower Helderberg Group, Albany county, New York.

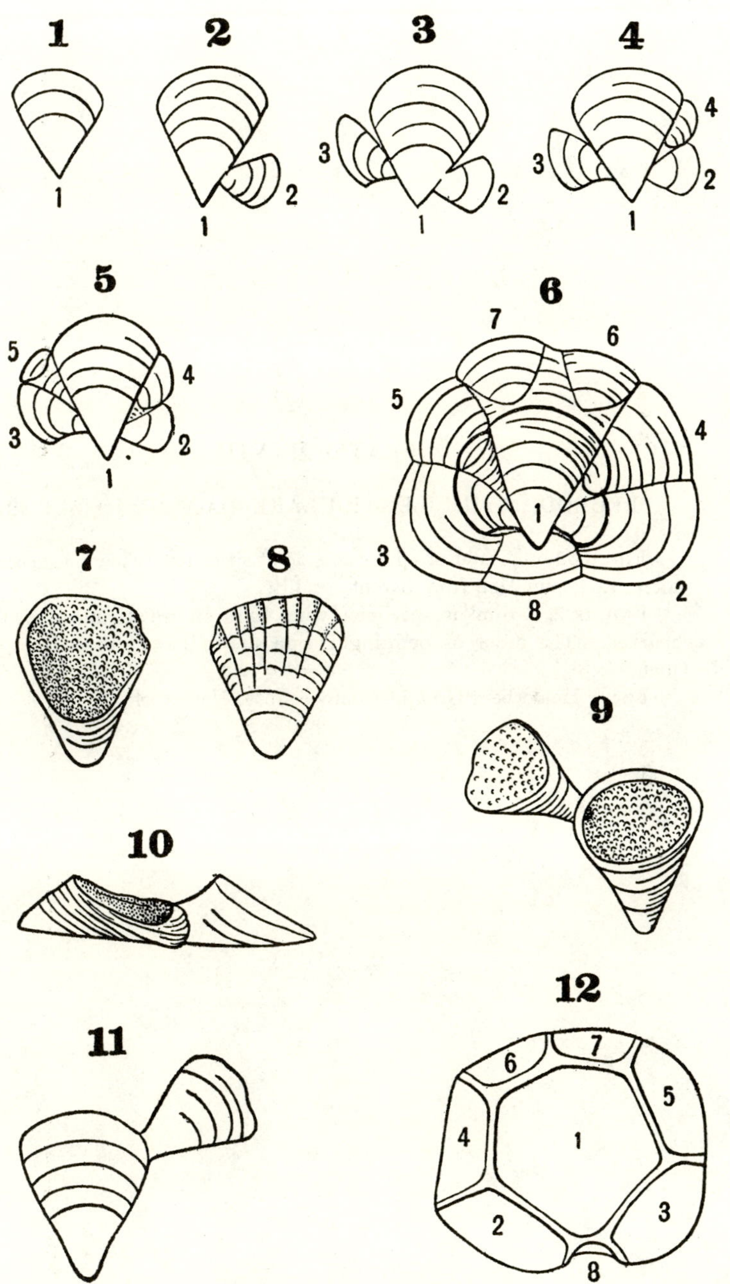

DEVELOPMENT OF PLEURODICTYUM

PLATE XXVIII

PLEURODICTYUM LENTICULARE Hall (Pages 423, 424)

Figure 1. — Lower or epithecal side of specimen ; showing successive alternate gemmation from parent corallite. $\times\ 3\frac{1}{2}$.

Figure 2. — Similar specimen with lines of growth more strongly marked.   The order of budding is opposite to that of preceding specimen. $\times\ 3\frac{1}{2}$.

Lower Helderberg Group, Albany county, New York.

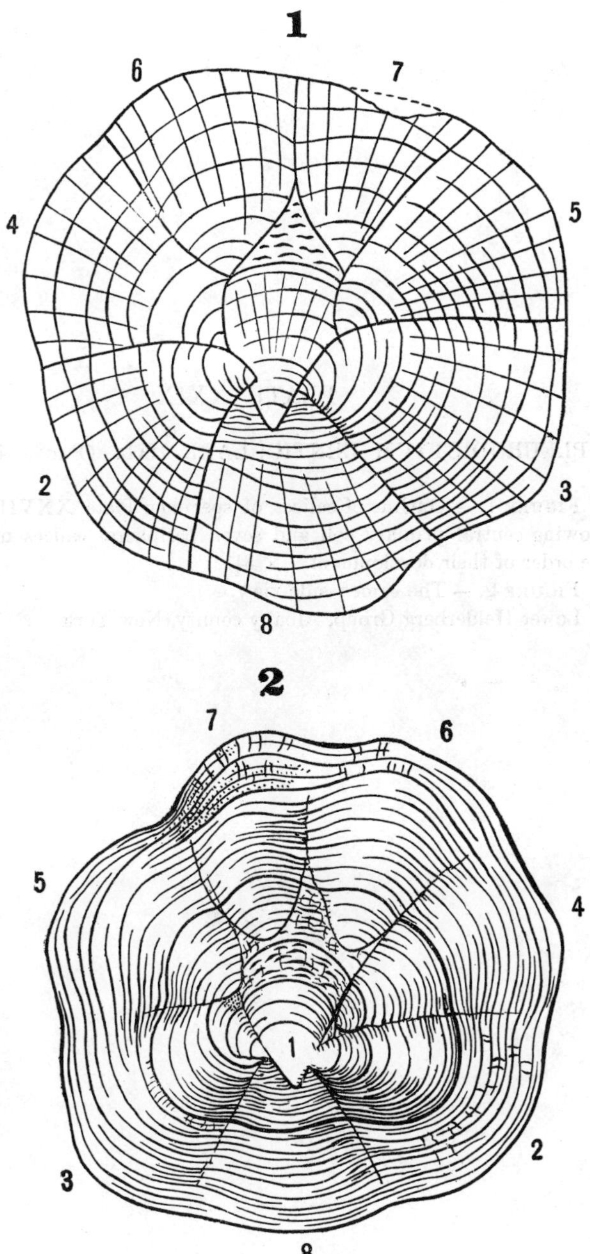

PLEURODICTYUM

PLATE XXIX

PLEURODICTYUM LENTICULARE Hall (Pages 423, 424)

FIGURE 1. — Outline of calices of specimen Plate XXVIII, figure 2; showing central primary cell and seven peripheral calices numbered in the order of their development. $\times\ 3\frac{1}{2}$.

FIGURE 2. — The same ; side view.

Lower Helderberg Group, Albany county, New York.

1

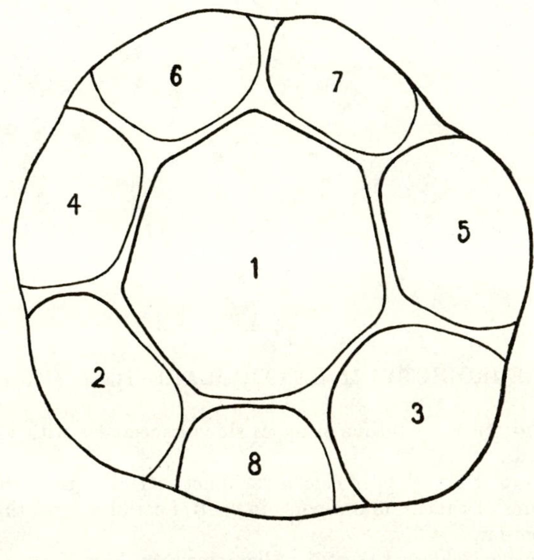

2

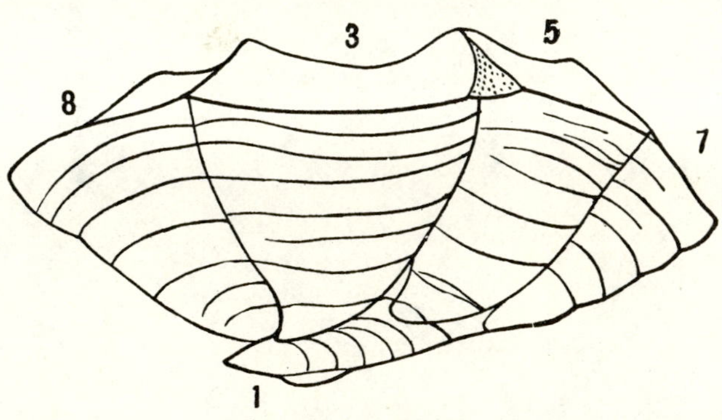

PLEURODICTYUM

PLATE XXX

PLEURODICTYUM LENTICULARE Hall (Pages 423, 424)

Figure 1. — Outline of upper side of specimen with eight peripheral calices. × 3½.

Figure 2. — Upper side of symmetrical specimen; showing general features of calices, mural pores in central corallite, and thickened epithecal border. × 3½.

Lower Helderberg Group, Albany county, New York.

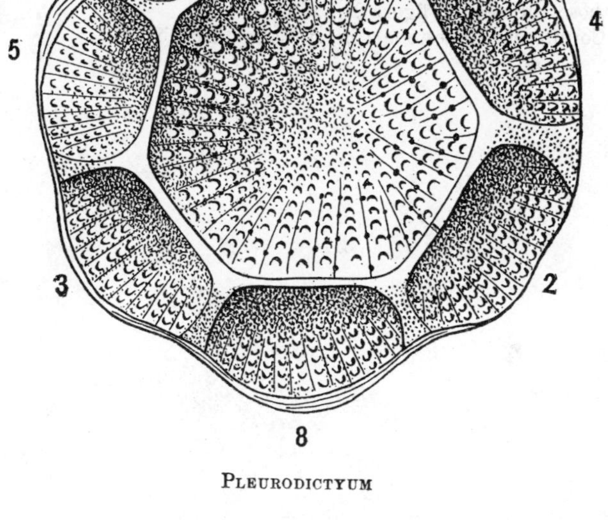

PLEURODICTYUM

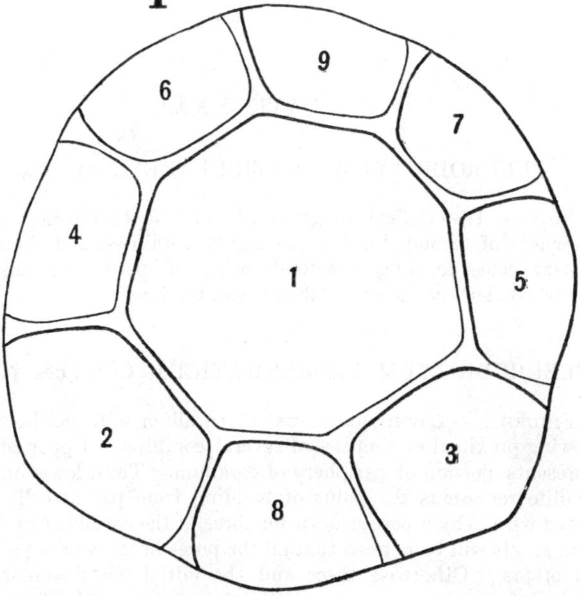

PLATE XXXI

PLEURODICTYUM LENTICULARE Hall (Page 424)

Figure 1. — Calical diagram of gerontic specimen; showing enlargement of second, fourth, and eighth corallites, and the addition of tertiary cells, forming a second series of peripheral calices. $\times 3\frac{1}{2}$. Lower Helderberg Group, Albany county, New York.

PLEURODICTYUM PROBLEMATICUM Goldfuss (Page 426)

Figure 2. — Lower side of cast of corallum with epitheca removed; showing proximal extremities of several corallites. Upper edge of figure represents portion of periphery of corallum. Thus, lower angle of each corallite represents the point of budding from parent cell, and is connected with it by a pore, shown for three of the corallites by dotted lines from p. It will be noticed that all the pores in the angles are larger than the others. Otherwise these and the initial pores cannot be distinguished from the ordinary mural pores between the flat sides of the corallites. $\times 7$. Devonian, Pelm, Germany.

FAVOSITES EPIDERMATUS? Rominger (Page 426)

Figure 3. — Side view of mature corallite with attached intermural bud. Specimen broken from interior of a large colony. $\times 3\frac{1}{2}$.

Figure 4. — The same, front view, with bud removed; showing pore or mural opening (p) at lower point of attachment of bud, corresponding to those indicated in figure 2. $\times 3\frac{1}{2}$.

Corniferous Limestone, Cherry Valley, New York.

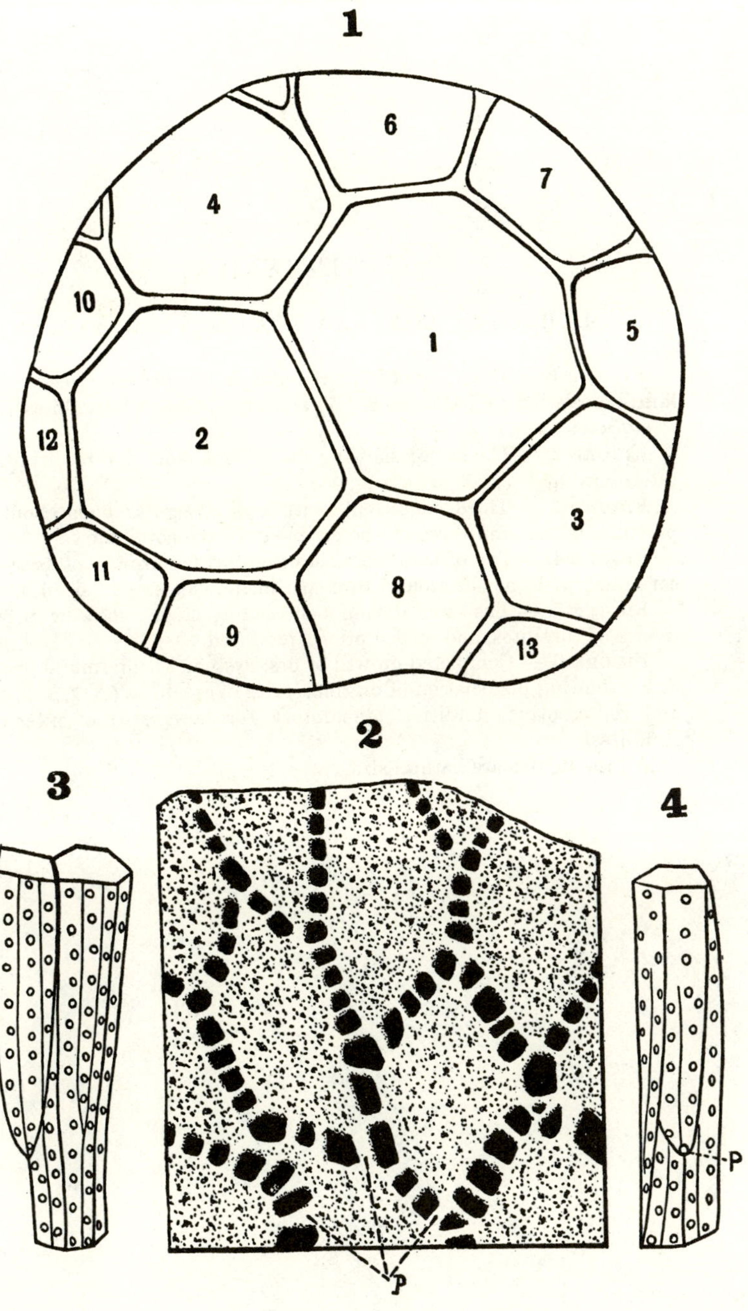

PLEURODICTYUM AND FAVOSITES

PLATE XXXII

MICHELINIA CONVEXA D'Orbigny (Pages 431, 432)

Figure 1. — Diagrammatic representation of upper surface of corallum; consisting of parent cell (A) and six peripheral corallites (*1, 2, 3,* etc.).

Figure 2. — The same; showing the introduction of three triangular intermural buds (*1', 2', 3',* etc.).

Figure 3. — Third condition; with six triangular buds about the parent corallite, and three on the periphery of the corallum.

Figure 4. — Top of corallum; showing further growth of preceding corallites, with the addition of three peripheral calices (*4'', 5'', 6''*).

Figure 5. — The same during a succeeding stage; showing increase in size of corallites, and modifications produced.

Figure 6. — Completed growth of first system of intermural gemmation; showing dissociation of original series of corallites (A, *1, 2, 3,* etc.), and representing condition preparatory for new series of interstitial corallites.

All the figures are natural size.

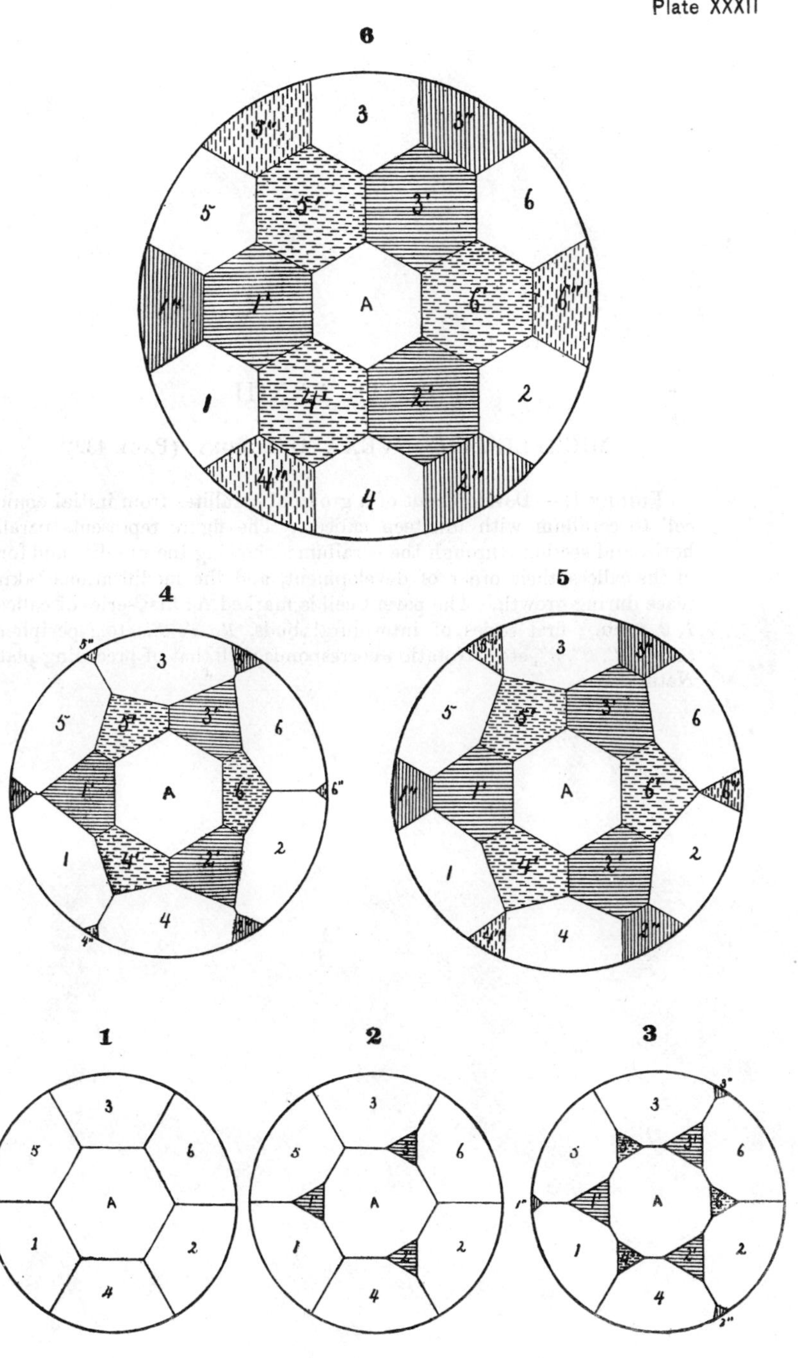

6
3
5
A
1
4
2
4
5
1
2
3
FAVOSITIDÆ

PLATE XXXIII

MICHELINIA CONVEXA D'Orbigny (Page 432)

Figure 1. — Development of a group of corallites from initial conical cell to corallum with nineteen calices. The figure represents parallel horizontal sections through the corallum ; showing the number and form of the calices, their order of development, and the modifications taking place during growth. The parent cell is marked A ; first series of calices, *1, 2, 3,* etc. ; first series of intermural buds, *1', 2', 3',* etc. ; peripheral series, *1'', 2'', 3''*, etc. Notation corresponds with that of preceding plate. Natural size.

1

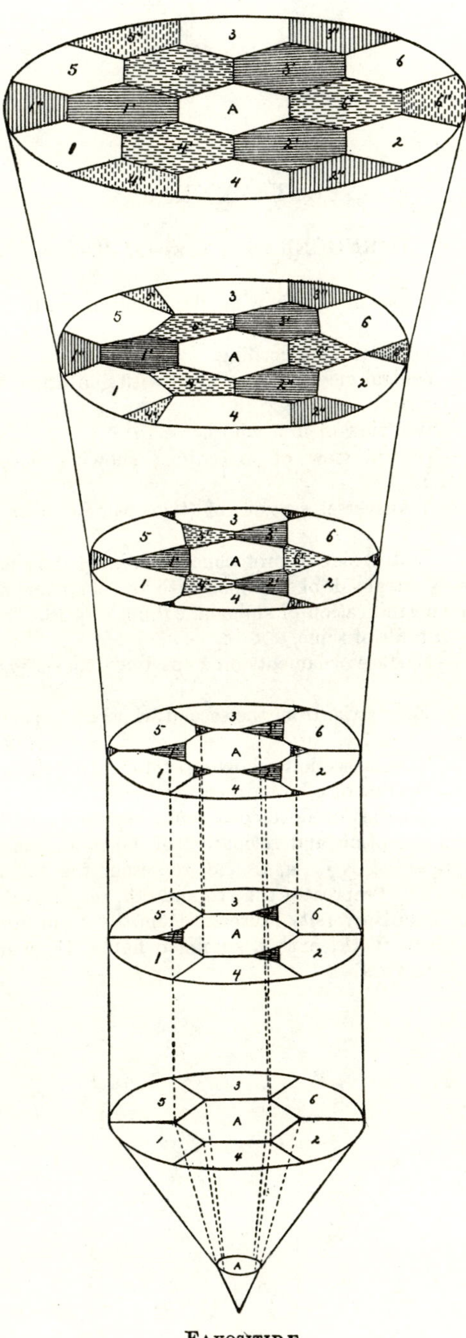

FAVOSITIDÆ

PLATE XXXIV

TORNOCERAS (Pages 437–439)

FIGURE 1. — Protoconch; showing first septum with lateral edges broken. × 18.

FIGURE 2. — Side view of preceding. × 18.

FIGURE 3. — Ventral view of protoconch with one attached air-chamber; showing siphonal cæcum. × 18.

FIGURE 4. — Side view of first whorl. × 18.

FIGURE 5. — Ventral view of preceding; showing development of ventral lobe. × 18.

FIGURE 6. — Transverse section of two whorls near the protoconch. × 18.

FIGURE 7. — Ventral side of protoconch, retaining the shell. × 18.

FIGURE 8. — Ventral side of specimen with first chamber; showing surface ornaments and indication of siphonal cæcum. × 18.

FIGURE 9. — Profile of same. × 18.

FIGURE 10. — Surface ornaments on a specimen consisting of a single whorl. × 18.

FIGURE 11. — Four striæ from the second whorl of a specimen; showing funnel sinus. × 18.

FIGURE 12. — Vertical section; showing septa and air-chambers. × 18.

FIGURE 13. — Outline of a half-grown specimen. × 3.

FIGURE 14. — A series of developed septa beginning with the first; showing gradual inception and formation of lobes and saddles to the adult period represented by j. a, b, c, d, represent the 1st, 2d, 3d, and 4th septa, while e and f represent the 7th and 8th, respectively. × 9.

Specimens in Yale University Museum, from the Hamilton Shales of Wende Station, New York, except specimen figure 13, which is from 18-Mile Creek, New York.

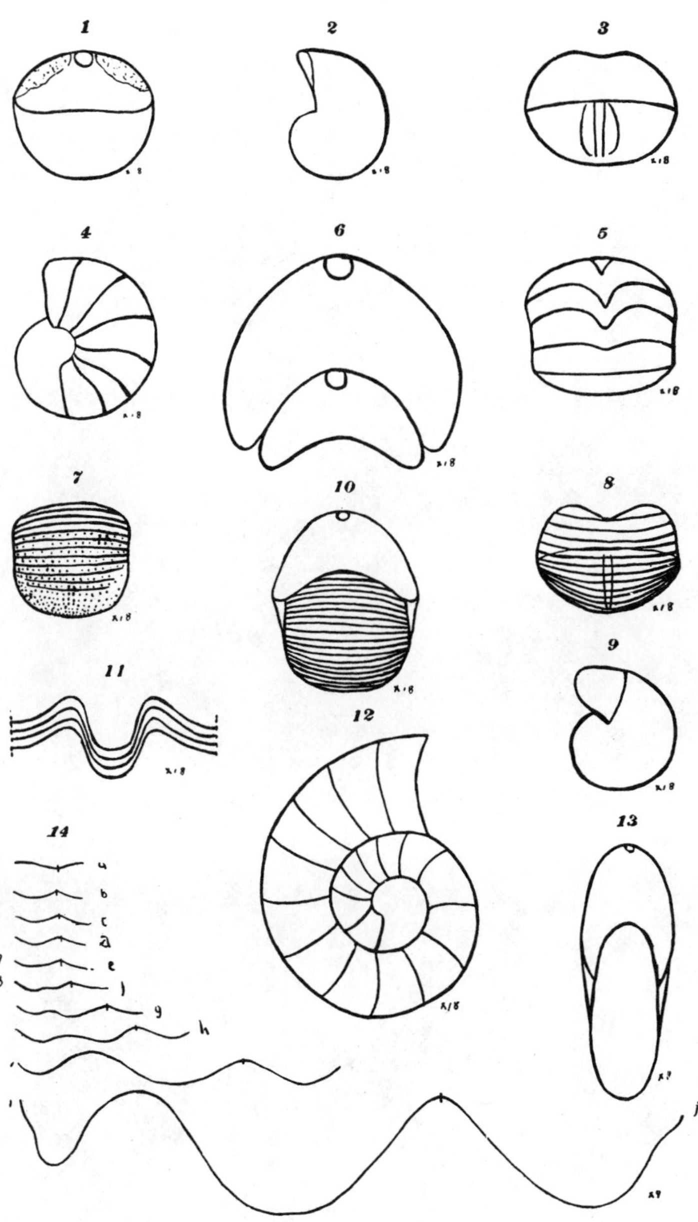

TORNOCERAS

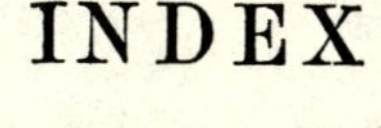# INDEX

INDEX

A

Acantharia, composite spine in, 49.

Acanthaster solaris, protective spines in, 56.

acanthogeny, 19.

Acanthosicyos horrida, cultivation of, Henslow, 74.

Acanthothyris, 245.
    Döderleini, spines in, 51.
    spinosa, spines in, 51.

Acaste, 129, 156.

acceleration in Brachiopoda, modifications from, 233.

Acerocare, 142, 145.

Achatinella, differentiation of, 65.
    variations in, 36, 37.

Achatinellæ, free variation in, Hyatt and Verrill, 36.
    specific differentiation in, Hyatt and Verrill, 36.

Achtheres, variation in anterior antennæ of, 192.

Acidaspidæ, 129, 131, 139, 157.
    defined, 151.
    extremes of spinosity in, 56.
    lobes in glabella of, 149.

Acidaspis, 70, 113, 129, 149, 151, 152, 157, 174, 176, 179, 210.
    accelerated development of spines in, 22.
    eyes in, 126.
    features of dorsal shield of, 182.
    glabella in larval, 126, 180.
    high specialization of, 174.
    no eye-line in, 180.
    spiniform processes in, 78.
    third segment in, 127.
    tuberculata, 173.

Acontaspis hastata, multiplication of spines in, 69.

Acontheus, cephalon in, 182.

Acrocephalites, 142.

Acrosoma arcuata, protective mimetic features in, 62.

Acrothele, 244.
    marginal upper beak in, 236.
    pedicle-opening in, 241.

Acrotreta, 244.
    marginal upper beak in, 236.

Actinocrinus lobatus, spines in, 51.
    pernodosus, spines in, 51.

Actinopeltis, 155, 156.

Actinopteria Boydi, spines in, 51.

Adouin's law regarding the Articulata, 83.

Æglina, 113, 132, 146, 188.
    armata, eyes in, 147.
        free-cheeks in, 117, 118.
    eyes in, 147.
    free-cheeks in, 124.
    princeps, eyes in, 147.
    rediviva, eyes in, 147.

Agassiz, A., homologous structures in Echinodermata, 94.
    L., foundation of orders, 242.
        interpretation of family characters, 290.
        ontogeny a standard of classification, 120.

Aglaspidæ, 132.

Aglaspis, 132.

Agnostidæ, 130, 131, 134.
    defined, 135.
    primitive head structure in, 135.
    primitive structure in, 135.

Agnostini, 134.

Agnostus, 130, 136, 176, 178.
    absence of eyes in, 183.
    Cambrian, 183.
    cephalon in, 127, 128.
    distinct plate in, 134.
    eye-spots lost in, 135.
    free-cheeks in, 124.

C

adult loop in, 291.
floridana, 297, 300.
  Grayi, 297.
    adult characters of loop in, 295.
  ismeniform stage of, 300.
  mühlfeldtiform stage in, 301.
  new genus, 297.
  Raphaelis, 297.
    adult characters of loop in, 295.
  septigera, 300, 409.
    adult characters of loop in, 295.
    brachial supports in, Friele,
      Deslongchamps, 406.
    development of loop in, Friele,
      Deslongchamps, 292.
    transformations in, Friele, 298.
Dallininæ, 295, 302.
  boreal distribution of, 304.
  defined, 307.
  median septum in, 296.
  ontogeny and morphology of, 300.
  small cardinal processes in, 296.
  stages of growth in, 288, 303.
Dalman, J. W., classification of trilobites, 111.
Dalmanella, 387, 388, 394.
  absence of embryonal sinus in, 388.
  elegantula, 315, 317, 318, 321, 322,
    352.
    discussion of, 317.
    earliest growth-stages in, 373.
  mature individuals in, 373.
Dalmanites, 113, 129, 156, 157, 179.
  eyes in, 126.
  features of dorsal shield of, 182.
  free-cheeks in, 117, 124, 181.
  genal angles in, 123.
  larvæ of, 128.
  nasutus, spinose cephalon in, 157.
  no eye-line in, 180.
  ontogeny of, 128, 129, 135.
  pleura from pygidium in, 127.
  pygidium in, 181.
  socialis, 175, 176.
  tridens, spinose cephalon in, 157.
Dana, J. D., principle of cephalization
  in Crustacea, 83.
Daphnia, absence of median eye in,
  193.
  nauplius appendages of, 192.
Darwin, C., horns in Chamæleontidæ,
  59.

origin of so-called rudimentary organs, 80.
sexual selection as affecting the
  growth of antlers in the deer, 58.
spines on shells of barnacles, 52.
Davidson, T., casts of pedicles of fossil
  Lingulæ and *Eichwaldia*, 257.
  relations of *Bilobites* with *Orthis*, 399.
  spines in Brachiopoda, 51.
Davidsonella, 245.
  schizolophus stage in, 279.
Davidsonia, 245, 284.
  hinge in, 236.
  spiral impressions on valves of 276.
Daviesiella, 245.
Dayia, 245, 413, 416.
  navicula, 413.
Decapoda, 200.
  spines on zoëa of, 46, 57.
  spiny forms of, 70.
Dechenella, 148.
  lobation of glabella in, 148.
deer, absence of antlers in ancestors of,
  24.
  antlers of, 9, 21.
  development of horns in, 14.
  early, antlers in both sexes of, 57.
  fallow, antler of, 53.
  family, compound antlers in, 68.
  growth in antlers of, 10.
  modern, shedding of antlers of, 97.
  sexual selection as affecting growth
    of antlers in, Darwin, 58.
  simple antlers of young, 24.
  spine production in, 13.
  Tertiary, simple antlers in, 52.
defences, mechanical and motor, in animals and plants, Morris, 97.
Deiphon, 153, 155, 156.
  Forbesi, spiniform processes in, 78,
    79.
  free-cheeks in, 118.
  glabella and pygidium in, 155.
Delthyris, 399.
delthyrium, 257.
deltidial plates, origin of, 257.
deltidium amplectens, 391.
  development of, Deslongchamps,
    Clarke, 263.
  discretum, 391.
  origin of, 257.
  sectans, 391.

Q

R

regressive, 6.

restrictions of, 7.

tendency of, 6,

variations in *Achatinella* and *Allorchestes*, 36, 37.

Vella, spiniform branches of, 75.

Verneuilina spinulosa, spines in, 44.

Vernon, H. M., reproductive divergence, 37.

Verrill, A. E., Achatinellæ, 36.

cannibalistic selection, 57.

characters in appendages of *Trinucleus*, 225.

importance of tabulæ in a natural classification of corals, 427.

vertebræ, spines of, 9.

vertebrates, hornless young of horned, 19.

Verworn, development of extinct Ostracoda, 310.

vitality, ebbing, Geddes, 12.

W

Waagen, W., classification of Brachiopoda, 242, 290.

septal and dental plates in the terebratuloids, 297.

Wachsmuth, C., and Springer, F., development of tubercles of Crinoidea and Asteroidea into spines, 50.

spines on ventral sacs of Crinoidea, 45.

Walcott, C. D., additional thoracic segments in *Triarthrus Becki*, 217.

biramous appendages in *Triarthrus*, 200.

casts of pedicles of fossil Lingulæ and *Eichwaldia*, 257.

classification of trilobites, 114.

development of trilobites, 310.

metamorphoses of trilobites, 166.

mouth in *Calymmene*, 209.

possible trilobite eggs, 169.

young of *Olenellus asaphoides*, 178.

Waldheimia, 411.

bicarinata, 413.

Mawii, 413.

recurved loop in, 413.

septigera, 297.

Wallace, A. R., on the origin of xerophilous plants, 74.

protective horns in males of Copridæ and Dynastidæ, 60.

Whitfieldella, absence of embryonal sinus in, 388.

nitida, 351, 352, 356, 373, 378, 392.

discussion of, 374.

early stages of growth in, 373.

tumida, 377.

Whitfieldia, 377.

Williams, H. S., variations in stock of *Atrypa reticularis*, 25.

Woodward, H., affinities of trilobites, 110.

Trilobita the earliest forms of Crustacea, 184.

X

Xanthidium armatum, spine growth from frustule of, 43.

Xiphogomium, 148.

Xiphosphæra pallas, spines in, 48.

Xylotrupes gideon, horns in, 60.

Y

Youngia, 155, 156.

Z

Zacanthoides, 113, 142, 143, 144.

pygidium in, 143.

Zaphrentis spinulosa, spinules in, 50.

Zellania, 246, 300, 304, 308.

schizolophus stage in, 279.

Zilla, branches of, transformed into spines, 74.

Zittel, K. A. von, classification of articulates, 112.

of Trilobita, 112, 132.

historical review of trilobites, 111.

oldest species of *Spondylus*, 23.

terminal spine in Nasellarian Radiolaria, 43.

zoëa of common crab, 56.

of Decapoda, spines in, 57.

Zoic maxima, Gratacap, 35.

zugolophus stage, 281, 284.

THE HISTORY OF PALEONTOLOGY

An Arno Press Collection

Abel, Othenio. **Paläobiologie und Stammesgeschichte**. 1929

Agassiz, L[ouis]. **Études Critiques sur les Mollusques Fossiles**. 1840-1845

Beecher, Charles Emerson. **Studies in Evolution**. 1901

Bernard, Félix. **Éléments de Paléontologie**. 1895

Buckland, William. **Geology and Mineralogy Considered with Reference to Natural Theology**. Two vols. 1836

Cuvier, Georges. **Memoirs on Fossil Elephants** *and* **On Reconstruction of the Genera Palaeotherium and Anoplotherium**. 1812

Dacqué, Edgar. **Vergleichende biologische Formenkunde der fossilen niederen Tiere**. 1921

Depéret, Charles. **The Transformations of the Animal World**. 1909

Gaudry, Albert. **Essai de Paléontologie Philosophique**. 1896

Gould, Stephen Jay, ed. **The Complete Works of Vladimir Kovalevsky**. 1980

Gould, Stephen Jay, ed. **The Evolution of** *Gryphaea*. 1980

Gould, Stephen Jay, ed. **Louis Dollo's Papers on Paleontology and Evolution**. 1980

Hatcher, John B. **The Ceratopsia**. 1907

Hyatt, Alpheus. **Phylogeny of an Acquired Characteristic**. 1893

Kummel, Bernhard, ed. **Status of Invertebrate Paleontology, 1953**. 1954

Mantell, Gideon Algernon. **The Medals of Creation**. Two vols. 1854

Matthew, W[illiam] D[iller]. **Outline and General Principles of the History of Life**. 1928

Miller, Hugh. **The Testimony of the Rocks**. 1857

Moore, Raymond C. and Lowell R. Laudon. **Evolution and Classification of Paleozoic Crinoids**. 1943

Newell, Norman D. **Late Paleozoic Pelecypods**. 1937/1942

Nicholson, H[enry] Alleyne. **The Ancient Life-History of the Earth**. 1877

d'Orbigny, Alcide [Dessalines]. **Cours Élémentaire de Paléontologie et de Géologie Stratigraphiques**. Two vols. in three. 1849/1851/1852

Osborn, Henry Fairfield. **The Origin and Evolution of Life**. 1917

Osborn, Henry Fairfield. **The Titanotheres of Ancient Wyoming, Dakota, and Nebraska**. Two vols. 1929

Owen, Richard. **Palaeontology.** 1860

Phillips, John. **Life on the Earth.** 1860

Pictet, F[rançois] J[ules]. **Traité Élémentaire de Paléontologie.** Four vols. 1844-1846

Romer, A[lfred] S[herwood] and L[lewellyn Ivor] Price. **Review of the Pelycosauria.** 1940

Schindewolf, Otto H. **Grundfragen der Paläontologie.** 1950

Schopf, Thomas J.M., ed. **Presidential Addresses of the Paleontological Society.** 1980

Scilla, [Agostino]. **De Corporibus Marinis.** 1747

Simpson, George Gaylord. **A Catalogue of the Mesozoic Mammalia** and **American Mesozoic Mammalia.** Two vols. in one. 1928/1929

Stensiö, Erik A[ndersson]. **The Downtonian and Devonian Vertebrates of Spitsbergen.** Two vols. 1927